Telecom Crash Course

Second Edition

Steven Shepard

McGraw-Hill/Osborne
New York   Chicago   San Francisco   Lisbon
London   Madrid   Mexico City   Milan
New Delhi   San Juan   Seoul   Singapore
Sydney   Toronto

McGraw-Hill

A Division of The **McGraw·Hill** *Companies*

5 6 7 8 9 0   IBT/IBT   1 9 8 7 6 5 4 3 2 1 0

ISBN 0-07-145143-9

The sponsoring editors for this book were Steve Chapman and Jane Brownlow and the production supervisor was David Zielonka. It was set in Century Schoolbook by MacAllister Publishing Services, LLC.

Printed and bound by IBT Global.

Throughout this book, trademarked names are used. Rather than put a trademark symbol after every occurrence of a trademarked name, we use names in an editorial fashion only, and to the benefit of the trademark owner, with no intention of infringement of the trademark. Where such designations appear in this book, they have been printed with initial caps.

 This book is printed on recycled, acid-free paper containing a minimum of 50 percent recycled de-inked fiber.

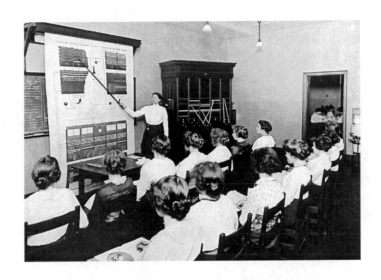

This book is for all the students,
In all the classes,
In all the companies,
In all the countries,
Who took the time to educate me,
While I had the pleasure of educating you.

This book is also for my brother Roger:
writer, musician, and all-round good guy.

ABOUT THE AUTHOR

Steven Shepard is the president of the Shepard Communications Group in Williston, Vermont. A professional author and educator with 22 years of varied experience in the telecommunications industry, he has written books and magazine articles on a wide variety of topics. He is the author of *Telecommunications Convergence: How to Profit from the Convergence of Technologies, Services, and Companies* (McGraw-Hill: New York, 2000); *A Spanish-English Telecommunications Dictionary* (Shepard Communications Group: Williston, Vermont, 2001); *Managing Cross-Cultural Transition: A Handbook for Corporations, Employees and Their Families* (Aletheia Publications: New York, 1997); *An Optical Networking Crash Course* (McGraw-Hill: New York, February 2001); *SONET and SDH Demystified* (McGraw-Hill: New York, 2001), *Telecomm Crash Course* (McGraw-Hill: New York, October 2001); *Telecommunications Convergence, Second Edition* (McGraw-Hill: New York, February 2002); *Videoconferencing Demystified* (McGraw-Hill: New York, April 2002); *Metro Networking Demystified* (McGraw-Hill: New York, October 2002); *The Shepard Report: Charting a Path in Uncertain Times* (SCG: March 2004); and *RFID* (McGraw-Hill: New York, July 2004). *VoIP Crash Course* will be released in mid-2005. Steve is also the Series Editor of the McGraw-Hill *Portable Consultant* book series.

Mr. Shepard received his undergraduate degree in Spanish and Romance Philology from the University of California at Berkeley and his masters degree in International Business from St. Mary's College. He spent 11 years with Pacific Bell in San Francisco in a variety of capacities including network analysis, computer operations, systems standards development, and advanced technical training, followed by 9 years with Hill Associates, a world-renowned telecommunications education company, before forming the Shepard Communications Group. He is a Fellow of the Da Vinci Institute for Technology Management of South Africa, a member of the Board of Directors of the Regional Educational Television Network, and a member of the Board of Trustees of Champlain College in Burlington, Vermont. He is also the resident director of the University of Southern California's Executive Leadership and Advanced Management Programs in Telecommunications and adjunct faculty member at the University of Southern California, The Garvin School of International Management (Thunderbird University), the University of Vermont, Champlain College, and St. Michael's College. He is married and has two children.

Mr. Shepard specializes in international issues in telecommunications with an emphasis on strategic technical sales, convergence and optical networking, the social implications of technological change, the development of multilingual educational materials, and the effective use of multiple delivery media. He has written and directed more than 40 videos and films, and written technical presentations on a broad range of topics for more than 70 companies and organizations worldwide. He is fluent in Spanish and routinely publishes and delivers presentations in that language. Global clients include major telecommunications manufacturers, service providers, software development firms, multinational corporations, universities, advertising firms, and regulatory bodies.

CONTENTS

Contents

Contents

Contents

Contents

ACKNOWLEDGMENTS

I have written many books—21 at last count, I believe—but none of them have been as satisfying or personally rewarding as the *Telecomm Crash Course*. The opportunity to write a second edition is a testimony to you, my readers, who give me a reason to keep researching, interviewing, observing, and writing. Thank you—you mean the world to me.

I also thank my small army of idea crafters, proofreaders, fact-checkers, sanity injectors, and workers in words. I am honored to call you friends. Specifically, thanks go to the following people who went beyond the techno-pale to make this second edition happen: Allison Ashcroft, Phil Asmundson, Kim Barker, Ian Bonde, Jane Brownlow, Joe Candido, Phil Cashia, Ann Cathers, Steve Chapman, Dick Dadamo, Bruce Degn, Jonathan Dunne, Donna Epps, Nicole Fahrer, Jack Gerrish, Jack Garrett, Steve Green, Dave Hill, Lisa Hoffmann, Alfred Hsi, Ron Hubert, Naresh Lakhanpal, Natasha Leger, Tracey Lewis, Viki MacMillan, Dee Marcus, Roy Marcus, Alvaro Marques, Gary Martin, John Martin, Bob Maurer, Dennis McCooey, Paul McDonagh-Smith, Alan Nurick, Katy O'Connor, Chris O'Gorman, Richard Parlato, Girish Pathak, Dick Pecor, Kenn Sato, Dave Stubbs, Elvia Szymanski, Philip Takken, Calvin Tong, Jack Tongue, Lisa Watson-Krause, Dave Whitmore, and Craig Wigginton.

I also thank my family. Sabine, you've put up with me for 24 years; thank you for all the insights, friendship, and love. Cristina and Steve, thanks for teaching me, constantly, to be a better person. Children really are the greatest gift.

PREFACE

I have always believed that the implications of telecommunications technologies are far more important than the technologies themselves. Technology is interesting, awe inspiring, and mystifying, but useless without a perceived application for its capabilities. As I write this book, I am also working on a parallel effort that will be published shortly called *The Social Implications of Technological Change*; it is a collection of essays that examines the impact that certain select technologies have on the social fabric. Telecommunications clearly falls squarely into this camp; ever since 1876, when Alexander Graham Bell spilled a beaker of acid into his lap and called out to his assistant Watson for help,[1] *The Network* has represented the ultimate disruptive force in human society. It and its ever more capable descendents have changed the way we conduct business, purchase goods and services, perform medical diagnoses, develop relationships, conduct research, engage in warfare, create art, and overcome physical and mental obstacles. It is a massive, inexorably powerful force to be reckoned with.

That was the force that drove my desire to write this book. I have studied telecommunications for more than 20 years, first working for Pacific Bell in California, and later as a professional writer and educator. I have always found the technology to be interesting, but far more interesting and inspiring to me is the way people use it. Innovation is a very powerful force. Let me explain by example.

In January 1998, following a long, wallowing flight down the crouched eastern spine of South America, I stepped off a plane in São Paulo, Brazil, hailed a driver, and headed into the city. São Paulo is one of the world's largest cities, a sprawling chaos of ultramodern skyscrapers nestled among lush city parks, skeletons of half-finished buildings, and an incongruously decaying inner city. The place lies across hundreds of square miles of gently rolling hills like a drying cowhide and is home to a staggering 17 million people.

The drive from the airport to the central business district of São Paulo takes about one and a half hours. Along the route, the wide, modern freeway passes through farmland, massive construction sites, rolling, scrub-covered hills, and a large, sad shantytown—a *favela*—which goes on for miles. The buildings that make up this slumlike extension of the modern

---

[1] According to the story he said, "Come here, Watson; I need you." I doubt it.

city of Saõ Paulo are shocking. They are cobbled together from whatever materials were handy at the time of construction—chicken wire stuffed with newspaper for walls, metal cans and barrels hammered flat for roofs, a scrap of burlap hung as a door. Open sewers are commonplace; naked children are everywhere. At one point during the drive to downtown, the traffic clotted up and the driver stopped the car. To the left, I looked out across the hills, each topped with billboards for companies I knew well: Lucent, Motorola, Nokia, Nortel, Ericsson, Alcatel. Looking out the other side of the car, sitting motionless in the hot air, I watched in horror as a young boy, no more than 10 years old, shimmied up a high voltage tower. Behind him, tied to his waist, he dragged a long black cable, on the end of which was a pair of clamps from an automobile battery cable. Without flinching, he expertly attached the clamps to the aluminum power cables overhead amidst a great flurry of sparks. Below him people clapped.

As we picked up speed again I noticed something else rather unusual about this place: Roughly one out of every six of these sad little hovels sported a small satellite dish on the roof. Shaking my head, I vowed to understand this seeming dichotomy before I left Brazil.

I was in the country for two purposes: to deliver a series of lectures on telecommunications convergence to a major manufacturer in the region and to write an article about the divestiture of Telebras, the monopoly-controlled telephone company, which was broken up on the very day I landed in Saõ Paulo. On the way into the city my driver had to divert around the normal path to the hotel because residents of the city were rioting over the breakup of Telebras. Imagine—a firebomb-wielding crowd had gathered to protest the divestiture of the phone company.

Several days later, when things had calmed down, I caught a cab and asked the driver to take me to the slum that I had driven through on my way into the city. It took me 10 minutes and a hefty tip to convince him that I had not been in the hot Brazilian sun too long and really did want to go where I asked him to take me. We drove for an hour before arriving at the area. I paid the driver to wait, told him I'd be back in a couple of hours, and wandered off into the less than fragrant shantytown in search of answers. As I walked away I could hear the driver muttering opinions about my sanity. I only hoped he would be there when I returned.

Fifteen minutes later I came to my first satellite dish, perched jauntily on a flattened oil barrel like a hat feather. With it came João Fernandez, a gregarious stonemason who was more than happy to talk with me about his way of life, his family, and his satellite dish. He invited me into his home, introduced me to his wife and four young children, offered me a beer from a bucket of cold water buried in the dirt floor of the shack, and began a most remarkable story that I will never forget.

"Here in Brazil there is a class system. We," he motioned around himself, taking in the shacks that surrounded his, "are at the bottom of the system. However, we have skills that people need. I am a stonemason; my neighbor over there is a plumber; the man behind me is a very skilled gardener. We need to work, and the upper classes need what we do." He stopped there, as if that explained away my perplexity. "It seems to me," I began, "that a satellite dish for television is an expensive luxury for you. You have told me yourself that you struggle to feed your family; how can you justify the cost of a satellite dish?" He looked at me blankly and then smiled. "Actually," he responded, "the dish isn't for television. It's for Internet access. And I don't pay for it alone; six families forego meals one day a week to be able to afford the connection." He pointed to a far corner of the shack. There, perched on a plastic crate, was an ancient personal computer. I was puzzled: Internet access? Well, *that* made a lot more sense, I thought sarcastically. He seemed to sense my confusion. "You see, in Brazil, there is a huge online jobs market. However, there is also a very deeply held sense of class that is unavoidable. If a potential employer were to know where I live, he would not hire me because of my social . . . condition. So, we create an online presence for ourselves at a special Web site, advertise our skills, and sell ourselves on the merits of our capabilities. 'Where I live' never enters the conversation. I simply agree to meet the employer at the jobsite." The little boy with the jumper cables, I later learned, was stealing electricity to power a stepdown transformer that provided juice to one whole area of the town. All this because of the Internet.

Several years later, I gave a 2-hour talk via satellite to 60 people in four countries in Africa about the impact of the Internet. Somewhere along the way we discovered that the 60 people had somehow become 2,500 participants, and instead of four countries we were broadcasting to 16. The 2-hour talk somehow became a 7-hour marathon that I conducted in English, French, and Swahili. In fact, it ended only because Comsat called and told us they needed their satellite back for another broadcast.

One year later I received the following e-mail from one of the participants:

> Dear Professor [sic] Shepard: I wish to thank you from the bottom of my heart for helping me start my Internet business in Kenya. It was not easy: I had to take out a large business loan for $100, but I believe that I will be able to pay it back within 18 months. Already I have made sales of my magazine in the

United States and Europe and because of your help my family is eating well and my children are smiling. God bless you and the Internet.

Asante Sana, rafiki wako William

It doesn't get any better than that, although in this case it did. That evening I left the broadcast facility in Washington and caught a cab to take me back to National Airport. I could tell from the driver's singsong accent that he was West African, so I asked him where he was from. "Abidjan, in Ivory Coast," he replied. I smiled and told him what I had been doing all day and that there had been quite a few people from the Ivory Coast in the broadcast. I also asked him how long he had been driving a cab and how long he had lived in the States. "Actually, I don't drive a cab," he replied. "I also don't live in Washington—at least not really." I observed that he *was* in Washington and *was* driving a cab, to which he laughed and replied, "This is my brother's car. His children are sick so I agreed to drive for him today because I have a day off." Nodding my understanding, I asked him where he worked. "CompUSA—have you heard of it? I build custom computers for customers. That's not really my main job, though."

By now I knew that this story was only going to get better. As we crossed the Arlington Memorial Bridge and passed the Pentagon on our way to the airport, he elaborated. "In my country, we have a fairly good telecommunications network, but it is only in the largest cities—Abidjan, Bouaké, Yamoussoukro, and then only for a few people who can afford the cost and are willing to wait the long time it takes to get the service—it can be years. There are many people who want it, however. So I started a business in Abidjan. You have heard of Kinko's, yes? Well, I am the Kinko's of Côte d'Ivoire. I rented a building, then went to CI-TELCOM (the local PTT), and asked them to install two E1 circuits in the building. It took me over a year to get them, but now I am the place where people come to make telephone calls, use computers, and surf the Web. Since the telephone company could not take the network to the people, I bring the people to the network. I have now three buildings, one in each of the largest cities. And I will build more."

I, of course, was speechless. Côte d'Ivoire has 14 million people, of which only 10,000 are Internet users on the country's 65,000 PCs. "So why are you in the States?" I asked him. "To ensure that my children get the best possible education. During the summer we live in Abidjan and run the business. But during the school year, we live in Washington so that my two children can go to school here. I want them to get an American education in an American university. And when I'm gone from Côte d'Ivoire, my family runs the businesses."

All of these experiences—the family in Brazil, the broadcast to Africa, the cab driver from Ivory Coast—caused something deep and profound to dawn on me. The message that was burning itself into my brain was this: If you aren't connected, you can't play. Years before I had laughed at the stories I heard from eastern Europe about the sad state of affairs of the telephone networks there. How the most popular birth gift in Romania was a service order for a phone to be installed, because it took 18 years from the point of order to the day of installation. How a burgeoning business had flourished in Mexico City, the principal employees dressed themselves in black like techno-ninjas, climbed apartment buildings in the deep darkness of moonless Mexican nights, and cut drop cables that they would then swing to the apartment of the highest bidder, knowing that the local telephone company's records were so inaccurate that they would never be able to determine the real owner of any cable pair. I heard unbelievable stories of West German corporations paying employees to come into the office at 3 AM to place calls to their East German counterparts so that they could ensure connectivity between the two offices—insurance against the fact that there were only about 20 circuits connecting West to East before the fall of the Berlin Wall, many of them nonfunctional. And Africa, achingly poor with 12 percent of the world's population and less than 1 percent of the world's telephones—a technology tragedy.

Today, of course, little of this is germane. Analysts predict that within the next 5 years Mexico will have one of the most advanced networks in the world. Romania and other "greenfield markets" have paid careful attention to the mistakes made by western industrialized nations with advanced telecommunications infrastructures, noted the things done well and poorly, and carefully built their own evolving networks around the success stories, avoiding the technology failures. Germany's reunification effort has been nothing short of miraculous; the network in the eastern part of Germany is as capable as that of the West. This is remarkable only when one considers that, prior to reunification, the last time the network in the eastern portion of the country was upgraded was 1921. And Africa, while still suffering from a paucity of service, now glows with the certainty of progress made possible by the shiny new optical rings and wireless infrastructures that surround the continent. Stanley's Dark Continent is about to be lit.

The volume of investment dollars flowing into these regions of the world is immense and is not slowing down. Why are these countries spending so much on telecommunications infrastructure? The answer is actually quite simple: *If you aren't connected, you can't play in the game, and you will* not *have a place on the world's economic stage.*

Telecommunications technology has become the interstitial tissue, the technological fabric, the nervous system that interconnects and holds the world together. In many ways, the telecommunications industry does not really exist—it is actually a part of every other industry, and every other industry is a part of it. It is inextricably intertwined with the heart and soul of every business on Earth. I defy you to show me a business that is not dependent upon connectivity for its very lifeblood. Show me a company that is not connected and I'll show you a company that is not long for this world.

Equally important is the critical role that information and knowledge play in today's business world. As Walter Wriston, the former chairman of Citibank observed, "The pursuit of wealth is now largely the pursuit of information and its application to the means of production. The rules, customs, skills, and talents necessary to uncover, capture, produce, preserve, and exploit information are now humankind's most important assets . . . . In the next few decades the attraction and management of intellectual capital will determine which institutions and nations will survive and prosper, and which will not."

In *The Lexus and the Olive Tree,* author Thomas Friedman says the following about a pervasive disease that is increasingly afflicting our world:

> [Microchip Immune Deficiency Syndrome (MIDS) is] a disease that can affect any bloated, overweight, sclerotic system in the post-Cold War era . . . . The symptoms of MIDS appear when a country or company exhibit a consistent inability to increase productivity, wages, living standards, knowledge use, and competitiveness and become too slow to respond to the challenges of the fast world. Countries and companies with MIDS tend to be those run on Cold War corporate models—where one or more people at the top hold all the information and make all the decisions, and all the people in the middle and the bottom simply carry out those decisions, using only the information they need to do their jobs. The only known cure for countries and companies with MIDS is "the fourth democratization." This is the democratization of decision making and information flows and the deconcentration of power, in ways that allow more people in a country or company to share knowledge and innovate faster. This enables them to keep up with a marketplace in which consumers are constantly demanding cheaper products and services tailored specifically for them. MIDS can be fatal for those companies and countries that do not get appropriate treatment in time.

His seminal book is a must-read for anyone interested in the inevitable impacts of globalization, particularly those derived from technology and information. In the book, Friedman discusses the concept of three levels of democratization, the third of which is described earlier. The other two are technology and finance. Democratization, as he describes it, is the free sharing of some capability or resource, in this case information, facilitative technologies, and financial resources and capabilities. There is no question that these three factors figure prominently in the development of nations and companies; together they represent a powerful force in the inexorable balance of global wealth and the development of nations and their people.

Why You Need to Read This Book

Anyone who strolls down the aisles of a local bookstore or the endless logical halls of the Amazon.coms of the world knows that booksellers now have a special place for technical books in their stores. There have always been *computer* books in bookstores and a few titles on such warm, rich topics as COBOL, UNIX™, and database management. Five years ago, however, there were *no* books on networking and precious few on applications. Today, they make up the bulk of the technical book section and cover a dizzying array of arcane topics. And they don't even look like technical books anymore: Their covers sport cartoon characters, strange-looking Australian lizards, and lots of color. Furthermore, they cover the technical details of the immense spectrum of communications technologies down to spectacularly granular detail—as my friend Gary Kessler likes to say, "including the first names of every bit in the bitstream."

That being said, I must also say this: This book is different. Yes, it covers a wide array of technologies to a reasonable depth. More importantly, however, the book focuses on the *implications* of the technologies—why they are important, whom they have an impact upon, and how they affect such critical factors as corporate competitive positioning, financial worth, globalization, and economic progress. The book is designed to not only help readers understand what the technologies are, how they work, where they fit in the network puzzle, and how they integrate with one another, but also how to choose among them, how to implement them as tools toward successful competitive advantage, and how to turn them from whizbang technologies into moneymaking applications.

Since this book was first published, it has been brought to my attention that it is widely used as a textbook in colleges and universities throughout the world, for which I am honored. I receive hundreds of e-mail messages every year from professors and instructors asking for advice on how best to teach the material in the book for maximum impact, so to help with that process I have added a list of questions at the end of each chapter. I hope you find them useful.

In short, this book is designed to get you started and give you a bird's-eye view of the most exciting industry on Earth today. I hope you have as much fun reading it as I have had writing it. As always, I welcome your comments and corrections. Enjoy!

Steven Shepard
Madrid, Johannesburg, Rimouski, Vancouver, Calgary, Williston
April 2005

First Things First

Telecommunications, like all highly visible and interesting fields, is full of apocryphal stories, technical myths, and fascinating legends. Everyone in the field seems to know someone who knows the outside-plant repair-person who found the poisonous snake in the equipment box in the man-hole,[1] or the person who was on the cable-laying ship when it pulled up the cable that had been bitten through by some species of deep water shark, or a collection of seriously evil hackers, or the backhoe driver who cut the cable that put Los Angeles off the air for 12 hours.

There is also a collection of technojargon that pervades the telecommunications industry and often gets in the way of the relatively straightforward task of learning how all this stuff actually works. To ensure that such things don't get in the way of absorbing what's in this book, I'd like to begin with a discussion of some of them. I am a writer, after all; words are important to me.

This is a book about telecommunications, which is the science of communicating over distance ("tele-", from the Greek *tēle*, meaning "far off"). It is, however, fundamentally dependent upon *data communications*, the science of moving traffic between computing devices so that this traffic can be manipulated in some way to make it useful. Data, in and of itself, is not particularly useful, consisting as it does of a stream of ones and zeroes that is meaningful only to the computing device that will receive and manipulate those ones and zeroes. The data does not really become useful until it is converted by some application into *information*, because a human can generally understand information. The human then acts upon the information using a series of intuitive processes that further convert the information into *knowledge*; at this point it becomes truly useful. Here's an example: A computer generates a steady stream of ones and zeroes in response to a series of business activities involving the computer that generates the ones and zeroes. Those ones and zeroes are fed into another computer, where an application converts them into a spreadsheet of sales figures (information) for the store from which they originated. A financial analyst studies the spreadsheet, calculates a few ratios, examines some historical data (including not only sales numbers but demographics, weather patterns, and political trends), and makes an informed prediction about future stocking requirements and advertising focal points for the store based on the knowledge that the analyst was able to create from the distilled information. That's knowledge. To take it one step further, once decisions have been made and the results of those

---

[1] I realize that this term has fallen out of favor, but I use it here for historical accuracy.

decisions have been seen, *wisdom* allows decision makers to learn from their decisions. So they can make them again if the results are positive, or choose differently if they're not. Data, information, knowledge, and wisdom—a critical continuum.

Data communications relies on a carefully designed set of rules that governs the manner in which computers exchange data. These rules are called *protocols*, and they are centrally important to the study of data communications. Dictionaries define "protocol" as "a code of correct conduct." From the perspective of data communications, they define it as "a standard procedure for regulating the transmission of data between computers," which is itself "a code of correct conduct." These protocols, which will be discussed in detail later in this book, provide a widely accepted methodology for everything from the pin assignments on physical connectors to the sublime encoding techniques used in secure transmission systems. Simply put, they represent the many rule sets that govern the game. Many countries play football, for example, but the rules are all slightly different. In the United States, players are required to weigh more than a car, yet be able to run faster than one. In Australian rules football, the game is declared forfeit if it fails to produce at least one body part amputation on the field or if at least one player doesn't eat another. They are both football, however. In data communications, the problem is similar: Many protocols out there accomplish the same thing. Data, for example, can be transmitted from one side of the world to the other in a variety of ways including T1, E1, microwave, optical fiber, satellite, coaxial cable, and even through the water. The end result is identical—the data arrives at its intended destination. Different protocols, however, govern the process in each case.

A discussion of protocols would be incomplete without a simultaneous discussion of *standards*. If protocols are the various sets of rules by which the game is played, standards govern which set of rules will be applied for a particular game. For example, let's assume that we need to move traffic between a PC and a printer. We agree that in order for the PC to be able to transmit a printable file to the printer, both sides must agree on a common representation for the zeroes and ones that make up the transmitted data. They agree, for example (and this is *only* an example), that they will both rely on a protocol that represents a zero as the absence of voltage and a one as the presence of a 3-volt pulse on the line, as shown in Figure 1-1. Since they agree on the representation, the printer knows when the PC is sending a one and when the PC is sending a zero. Imagine what would happen if they failed to agree on such a simple thing beforehand. If the transmitting PC decides to represent a one

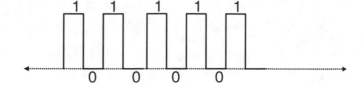

Figure 1-1
Voltage representations of data

as a 300-volt pulse, and the printer expects a 3-volt pulse, the two devices will have a brief (but exciting!) conversation, the ultimate result of which will be the release of a small puff of silicon smoke from the printer.

Now they have to decide on a standard that they will use for actually originating and terminating the data that they will exchange. They are connected by a cable (Figure 1-2) that has nine pins on one end and nine jacks on the other. Logically, the internal wiring of the cable would look like Figure 1-3. However, when we stop to think about it, this one-to-one correspondence of pin-to-socket will not work. If the PC transmits on pin 2, which in our example is identified as the *send data lead*, it will arrive at the printer on pin 2—the send data lead. This would be analogous to holding two telephone handsets together so that two communicating parties can talk. It won't work without a great deal of hollering, since the handsets are oriented microphone-to-microphone and speaker-to-speaker! Instead, some agreement has to be forged to ensure that the traffic placed on the *send data lead* somehow arrives on the *receive data lead*, and vice versa. Similarly, the other leads must be able to convey information to the other end so that normal transmission can be started and stopped. For example, if the printer is ready to receive the print file, it might put voltage on the *data terminal ready* (DTR) *lead*, which signals to the PC that it is ready to receive traffic. The PC might respond by setting its own DTR lead high, as a form of acknowledgement, and then transmitting the file that is to be printed. The printer will keep its DTR lead high until it wants the PC to stop sending. For example, if the printer senses that it is running out of buffer space because the PC is transmitting faster than the slower printer can print, it will drop the DTR lead, causing the PC to temporarily halt its transmission of the print file. As soon as the printer is ready to receive again, it sets the DTR lead high, and printing resumes. As long as both the transmitter and the receiver abide by this standard set of rules, data communications will work properly. This process of swapping the data on the various leads of a cable, incidentally, is done by the modem—or by a *null modem cable*

Figure 1-2
Pin
assignments
on a cable
connector

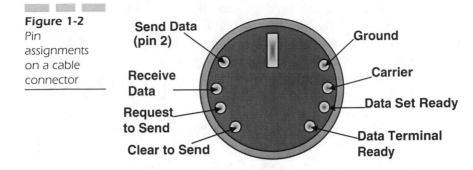

Send Data
(pin 2)

Receive
Data

Request
to Send

Clear to Send

Ground

Carrier

Data Set Ready

Data Terminal
Ready

Figure 1-3
Logical wiring
scheme

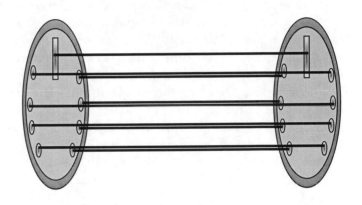

that makes the communicating devices think they are talking to a modem. The null modem cable is wired so that the send data lead on one end is connected to the receive data lead on the other end, and vice-versa. Similarly, a number of control leads—such as the *carrier detect lead*, the DTR lead, and the *data set ready* (DSR) *lead*—are wired together so that they give false indications to each other to indicate that they are ready to proceed with a transmission, when in fact no response from the far end modem has been received.

Standards: Where Do They Come From?

Physicists, electrical engineers, and computer scientists generally design data communications protocols. For example, the Transmission Control

Protocol (TCP) and the Internet Protocol (IP), discussed in detail later in the book, were written during the heady days of the Internet, back in the 1960s, by such early pioneers as Vinton Cerf and the late John Postel (I want to say, "back in the last century," to make them seem like *real* pioneers!). Standards, on the other hand, are created as the result of a consensus-building process that can take years to complete. By design, standards must meet the requirements of the entire data and telecommunications industry, which is of course global. It makes sense, therefore, that some international body be responsible for overseeing the creation of international standards. One such body is the United Nations (UN). Its 150+ member nations work together in an attempt to harmonize whatever differences they have at various levels of interaction, one of which is international telecommunications. The International Telecommunications Union (ITU), a suborganization of the UN, is responsible for not only coordinating the creation of worldwide standards but also publishing them under the auspices of its *own* suborganizations. These include the Telecommunications Standardization Sector (TSS, sometimes called the ITU-T, and formerly called the Consultative Committee on International Telegraphy and Telephony, or the CCITT), the Telecommunications Development Sector (TDS), and the Radio Communication Sector (RCS, formerly the Consultative Committee on International Radio, or the CCIR). The organizational structure is shown in Figure 1-4.

Of course, the UN and its suborganizations cannot perform this task alone, nor should they. Instead, they rely upon the input of hundreds of industry-specific organizations as well as local, regional, national, and

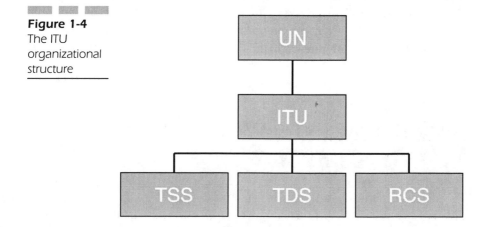

Figure 1-4
The ITU
organizational
structure

international standards bodies that feed information, perspectives, observations, and technical direction to the ITU, which serves as the coordination entity for the overall international standards creation process. These include the American National Standards Institute (ANSI), the European Telecommunications Standards Institute (ETSI, formerly the Conference on European Post and Telegraph, or CEPT), Telcordia (formerly Bellcore, now part of Science Applications International Corporation [SAIC] but as of this writing, about to be sold yet again), the International Electrotechnical Commission (IEC), the European Computer Manufacturers Association (ECMA), and a host of others.

It is worthwhile to mention a bit about the ITU as a representative standards body. Founded in 1947 as part of the United Nations, it descended from a much older body called the Union Telegraphique, founded in 1865 and chartered to develop standards for the emerging telegraph industry. Over the years since its creation, the ITU and its three principal member bodies have developed three principal goals:

- To maintain and extend international cooperation for the improvement and interconnectivity of equipment and systems through the establishment of technical standards
- To promote the development of the technical and natural facilities (read spectrum) for most efficient applications
- To harmonize the actions of national standards bodies to attain these common aims, and most especially to encourage the development of communications facilities in developing countries

The Telecommunications Standardization Sector (TSS)

The goals of the TSS (or ITU-T), according to the ITU, are as follows:

- To fulfill the purposes of the Union relating to telecommunication standardization by studying technical, operating, and tariff questions and adopting formal recommendations on them with a view to standardizing telecommunications on a world-wide basis.
- To maintain and strengthen its pre-eminence in international standardization by developing Recommendations rapidly.

- To develop Recommendations that acknowledge market- and trade-related consideration.

- To play a leading role in the promotion of cooperation among international and regional standardization organizations and forums and consortia concerned with telecommunications.

- To address critical issues that relate to changes due to competition, tariff principles, and accounting practices.

- To develop Recommendations for new technologies and applications such as appropriate aspects of the GII [Global Information Infrastructure] and Global Multimedia and Mobility.

The Telecommunications Standardization Bureau (TSB)

The Telecommunication Standardization Bureau (TSB) provides secretarial support for the work of the ITU-T Sector and services for the participants in ITU-T work, diffuses information on international telecommunications worldwide, and establishes agreements with many international standards development organizations. These functions include the following:

- *Study group management*: The management team of the study groups is composed of the chairman, vice-chairmen of the study group, chairmen of the working parties and the TSB counselor/engineer.

- *Secretarial support and meeting organization*: TSB provides secretarial services for ITU-T assemblies and study group meetings. TSB counselors and engineers coordinate the work of their study group meeting, and their assistants ensure the flow of meeting document production.

- *Logistics services*: The TSB provides services such as meeting room allocation, registration of participants, document distribution, and facilities for meeting participants.

- *Approval of recommendations and other texts*: The TSB organizes and coordinates the approval process of recommendations.

- *Access to ITU-T documents for ITU-T members*: The TSB organizes and controls the dispatch of documents in paper form to participants in ITU-T work and provides electronic document handling services (EDH) that allow easy and rapid exchange of documents, information, and ideas among ITU-T participants in order to facilitate standards development. The ITU-T participants can have electronic access, via Telecom Information Exchange Services (TIES), to study group documents such as reports, contributions, delayed contributions, temporary and liaison documents, and so on.

The TSB also provides the following services:

- Maintenance of the ITU-T Web site and distribution of information about the activities of the Sector including the schedule of meetings, TSB Circulars, Collective Letters, and all working documents
- Update of services for the List of ITU-T Recommendations, the ITU-T Work Program Database, the ITU-T Patent Statements Database, and the ITU-T Terms and Definitions Database (SANCHO), as well as update services for other databases as required
- Country code number assignment for telephone, data, and other services
- Registrar services for universal international freephone numbers (UIFN)
- Technical information on international telecommunications and close collaboration with the ITU Radiocommunication Sector and the ITU Telecommunication Development Sector for matters of interest to developing countries
- Administrative and operational information through the *ITU Operational Bulletin*
- Coordination of the editing, publication, and posting of the recommendations

The Radio Bureau

The functions of the Radio Bureau include:

- Administrative and technical support to radiocommunication conferences, radiocommunication assemblies, and study groups, including working parties and task groups
- Application of the provisions of the radio regulations and various regional agreements

- Recording and registration of frequency assignments and also orbital characteristics of space services, and maintenance of the Master International Frequency Register

- Consulting services to member states on the equitable, effective, and economical use of the radio-frequency spectrum and satellite orbits, and investigation and assistance in resolving cases of harmful interference

- Preparation, editing, and dispatch of circulars, documents, and publications developed within the Sector

- Delivery of technical information and seminars on national frequency management and radiocommunications, and close work with the Telecommunication Development Bureau to assist developing countries

The Standards

A word about the publications of the ITU. First of all, the "standards" are referred to as *recommendations* because the ITU has no enforcement authority over the member nations that use them. Its strongest influence is exactly that—the ability to *influence* its member telecommunications authorities to use the standards because it makes sense to do so on a global basis.

The standards are published every four years, following enormous effort on the part of the representatives that sit on the organization's task forces. These representatives hail from all corners of the industry; most countries designate their national telecommunications company (where they still exist) as their representative to the ITU-T, while others designate an alternate, known as a recognized private operating agency (RPOA). The United States, for example, has designated the Department of State as its duly elected representative body. Other representatives may include manufacturers (Lucent, Cisco, Nortel, Fujitsu), research and development organizations (Bell Northern Research, Bell Laboratories, Xerox PARC), and other international standards bodies.

The efforts of these organizations, companies, governments, and individuals result in the creation of a collection of new and revised standards recommendations published on a four-year cycle. Historically, the standards were color-coded, published in a series of large-format, soft-cover books, differently colored on a rotating basis. For example, the 1984 books were red; the 1988 books, blue; the 1992 books, white. It is common to hear network people talking about "going to the blue book." They are referring (typically) to the generic standards published by the ITU for that particular year. It is also common to hear people talk about the CCITT. Old habits die hard: The organization ceased to exist in the early 1990s, replaced by the ITU-T. The previous name is still used, however.

The activities of the ITU-T are parceled out according to a cleverly constructed division of labor. Three efforts result: (1) study groups create the actual recommendations for telecomm equipment, systems, networks, and services (there are currently 16 study groups); (2) plan committees develop plans for the intelligent deployment and evolution of networks and network services; and (3) specialized autonomous groups (currently three) produce resources that support the efforts of developing nations. The study groups and their responsibilities are as follows:

- SG 2—Operational aspects of service provision, networks and performance
- SG 3—Tariff and accounting principles including related telecommunications economic and policy issues
- SG 4—Telecommunication management, including Telecommunications Managed Network (TMN)
- SG 5—Protection against electromagnetic environment effects
- SG 6—Outside plant
- SG 7—Data networks and open system communications
- SG 9—Integrated broadband cable networks, and television and sound transmission
- SG 10—Languages and general software aspects for telecommunication systems
- SG 11—Signaling requirements and protocols
- SG 12—End-to-end transmission performance of networks and terminals
- SG 13—Multiprotocol and IP-based networks and their internetworking
- SG 15—Optical and other transport networks
- SG 16—Multimedia services, systems, and terminals
- SG 17—Security, languages, and telecommunication software
- SG 18—Mobile telecommunication networks
- SSG—Special Study Group, IMT-2000 and Beyond

The standards are published in a series of alphabetically arranged documents and are available as books, online resources, and CDs. They are functionally arranged according to the alphabetical designator of the standard as follows:

A—Organization of the work of ITU-T

B—Means of expression: definitions, symbols, and classification

C—General telecommunication statistics

D—General tariff principles

E—Overall network operation, telephone service, service operation, and human factors

F—Nontelephone telecommunication services

G—Transmission systems and media, digital systems, and networks

H—Audiovisual and multimedia systems

I—Integrated services digital network

J—Transmission of television, sound program, and other multimedia signals

K—Protection against interference

L—Construction, installation, and protection of cables and other elements of outside plant

M—TMN and network maintenance: international transmission systems, telephone circuits, telegraphy, facsimile, and leased circuits

N—Maintenance: international sound program and television transmission circuits

O—Specifications of measuring equipment

P—Telephone transmission quality, telephone installations, and local line networks

Q—Switching and signaling

R—Telegraph transmission

S—Telegraph services terminal equipment

T—Terminals for telematic services

U—Telegraph switching

V—Data communication over the telephone network

X—Data networks and open system communication

Y—Global information infrastructure and Internet protocol aspects

Z—Languages and general software aspects for telecommunication
systems

Within each letter designator can be found specific, numbered recommendations. For example, recommendation number 25 in the "X" book
contains the specifications for transporting packet-based data across a
public network operating in packet mode. This, of course, is the now
famous X.25 packet-switching standard. Similarly, Q.931 provides the
standard for signaling in integrated services digital networks (ISDN)
networks, and so on. The documents are remarkably easy to read and
contain vast amounts of information. I am always surprised to discover
how many people who work in telecommunications have never read the
ITU standards. Take this, then, as *my* recommendation: Find some of
them and flip through them. They can be very useful.

I remember very well an experience I had years ago with the ITU
standards. My family was visiting friends in California, but because of
work commitments I was in Vermont preparing to travel. It was a dark,
snowy January night and I was sitting in front of a roaring fire, drinking
a warm brandy and reading Q.931. The worst part was that I was really
into it, enjoying the fact that I was understanding the technojargon. And
then it hit me: I had become . . . *one of them*. I was a propellerhead.

I spent some time writing about the ITU and its standards activities
simply to explain the vagaries of the process (one of my favorite
telecomm jokes goes like this: "There are two things you never want to
watch being made. One of them is sausage; the other is standards.") and
the role of these bodies. The ITU is representative of the manner in
which all standards are developed, although the frequency of update, the
cycle time, the relative levels of involvement of the various players, and
the breadth of coverage of the documents vary dramatically.

Other Important Organizations

In addition to the formal standards bodies, some industry groups are
worth paying attention to. Some of them, like the United States Telecom
Association (USTA) and the National Telecommunications and Information Administration (NTIA), are trade organizations that promote the

efficient and effective operation of the industry. Others, like the Cellular Telecommunications and Internet Association (CTIA), focus on specific sectors of the marketplace. Still others, like SuperComm, are industry groups that sponsor annual trade shows where vendors gather to display their wares. These shows are worth attending, particularly if you are running low on keychains, T-shirts, and pens (only kidding).

There are also technology specific groups that focus specifically on optical, wireless, component, switching, ATM, and so on. All are reachable online; depending on the nature of your interest in the industry, they are worth contacting.

The Network

For years now, communications networks have been functionally depicted as shown in Figure 1-5: a big, fluffy, opaque cloud into which disappear lines representing circuits that magically reappear on the other side of the cloud. I'm not sure why we use clouds to represent networks; knowing what I know about their complex innards and how they work, a hairball would be a far more accurate representation.

In truth, clouds are pretty good representations of networks from the point of view of the customers who use them. Internally, networks are remarkably complex assemblages of hardware and software, as you will see in the chapter on telephony. Functionally, however, they are straightforward: Customer traffic goes into the network on the *Gozinta;* the traf-

Figure 1-5
The network
cloud

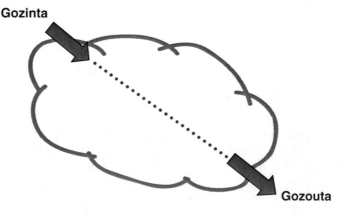

Gozinta

Gozouta

fic then emerges, unchanged, on the *Gozouta*. How it happens is unimportant to the customer; all he or she cares about—or *wants* to care about—is that the network receives, interprets, transports, and delivers his or her voice/video/images/data/music to the destination in a correct, timely, and cost-effective fashion. Later in the book we will discuss the various technologies that live within the network, but for now suffice it to say that its responsibilities fall into two categories: access and transport, as illustrated in Figure 1-6.

Network Access

As the illustration shows, network access is exactly that: the collection of technologies that support connectivity between the customer and the transport resources of the network. At its most common level, access is the local loop—the (typically) two-wire circuit that connects a customer's telephone to the local switch, which provides telephony service to that customer. As the network has become more data-aware, other solutions have emerged that provide greater bandwidth as well as multiservice capability. ISDN, which uses the two-wire local loop, provides greater bandwidth than the traditional analog local loop through digitization and time-division multiplexing (both explained shortly). Digital subscriber line (DSL) is also a local loop-based service, but offers even more diverse service than ISDN in the areas where it is available. Cable

Figure 1-6
Access vs.
transport
regions of the
network

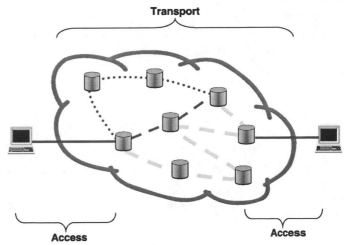

modem service, which does *not* use the telephony local loop, offers high downstream (toward the customer) bandwidth and smaller upstream (from the customer) capacity. Wireless services, including Wi-Fi, WiMAX, LMDS, MMDS, satellite, cellular, microwave, and others represent alternative options for access connectivity. All of these will be discussed in greater detail later in the book.

Miscellaneous Additional Terms   A number of other terms need to be introduced here as well, beginning with data terminal equipment (DTE) and data circuit terminating equipment (DCE). DTE is exactly that—it is the device that a user employs to gain access to the network. A DCE is the device that actually terminates the circuit at the customer's premises, typically a modem. The relationship between these network elements is shown in Figure 1-7. One important point: Because the bulk of the usage is over the public switched telephone network (PSTN), which is optimized for the transport of voice, the primary role of the DCE is to make the customer's DTE look, smell, taste, and feel like a telephone to the network. For example, if the DTE is a PC, then the modem's job is to collect the high-frequency digital signals being produced by the PC and modulate them into a range of frequencies that are acceptable to the bandwidth-limited voiceband of the telephone network. That's where the name comes from, incidentally—modulate/demodulate (MO-DEM).

Another pair of terms that must be introduced here is *parallel* and *serial*. You have undoubtedly seen the ribbon cables that are used to transport data inside a PC (Figure 1-8) or the parallel wires etched into the motherboard inside the PC (Figure 1-9). These parallel conductors are called a *bus*, and are used for the high-speed transport of multiple simultaneous bits in parallel fashion from one device inside the computer to another. Serial transmission, on the other hand, is used for the "single-file" transport of multiple bits, one after the other, usually deployed *outside* a computer.

Figure 1-7
The relationship between the DTE and DCE in a typical network

DTE   DCE
(Computer) (Modem)

DCE   DTE
(Modem) (Computer)

Figure 1-8
The parallel
bus ribbon
cable in a PC

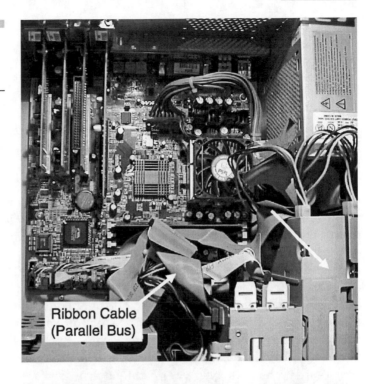

Ribbon Cable
(Parallel Bus)

Figure 1-9
The parallel
bus, this time
etched (inside
the circle) on
the mother-
board of the
PC

Finally, we offer *simplex, half-duplex,* and *full-duplex transmission.*
Simplex transmission means "one-way only," like a radio broadcast. Half-
duplex transmission means "two-way, but only one way at a time," like

CB radio. Finally, full-duplex means "two-way simultaneous transmission," like telephony or two-way data transmission.

Network Transport

The fabric of the network cloud is a rich and unbelievably complex collection of hardware and software that moves customer traffic from an ingress point to an egress point, essentially anywhere in the world. It's a function that we take entirely for granted because it is so ingrained in day-to-day life. But stop for a moment to think about what the network actually does. Not only does it deliver voice and data traffic between end points, but it does so easily and seamlessly, with various levels of service quality, as required, to any point on the globe (and in fact beyond!) in a matter of seconds—and with *zero* human involvement. It is the largest fully automated machine on the planet and represents one of the greatest technological accomplishments of all time. Think about that: I can pick up a handset here in Vermont, dial a handful of numbers, and seconds later a telephone rings in Ouagadougou, Burkina Faso, in North Central Africa. How that happens borders on the miraculous. We will explore it in considerably greater detail later in the book.

Transport technologies within the network cloud fall into two categories: fixed transport and switched transport. *Fixed transport*, sometimes called private line or dedicated facilities, includes such technologies as T1, E1, DS3, SONET, SDH, dedicated optical channels, and microwave. *Switched transport technologies* include modem-based telephone transport, X.25 packet switching, frame relay, switched Ethernet, and asynchronous transfer mode (ATM). Together with the access technologies described previously and customer premises technologies such as Ethernet, transport technologies offer the infrastructure components required to craft an end-to-end solution for the transport of customer information.

The Many Flavors of Transport

Over the last few years the network has been functionally segmented into a collection of loosely defined "regions" that define unique service types. These include the local area (sometimes referred to as the premises region), the metropolitan area, and the wide area (sometimes known as the core). *Local area networking* has historically defined a network that provides services to an office, a building, or even a campus. *Metro networks* generally provide connectivity within a city, particularly to multiple physical locations of the same company. They are usually deployed across a ring architecture. *Wide area networks*, often called core transport, provide long-distance transport and are typically deployed using a mesh networking model.

Transport Channels

The physical circuit over which customer traffic is transported in a network is often referred to as a *facility*. Facilities are characterized by a number of qualities such as distance, quality (signal level, noise, and distortion coefficients), and bandwidth. *Distance* is an important criterion because it places certain design limitations on the network, making it more expensive as the circuit length increases. Over distance, signals tend to weaken and become noisy, and specialized devices (amplifiers, repeaters, regenerators) are required to periodically clean up the signal quality and maintain the proper level of "loudness" to ensure intelligibility and recognizability at the receiving end.

Quality is related to distance in the sense that they share many of the same affecting factors. Signal level is clearly important, as is noise, both of which were just discussed. Distortion is a slightly different beast and must be dealt with equally as carefully. Noise is a random event in networks caused by lightning, fluorescent lights, electric motors, sunspot activity, and squirrels chewing on wires; it is unpredictable and largely random. Noise, therefore, cannot be anticipated with any degree of accuracy; its effects can only be recovered from.

Distortion, on the other hand, is a measurable, predictable characteristic of a transmission channel and is usually frequency-dependent. For example, certain frequencies transmitted over a particular channel will be weakened, or attenuated, more than other frequencies. If we can measure this, then we can condition the channel to equalize the treatment that all frequencies receive as they are transmitted down that channel. This process is indeed known as *conditioning,* and is part of the higher cost involved in buying a dedicated circuit for data transmission. For example, think about the last time you attended a parade. How did you know the band was approaching? You heard the low-frequency tones of the bass drums and tubas long before you saw the band—or you heard the higher frequencies of the trumpets and trombones. Higher frequency signals weaken faster than lower frequencies and, therefore, don't travel as far without amplification and regeneration.

Bandwidth is the last characteristic that we will discuss here, and the quest for more of it is one of the great challenges of telecommunications. Bandwidth is a measure of the number of bits that can be transmitted down a facility in any one-second period. In most cases it is a fixed characteristic of the facility and is the characteristic that most customers pay for. The measure of bandwidth is bits-per-second, although today the measure is more typically thousands (kilobits), millions (megabits), or billions (gigabits) per second.

Facilities are often called *channels* because physical facilities are often used to carry multiple streams of user data through a process called *multiplexing*. Multiplexing is the process of allowing multiple users to share access to a transport facility, either by taking turns or using separate frequencies within the channel. If the users take turns, as shown in Figure 1-10, the multiplexing process is known as *time-division multiplexing* (TDM), because time is the variable that determines when each user gets to transmit through the channel. If the users share the channel by occupying different frequencies, as shown in Figure 1-11, the process is called *frequency-division multiplexing* (FDM), because frequency is the variable that determines who can use the channel. It is often said that in TDM, users of the facility are given *all* of the frequency *some* of the time, because they are the only one using the channel during their timeslot. In FDM, users are given *some* of the frequency *all* of the time, because they are the only one using their particular frequency "band" at any point.

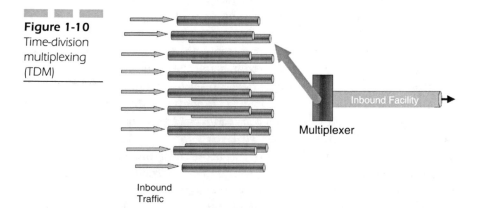

Figure 1-10
Time-division
multiplexing
(TDM)

Inbound
Traffic

Multiplexer

Inbound Facility

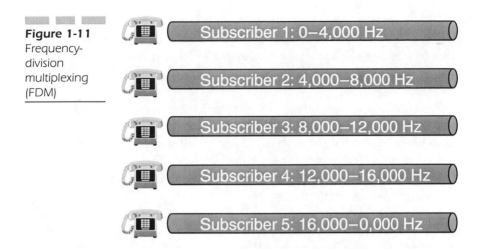

Figure 1-11
Frequency-division multiplexing (FDM)

Analog versus Digital Signaling: Dispensing with Myths

FDM is normally considered to be an *analog technology,* while time-division multiplexing is a *digital technology.* The word "analog" means "something that bears a similarity to something else," while the word "digital" means "discrete." Analog data, for example, typically illustrated as some form of sine wave such as that shown in Figure 1-12, is an exact representation of the values of the data being transmitted. The process of using manipulable characteristics of a signal to represent data is called *signaling.*

Figure 1-12
Sine wave

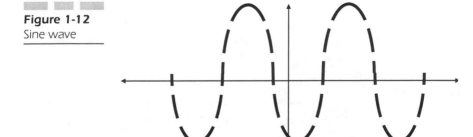

We should also introduce a few terms here just to keep things marginally confusing. When speaking of signaling, the proper term for digital is *baseband,* while the term for analog signaling is *broadband.* When talking about data (not signaling), the term "broadband" means "big channel."

The sine wave, undulating along in real time in response to changes in one or more parameters that control its shape, represents the exact value of each of those parameters at any point in time. The parameters are amplitude, frequency, and phase. We will discuss each in turn. Before we do, though, let's relate analog waves to the geometry of a circle. Trust me—this helps.

Consider the diagram shown in Figure 1-13. As the circle rolls along the flat surface, the dot will trace the shape shown by the dotted line. This shape is called a sine wave. If we examine this waveform carefully, we notice some interesting things about it. First of all, every time the circle completes a full revolution (360 degrees), it draws the shape shown in Figure 1-14. Thus, halfway through its path, indicated by the zero point on the graph, the circle has passed through 180 degrees of travel. This makes sense, since a circle circumscribes 360 degrees.

The reason this is important is that we can manipulate the characteristics of the wave created in this fashion to cause it to carry varying amounts of information. Those characteristics—amplitude, frequency, and phase—can be manipulated as follows.

Amplitude Modulation   *Amplitude* is a measure of the loudness of a signal. A loud signal, such as that currently thumping through the window of my office from the subwoofer in the back window of the car that belongs to the kid across the street, has high-amplitude components (but very low-frequency components, as evidenced by the fact that I can hear him coming when he's still in southern Vermont), while lower volume signals are lower in amplitude. Examples are shown in Figure 1-15. The

Figure 1-13
Creating a sine
wave

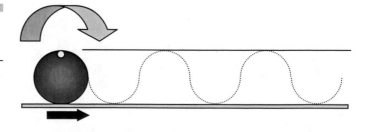

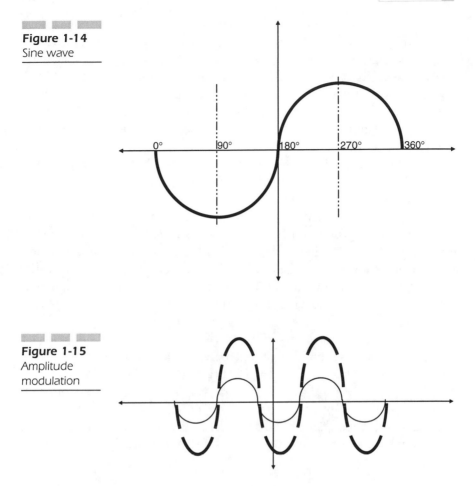

Figure 1-14
Sine wave

Figure 1-15
Amplitude
modulation

dashed line represents a high-amplitude signal, while the solid line represents a lower amplitude signal. So, how could this be used in the data communications realm? Simple: Let's let a high-amplitude signal represent a digital zero and a low-amplitude signal represent a digital one. If I then send four high-amplitude waves followed by four low-amplitude waves, I have actually transmitted the series 00001111. This technique is called amplitude modulation (AM); modulation simply means "variation." This is how AM radio works—and why it's called "AM."

Frequency Modulation Frequency modulation (FM) is similar to amplitude modulation except that instead of changing the loudness of the signal, we change the number of signals that pass a point in a given second (see Figure 1-16). The left side of the graph contains a lower frequency signal component, while a higher frequency component appears to its right. We can use this technique in the same way we used AM: If we let a high-frequency component represent a zero and a low-frequency component represent a one, then I can transmit our 00001111 series by transmitting four high-frequency signals followed by four low-frequency signals.

An interesting historical point about FM: The technique was invented by radio pioneer Edwin Armstrong in 1933. Armstrong created FM as a way to overcome the problem of noisy radio transmission. Prior to FM's arrival, AM was the only technique available and it relied on modulation of the loudness of the signal *and* the inherent noise to make it stronger. FM did not rely on amplitude, but rather on frequency modulation; therefore, it was much cleaner and offered significantly higher fidelity than AM radio. Keep in mind that signals pick up noise over distance; when an amplifier amplifies a signal, it also amplifies the noise.

Many technical historians of World War II believe that Armstrong's invention of FM transmission played a pivotal role in the outcome of the war. When WWII was in full swing, FM technology was only available to Allied forces. AM radio, the basis for most military communications at the time, could be jammed by simply transmitting a more powerful signal that overloaded the transmissions of military radios. FM, however, was not available to the Axis powers and, therefore, could not be jammed as easily.

Phase Modulation   Phase modulation (PM) is a little more difficult to understand than the other two modulation techniques. Phase is defined

Figure 1-16
Frequency
modulation

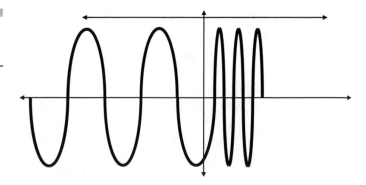

mathematically as "the fraction of a complete cycle elapsed as measured from a particular reference point." Any questions? OK, Now let's make that definition make sense. Consider the drawing shown in Figure 1-17. The two waves shown in the diagram are exactly 90 degrees "out of phase" of each other, because they do not share a common start point—wave B begins 90 degrees later than wave A. In the same way that we used amplitude and frequency to represent zeroes and ones, we can manipulate the phase of the wave to represent digital data.

A few years ago I saw a very graphic (and way too cool) example of how phase modulation can be used in a practical way. I was in Arizona with a film crew, shooting a video for one of my corporate clients. The theme of the video was based on the statement that "if we don't intelligently deploy technology, the competition will leave us in the dust." Building on that phrase about dust, we rented a ghost town in Arizona and created a video metaphor around it. In the last scene of the video, a horse-drawn wagon loaded with boxes of technology products disappears into a bloody Arizona sunset, and just before the wagon disappears over the hill, it blasts into the sky on a digital effect that looks like a rocket trail. We loved it. The only problem was that when we got back into the postproduction studio and began to assemble the final show, we discovered—to our horror—that the sound of an airplane could be heard

Figure 1-17
Phase
modulation

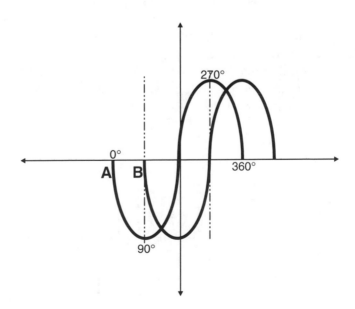

behind our narrator. Since airplanes didn't exist in the old west, we had a big problem. Not to be bested, our audio engineer asked us to pipe the sound into the audio booth. Listening to the recording, he went to his wall of audio CDs and selected a collection of airplane sounds. He listened to several of them until he found one that was correct. Setting the levels so that they matched those of the video soundtrack, he inverted the CD signal (180 degrees out of phase with the soundtrack signal) and electronically added it to the soundtrack. The airplane noise disappeared from the narration.

Digital Signaling

Data can be transmitted in a digital fashion as well. Instead of a smoothly undulating wave crashing on the computer beach, we can use an approximation of the wave to represent the data. This technique is called *digital signaling*. In digital signaling, an interesting mathematical phenomenon called the *Fourier Series* is called into play to create what most people call a square wave, shown in Figure 1-18. In the case of digital signaling, the Fourier Series is used to approximate the square nature of the waveform. The details of how the series actually works are beyond the scope of this book, but suffice it to say that, by mathematically combining the infinite series of odd harmonics of a fundamental wave, the ultimate result is a squared-off shape that approximates the square wave that commonly depicts digital data transmission. This technique is called digital signaling, as opposed to the amplitude-, frequency-, and phase-dependent signaling techniques used in analog systems.

In digital signaling, zeroes and ones are represented as either the absence or presence of voltage on the line and, in some cases, by either positive or negative voltage—or both. Figure 1-19, for example, shows a

Figure 1-18
Square wave

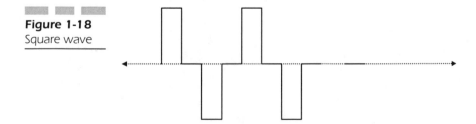

technique in which a zero is represented by the presence of positive voltage, while a one is represented as zero voltage. This is called a *unipolar signaling scheme*. Figure 1-20 shows a different technique, in which a zero is represented as positive voltage, while a one is represented as negative voltage. This is called a *nonreturn to zero (NRZI) signaling scheme*, because zero voltage has no meaning in this technique. Finally, Figure 1-21 demonstrates a *bipolar signaling system*. In this technique, the presence of voltage represents a one, but notice that every other one is opposite in polarity from the one that precedes it *and* the one that follows it. Zeroes, meanwhile, are represented as zero voltage. This technique, called *alternate mark inversion* (AMI), is commonly used in T- and E-Carrier systems for reasons that will be discussed later.

There are other techniques in use, but these are the most common.

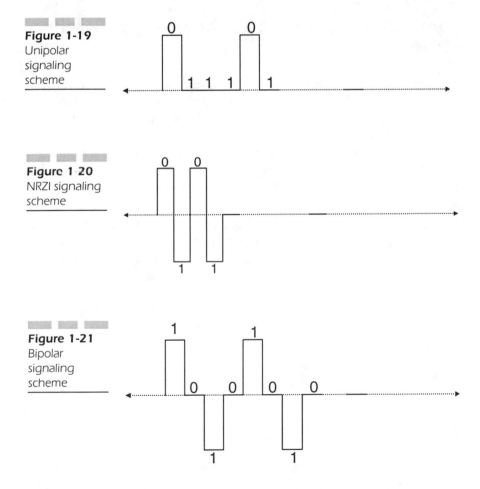

Figure 1-19
Unipolar
signaling
scheme

Figure 1-20
NRZI signaling
scheme

Figure 1-21
Bipolar
signaling
scheme

Clearly, both analog and digital signaling schemes can be used to represent digital data depending upon the nature of the underlying transmission system. It is important to keep the difference between *data* and *signaling* clearly separate. Data is the information that is being transported, and it can be either analog or digital in nature. For example, music is a purely analog signal because its values constantly vary over time. It can be represented, however, through both analog and digital signaling techniques. The zeroes and ones that spew forth from a computer are clearly digital information, but they, too, can be represented either analogically or digitally. For example, the broadband access technology known as DSL is not digital at all: There are analog modems at each end of the line, which means that analog signaling techniques are used to represent the digital data that is being transmitted over the local loop.

Combining Signaling Techniques for Higher Bit Rates

Let's assume that we are operating in an analog network. Under the standard rules of the analog road, one signaling event represents one bit. For example, a high-amplitude signal represents a one, and a low-amplitude signal represents a zero. But what happens if we want to increase our bit rate? One way is to simply signal faster. Unfortunately, the rules of physics limit the degree to which we can do that. In the 1920s, a senior researcher at Bell Laboratories, who has now become something of a legend in the field of communications, came to the realization that the bandwidth of the channel over which the information is being transmitted has a direct bearing on the speed at which signaling can be accomplished across that channel. According to Harry Nyquist, the broader the channel, the faster the signaling rate can be. In fact, put another way, the signaling rate can never be faster than two times the highest frequency that a given channel can accommodate. Unfortunately, the telephone local loop was historically engineered to support the limited bandwidth requirements of voice transmission. The traditional voice network was engineered to deliver 4 KHz of bandwidth to each local loop,[2] which means that the fastest signaling rate achievable over a tele-

---

[2]One way in which this was done was through the use of load coils. Load coils are electrical traps that tune the local loop to a particular frequency range, only allowing certain frequencies to be carried. This created a problem later for digital technologies, as we will discuss.

phony local loop is 8,000 baud. Yet, during the late 1980s and the early 1990s, it was common to see advertisements for 9,600 baud modems. This is where the confusion of terms becomes obvious: As it turns out, these were *9,600 bit-per-second modems*—a big difference. Modems with 9,600 baud were patently impossible. This, however, introduces a whole new problem: How do we create higher bit rates over signaling rate-limited (and therefore bandwidth limited) channels? To achieve higher signaling rates, one of two things must be done: Either broaden the channel, which is not always feasible, or figure out a way to have a single signaling event convey more than a single bit.

It's time to introduce a new word: *Baud*. Baud is the signaling rate. It may or may not be the same as the bit rate, depending on the scheme being used.

Consider the following example. We know from our earlier discussion that we can represent two bits by sending a high-amplitude signal followed by a low-amplitude signal (high-amplitude signal represents a zero, low-amplitude signal represents a one). What would happen though, if we were to combine amplitude modulation with frequency modulation? Consider the four waveforms shown in Figure 1-22. By combining the two possible values of each characteristic (high or low, frequency or amplitude), we create four possible states, each of which can actually represent two bits as shown in Figure 1-23. We can have a high-bandwidth, high-frequency signal; a high-bandwidth, low-frequency signal; a low-bandwidth, low-frequency signal; and a low-bandwidth, high-frequency signal. Consider what we have just done: We have created a system in which each signaling event represents two bits, which means that our bit rate is twice our signaling rate.

Let's take our concept one step farther. Figure 1-24 shows a system in which we are encoding four bits for each signal, a technique known as

Figure 1-22
Di-bit encoding
scheme

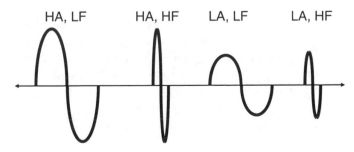

HA, LF   HA, HF   LA, LF   LA, HF

Figure 1-23
Di-bit values

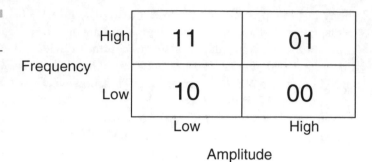

	Low	High
High	11	01
Low	10	00

Frequency

Amplitude

Figure 1-24
Quadrature
amplitude
modulation
(QAM)

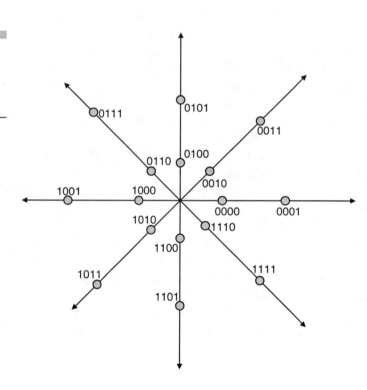

quad-bit encoding. This scheme, sometimes called *quadrature amplitude modulation* (QAM) (pronounced "Kwăm"), permits a single signal to represent four bits, which means that there is a 4:1 ratio between the bit rate and the signaling rate. Thus, it is possible to achieve higher bit rates in the bandwidth-limited telephony local loop by using multibit encoding techniques such as QAM. The first "high bit rate modems" (9,600 bits-per-second) used this technique or a variation of it to overcome the design limitations of the network. In fact, these multibit schemes are also used by the cable industry to achieve the high bit rates they need to operate their multimedia broadband networks.

There is one other limitation that must be mentioned: noise. Look at Figure 1-25. Here we have a typical QAM graph, but now we have added noise in the form of additional points on the graph that have no implied value. When a receiver sees them, however, how does it know which points are noise and which are data? Similarly, the oscilloscope trace shown in Figure 1-26 of a high-speed transmission would be difficult to interpret if there were noise spikes intermingled with the data. There is, therefore, a well-known relationship between the noise level in a circuit and the maximum bit rate that is achievable over that circuit, a relationship that was first described by Bell Labs researcher Claude Shannon, who is widely known as the father of information theory. In 1948, Shannon published *A Mathematical Theory of Communication,* which is now universally accepted as the framework for modern communications. We won't delve into the complex (but fascinating) mathematics that underlie Shannon's Theorem, but simply know that

Figure 1-25
QAM, this time with noise added to the constellation

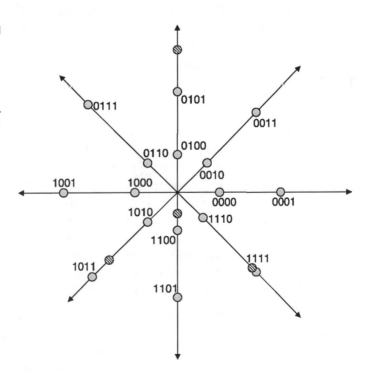

Figure 1-26
Oscilloscope
trace

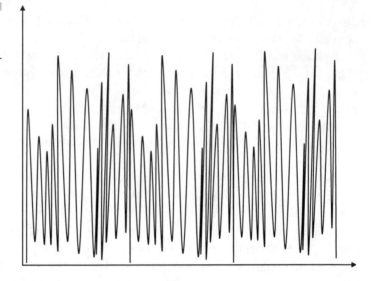

his conclusions are seminal: The higher the noise level in a circuit, the lower the achievable bandwidth. The bottom line? *Noise matters.* It matters so much, in fact, that network designers and engineers make its elimination the first order of business in their overall strategies for creating high-bandwidth networks. This is one of the reasons that optical fiber-based networks have become so critically important in modern transport systems—they are far less subject (but not immune!) to noise, and absolutely immune to the electromagnetic interference that plagues copper-based networks. Cable companies that now offer data services have the same issues and concerns. Every time a customer decides to play installer by adding a cable spur for a new television set in his or her home and crimping the connector on the end of the cable with a pair of pliers instead of a tool specifically designed for the purpose, he or she creates a point where noise can leak into the system, causing problems for everyone. And it gets even more melodramatic than that: According to John Judson, a cable systems maintenance manager in the Midwest, unauthorized connection to the cable network can cause problems that go way beyond unauthorized access to service. He observes:

> Cable networks are high-frequency systems. Some of the harmonics created in cable networks just happen to fall within the range of frequencies used in avionics, and therefore have the potential to affect aviation communications and navigation. So when you see the cable truck that looks like a commercial fishing boat cruising the neighborhood with

all the antennas on top, they're looking for signal leakage from unauthorized taps. They *will* find them and they *will* come in and fix them, and you *will* get a bill for it. So if you want to add a connection in the house, call us. It's cheaper.

That completes our introduction of common terms, with one exception: the Internet.

The Internet: What Is It?

The Internet is a vast network of networks, recognized as the fastest growing phenomenon in human history. In the words of Douglas Adams, author of *A Hitchhiker's Guide to the Galaxy,* the Internet is "Big. Really big. Vastly, hugely, mind-bogglingly big." And, it is getting bigger: The Internet doubles in size roughly every year, and that growth rate is expected to continue.

Not only is the Internet global in physical scope, it is universally recognized. *Everybody* knows about the Internet. In 1993, it came booming into the public consciousness, put down roots, spread like a biological virus, and flourished. Like other famous public figures, it has been on the cover of every major magazine in the world, has been the star of books, articles, TV shows, and movies, and has been praised as the most significant social force in centuries and debased as the source of a plethora of worldwide ills. Yet, for all this fame and notoriety, little is actually known about the Internet itself—at least, its private side. It is known to be a vast network of interconnected networks, with new appendages connecting approximately every ten minutes. According to the Network Wizards' Internet Domain Survey at www.nw.com, it connects approximately 300 million host computers, provides services to approximately 800 million users, and comprises roughly two million interconnected networks worldwide.

The World Wide Web

The World Wide Web (WWW) was first conceived by Tim Berners-Lee, who is considered the father of the World Wide Web. A physicist by training, Berners-Lee began his career in the computer and telecommunications industries following graduation from Oxford, before accepting a consulting position as a software engineer with the European Organization for Nuclear Research (CERN) during the late 1970s.

During his stint in Geneva, Berners-Lee observed that CERN suffers from the problems that plague most major organizations: information location, management, and retrieval. CERN is a research organization with large numbers of simultaneous ongoing projects, a plethora of internally published documentation, and significant turnover of people. Much of the work conducted at CERN revolves around large-scale high-energy physics collaborations that demand instantaneous information sharing between physicists all over the world. Berners-Lee found that his ability to quickly locate and retrieve specific information was seriously impaired by the lack of a single common search capability and the necessarily dispersed nature of the organization. To satisfy this need, he collaborated with Robert Cailliau to write the first WWW client, a search and archive program that they called *Enquire*. Enquire was never published as a product, although Berners-Lee, Cailliau, and the CERN staff used it extensively. It did, however, prove to be the foundation for the World Wide Web.

In May of 1990, Berners-Lee published *Information Management: A Proposal*, in which he described his experiences with hypertext systems and the rationale for Enquire. He described the system's layout, feel, and function as being similar to Apple's Hypercard, or the old adventure games in which players moved from "page to page" as they navigated through the game.[3] Remember this? Some of you will:

>YOU FIND YOURSELF IN A SMALL ROOM. THERE IS A
   DOOR TO THE LEFT.
>>OPEN DOOR

Enquire had no graphics, and was therefore rudimentary compared to modern Web browsers. To its credit, the system ran on a multiuser platform and could therefore be accessed simultaneously by multiple users. To satisfy the rigorous demands of the CERN staff, Berners-Lee and Cailliau designed the system around the following parameters:

- It had to offer remote access from across a diversity of networks.
- It had to be system and protocol independent, since CERN was home to a wide variety of system types—VM/CMS, Mac, VAX/VMS, and Unix.

[3] You'll love this: I recently discovered that some kind soul has ported Adventure to the PC platform, and the game, called WinFrotz, is downloadable from www.pcworld.com/downloads/file_description/0,fid,22456,00.asp.

- It had to run in a distributed processing environment.
- It had to offer access to all existing data types as well as to new types that would follow.
- It had to support the creation of personal, private links to new data sources as each user saw fit to create them.
- It had to support, in the future, diverse graphics types.
- It (ideally) had to support a certain amount of content and data analysis.

In November 1990, Berners-Lee wrote and published, with Robert Cailliau, *WorldWide Web: A Proposal for a HyperText Project*. In it, the authors described an information retrieval system in which large and diverse compendia of information could be searched, accessed, and reviewed freely, using a standard user interface based on an open, platform-independent design. This paper relied heavily on Berners-Lee's earlier paper.

In *WorldWide Web: A Proposal for a HyperText Project*, Berners-Lee and Cailliau proposed the creation of a "world wide web" of information that would allow the various CERN entities to access the information they need based on a common and universal set of protocols, file exchange formats, and keyword indices. The system would also serve as a central (although architecturally distributed) repository of information and would be totally platform independent. Furthermore, the software would be available to all and distributed free of charge.

Once the paper had been circulated for a time, the development of what we know today as the World Wide Web occurred with remarkable speed. The first "system" was developed on a NeXT platform. The first general release of the WWW inside CERN occurred in May of 1991, and in December the world was notified of the existence of the World Wide Web (known then as W3) thanks to an article in the CERN computer newsletter.

Over the course of the next few months, browsers began to emerge. Erwise, a GUI client, was announced in Finland, and Viola was released in 1992 by Pei Wei of O'Reilly & Associates. The National Center for Supercomputing Applications (NCSA) joined the W3 consortium, but didn't announce their Mosaic browser until February of 1993.

Throughout all of this development activity, W3 servers, based on the newly released Hypertext Transfer Protocol (HTTP) that allowed diverse sites to exchange information, continued to proliferate. By January of 1993, there were 50 known HTTP servers; by October there were over 200 servers, and WWW traffic comprised 1 percent of aggregate NSF backbone traffic. Very quietly, the juggernaut had begun.

In May 1994, the first international WWW conference was held at CERN in Geneva, and from that point on they were organized routinely, always to packed houses and always with a disappointed cadre of over-subscribed would-be attendees left out in the cold.

And, from that point on, the lines that clearly define "what happened when" begin to blur. NCSA's Mosaic product, developed largely by Marc Andreessen at the University of Illinois in Chicago, hit the mainstream and brought the WWW to the masses. Andreessen, together with Jim Clark, would go on to found Netscape Corporation shortly thereafter.

The following timeline, included courtesy of PBS, shows the highlights of the Internet's colorful history (as well as a few other great unrelated moments).

Internet Timeline (1960–1997)

1960   There is no Internet . . .

1961   Still no Internet . . .

1962   The RAND Corporation begins research into robust, distributed communication networks for military command and control.

1962–1969   The Internet is first conceived in the early 1960s. Under the leadership of the Department of Defense's (DOD's) Advanced Research Projects Agency (ARPA), it grows from a paper architecture into a small network (ARPANET) intended to promote the sharing of supercomputers amongst researchers in the United States.

1963   The Beatles play for the Queen of England.

1964   *Dr. Strangelove* portrays a nuclear holocaust, which the new network must survive.

1965   The DOD's Advanced Research Projects Agency begins work on ARPANET. ARPA sponsors research a "cooperative network of time-sharing computers."

1966   The U.S. Surveyor probe lands safely on moon.

1967   The first ARPANET papers are presented at the Association for Computing Machinery Symposium. Delegates at a symposium for the Association for Computing Machinery in Gatlinberg, Tennessee, discuss the first plans for the ARPANET.

1968   The first generation of networking hardware and software are designed.

1969   ARPANET connects the first four universities in the United States. Researchers at four U.S. campuses create the first hosts of the ARPANET, connecting Stanford Research Institute, UCLA, UCSB, and the University of Utah.

1970   ALOHANET is developed at the University of Hawaii.

1970–1973   The ARPANET is a success from the very beginning. Although originally designed to allow scientists to share data and access remote computers, email quickly becomes the most popular application. The ARPANET becomes a high-speed digital post office as people use it to collaborate on research projects and discuss topics of various interests.

1971   The ARPANET grows to 23 hosts connecting universities and government research centers around the country.

1972   The InterNetworking Working Group (INWG) becomes the first of several standards-setting entities to govern the growing network. Vinton Cerf is elected the first chairman of the INWG, and later becomes known as a father of the Internet.

1973   The ARPANET goes international with connections to University College in London, England, and the Royal Radar Establishment in Norway.

1974–1981   Bolt, Beranek & Newman opens Telenet, the first commercial version of the ARPANET. The general public gets its first vague hint of how networked computers can be used in daily life as the commercial version of the ARPANET goes online. The ARPANET starts to move away from its military/research roots.

1975   Internet operations are transferred to the Defense Communications Agency.

1976   Queen Elizabeth goes online with the first royal email message.

1977   Unix-to-Unix Copy Program (UUCP) provides email on THEORYNET.

1978   TCP's checksum design is finalized.

1979   Tom Truscott and Jim Ellis, two grad students at Duke University, and Steve Bellovin, at the University of North Carolina, establish the first USENET newsgroups. Users from all over the world join these discussion groups to talk about the net, politics, religion, and thousands of other subjects.

1980   Mark Andreessen turns 8 years old. In 14 more years he will revolutionize the Web with the creation of Mosaic.

1981   ARPANET has 213 hosts. A new host is added approximately once every 20 days.

1982–1987   The term "Internet" is used for the first time. Bob Kahn and Vinton Cerf are key members of a team that creates TCP/IP, the common language of all Internet computers. For the first time, the loose collection of networks that made up the ARPANET is seen as an "internet," and the Internet as we know it today is born. The mid-1980s marks a boom in the personal computer and superminicomputer industries. The combination of inexpensive desktop machines and powerful, network-ready servers allows many companies to join the Internet for the first time. Corporations begin to use the Internet to communicate with each other and with their customers.

1983   TCP/IP becomes the universal language of the Internet.

1984   William Gibson coins the term "cyberspace" in his novel *Neuromancer*. The number of Internet hosts exceeds 1,000.

1985   Internet e-mail and newsgroups are now part of life at many universities.

1986   Case Western Reserve University in Cleveland, Ohio, creates the first "Freenet" for the Society for Public Access Computing.

1987   The number of Internet hosts exceeds 10,000.

1988–1990   The Internet Worm is unleashed. The Computer Emergency Response Team (CERT) is formed to address security concerns raised by the Worm. By 1988, the Internet is an essential tool for communications; however, it also begins to create concerns about privacy and security in the digital world. New words such as "hacker," "cracker," and "electronic break-in" are created. These new worries are dramatically demonstrated on November 1, 1988 when a malicious program called the "Internet Worm" temporarily disables approximately 6,000 of the 60,000 Internet hosts. System administrator turned author, Clifford Stoll, catches a group of

Cyberspies and writes the best-seller, *The Cuckoo's Egg*. The number of Internet hosts exceeds 100,000. A happy victim of its own unplanned, unexpected success, the ARPANET is decommissioned, leaving only the vast network-of-networks called the Internet. The number of hosts exceeds 300,000.

1991 The World Wide Web is born!

1991–1993 Corporations wishing to use the Internet face a serious problem: Commercial network traffic is banned from the National Science Foundation's NSFNET, the backbone of the Internet. In 1991 the NSF lifts the restriction on commercial use, clearing the way for the age of electronic commerce. At the University of Minnesota, a team led by computer programmer Mark MaCahill releases "gopher" in 1991; it is the first point-and-click way of navigating the files of the Internet. Originally designed to ease campus communications, gopher is freely distributed on the Internet. MaCahill calls it "the first Internet application my mom can use." 1991 is also the year in which Tim Berners-Lee, working at CERN in Switzerland, posts the first computer code of the World Wide Web in a relatively innocuous newsgroup, "alt.hypertext." The ability to combine words, pictures, and sounds on Web pages excites many computer programmers who see the potential for publishing information on the Internet in a way that can be as easy as using a word processor. Marc Andreessen and a group of student programmers at NCSA (located on the campus of the University of Illinois at Urbana Champaign) will eventually develop a graphical browser for the WWW called Mosaic. Traffic on the NSF backbone network exceeds one trillion bytes per month. The first audio and video broadcasts take place over a portion of the Internet known as the "MBONE." One million hosts have multimedia access to the Internet over the MBONE. More than one million hosts are part of the Internet. Mosaic, the first graphics-based Web browser, becomes available. Traffic on the Internet expands at a 341,634 percent annual growth rate.

1994 The Rolling Stones broadcast the Voodoo Lounge tour over the MBONE. Marc Andreessen and Jim Clark form Netscape Communications Corp. Pizza Hut accepts orders for a mushroom, pepperoni with extra cheese over the net, and Japan's Prime Minister goes online at www.kantei.go.jp. Backbone traffic exceeds ten trillion bytes per month.

1995   NSFNET reverts to a research project, leaving the Internet in commercial hands. The Web now comprises the bulk of Internet traffic. The Vatican launches www.vatican.va. James Gosling and a team of programmers at Sun Microsystems release an Internet programming language called Java, which radically alters the way applications and information can be retrieved, displayed, and used over the Internet.

1996   Nearly ten million hosts are online. The Internet covers the globe. As the Internet celebrates its 25th anniversary, the military strategies that influenced its birth become historical footnotes. Approximately forty million people are connected to the Internet. More than $1 billion per year changes hands at Internet shopping malls, and Internet-related companies like Netscape are the darlings of high-tech investors. Users in almost 150 countries around the world are now connected to the Internet. The number of computer hosts approaches ten million (interesting, considering that today there are more than three hundred million!).

Within 30 years, the Internet has grown from a Cold War concept for controlling the tattered remains of a postnuclear society to the Information Superhighway. Just as the railroads of the nineteenth century enabled the Machine Age and revolutionized the society of the time, the Internet takes us into the Information Age and profoundly affects the world in which we live.

The Age of the Internet arrives.

1997   Some people telecommute over the Internet, allowing them to choose where to live based on quality of life, not proximity to work. Many cities view the Internet as a solution to their clogged highways and fouled air. Schools use the Internet as a vast electronic library, with untold possibilities. Doctors use the Internet to consult with colleagues half a world away. And even as the Internet offers a single Global Village, it threatens to create a second-class citizenship among those without access. As a new generation grows up as accustomed to communicating through a keyboard as in person, life on the Internet will become an increasingly important part of life on Earth.

1998–2004   The Internet bubble climbs to insane levels and money flows into the telecom and IT industries like water over Niagara. The bubble rises, and falls, and $7 trillion dollars in market value evaporate. Recovery begins in early 2004, just in time for the next big bubble to begin . . .

Chapter Summary

This chapter is designed to acquaint the reader with the fundamental terms and concepts that characterize the data and telecommunications worlds today. Now we can move deeper into the magic. In the next chapter, we introduce the design, philosophy, structure, and use of data communications protocols.

Chapter 1 Questions

1. What is the difference between telecommunications and data communications?

2. What is the difference between a protocol and a standard?

3. Why do you suppose the United Nations has global responsibility for the development of telecommunications standards?

4. Why are there so many different standards bodies? How do they work together to ensure global acceptance of new standards?

5. Give examples of each of the following: simplex, half-duplex, and full-duplex services. Which are most common?

6. What is the difference between intelligibility and recognizability?

7. Define the following terms: amplitude modulation, frequency modulation, and phase modulation.

8. Explain the difference between data encoding and signaling.

9. It is possible to transport digital data across an analog facility. Explain.

10. In simple terms, explain the difference between the work done by Harry Nyquist and Claude Shannon.

CHAPTER 2

Protocols

Click. One simple action that kicks off a complex series of events that results in the transmission of an e-mail message, or in the creation of a digital medical image, or in the establishment of a videoconference between a child and a grandmother. The process through which this happens is a remarkable symphony of technological complexity, and it is all governed by a collection of rules called protocols. This chapter is dedicated to them.

Data Communications Systems and Functions

If I were to walk up to you on the street in pretty much any Western country, and extend my hand in greeting, you would quite naturally reach your hand out, grab mine, and shake it. There is a commonly accepted set of social rules that we agree to abide by, one of which is shaking hands as a form of greeting. It doesn't work everywhere: In Tibet, it is customary to extend one's tongue as far as it can be extended as a form of greeting (clearly a sign of a great culture!). In China, unless you are already friends with the person you are greeting, it is not customary to touch in any fashion. You, of course, had a choice when I extended my hand. You could have hit it, licked it, spit in it. But, because of the accepted rules that govern Western society, you took my hand in yours and shook it. These rules that govern communication—*any form of communication*—are called protocols. And the process of using protocols to convey information is called data communications. It's no accident, incidentally, that the obnoxious racket that analog modems make when they are attempting to connect to each other is called a handshake. The noise they make is their attempt to negotiate a common set of rules that works for both of them for that particular session.

The Science of Communications

Data communications is the procedure required to collect, package, and transmit data from one computing device to another, typically (but not always) over a wide area network. It is a complex process with many layers of functionality. To understand data communications, we must break it into its component parts and examine each part individually, relying on the old adage that "the only way to eat an elephant is one bite at a time." Like a Russian Matreshka doll, data communications are made up of layer upon layer of operational functionality that work together to accomplish the task at hand—namely, the communication of data. These component parts are known as *protocols*, and they have one responsibility: to ensure the integrity of the data that they transport from the source device to the receiver. This integrity is measured in a variety of ways including *bit-level integrity*, which ensures that the bits themselves are not changed in value as they transit the network; *data integrity*, which guarantees that the bits are recognizable as packaged entities called frames or cells; *network integrity*, which provides for the assured delivery of those entities, now in the form of packets, from a source to a destination; *message integrity*, which not only guarantees the delivery of the packets, but, in fact, their *sequenced* delivery to ensure the proper arrival of the entire message; and finally, *application integrity*, which provides for the proper execution of the responsibilities of each application. This is shown graphically in Figure 2-1.

Protocols exist in a variety of forms and are not limited to data communications applications. Military protocols define the rules of engagement that modern armies agree to abide by; diplomatic protocols define the manner in which nations interact and settle their political and geographic differences; in addition, medical protocols document the manner in which medications are used to treat illness. The word "protocol" is defined as "a set of rules that facilitates communication." Data communications, then, is the science built around the protocols that govern the exchange of digital data between computing systems.

Data Communications Networks

Data communications networks are often described in terms of their architectures, as are protocols. Protocol architectures are often said to be *layered* because they are carefully divided into highly related but nonoverlapping functional entities. This "division of labor" not only makes it easier to understand how data communications works, but also makes the deployment of complex networks far easier.

The amount of code (lines of programming instructions) required to successfully carry out the complex task of data transmission is quite large. If the program that carries out all of the functions in that process

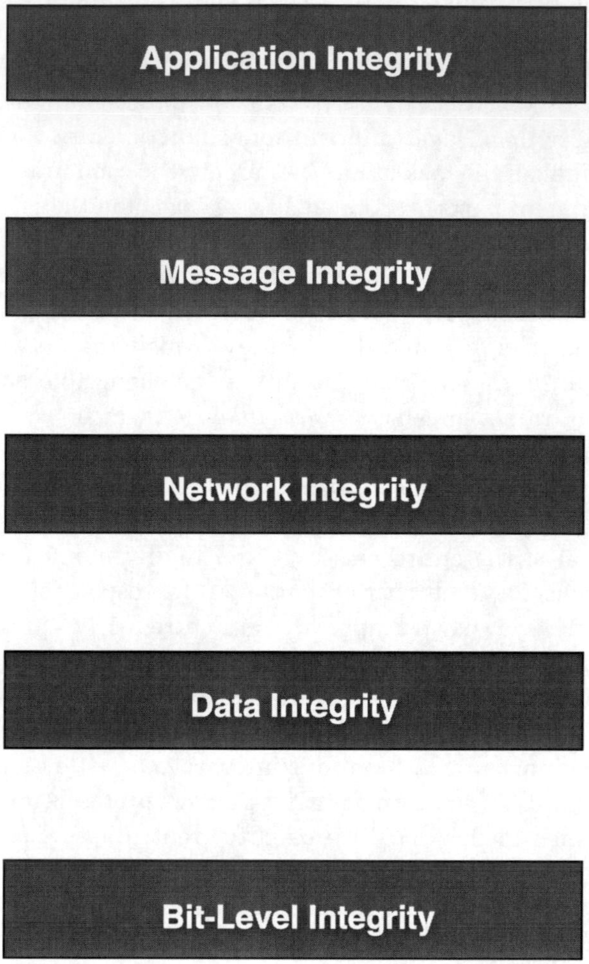

Figure 2-1
The various
integrity levels
of the OSI
Model

Application Integrity

Message Integrity

Network Integrity

Data Integrity

Bit-Level Integrity

were written as a single, large, monolithic chunk of code, then it would be difficult to make a change to the program when updates were required, simply because of the monolithic nature of the program. Now imagine the following: Instead of a single set of code, we break the program into functional pieces, each of which handles a particular specific function required to carry out the transmission task properly. With this model, changes to a particular module of the overall program can be accomplished in a way that only affects that particular module, making the process far more efficient. This modularity is one of the great advantages of layered protocols.

Consider the following simple scenario, shown in Figure 2-2: A PC-based e-mail user in Madrid with an account at ISP Terra Networks wants to send a large, confidential message to another user in Marseilles. The Marseilles user is attached to a mainframe-based corporate e-mail system. In order for the two systems to communicate, a complex set of challenges must first be overcome. Let's examine them a bit more closely.

The first and most obvious challenge that must be overcome is the difference between the actual user interfaces on the two systems. The PC-based system's screen presents information to the user in a Graphical User Interface (GUI, pronounced "gooey") format that is carefully designed to make it intuitively easy to use. It eliminates the need to rely on the old command line syntax that was used in DOS environments.

The mainframe system was created with intuitive ease of use in mind, but because a different company designed the interface for a mainframe host, under a different design team, it bears minimal resemblance to the PC system's interface. Both are equally capable, but completely different.

As a result of these differences, if we were to transmit a screen of information from the PC directly to the mainframe system, it would be unreadable simply because the two interfaces do not share common field names or locations.

The next problem that must be addressed is security, illustrated in Figure 2-3. We mentioned earlier that the message that is to be sent from the user in Madrid is confidential, which means that it should probably be encrypted to protect its integrity. Also, because the message is large, the sender will probably compress it to reduce the time it takes to transmit it. Compression, which will be discussed in more detail later, is simply the process of eliminating redundant information from a file before it is transmitted or stored to make it easier to manage.

Another problem has to do with the manner in which the information being transmitted is represented. The PC-based Eudora message encodes its characters using a seven-bit character set called the American Standard Code for Information Interchange (ASCII). A sample of the

Figure 2-2
PC to mainframe communications

Madrid

Marseilles

Figure 2-3
Managing
security

Madrid

Marseilles

ASCII codeset is shown in Table 2-1. Mainframes, however, use a different codeset called the Extended Binary Coded Decimal Interchange Code (EBCDIC); the ASCII traffic must be converted to EBCDIC if the mainframe is to understand it, and vice versa, as shown in Figure 2-4.

Table 2-1

ASCII-Decimal
Conversion
Table

Character	ASCII Value	Decimal Value
0	0110000	48
1	0110001	49
2	0110010	50
3	0110011	51
4	0110100	52
5	0110101	53
6	0110110	54
7	0110111	55
8	0111000	56
9	0111001	57
A	1000001	65
B	1000010	66
C	1000011	67
D	1000100	68
E	1000101	69
F	1000110	70
G	1000111	71

Character	ASCII Value	Decimal Value
H	1001000	72
I	1001001	73
J	1001010	74
K	1001011	75
L	1001100	76
M	1001101	77
N	1001110	78
O	1001111	79
P	1010000	80
Q	1010001	81
R	1010010	82
S	1010011	83
T	1010100	84
U	1010101	85
V	1010110	86
W	1010111	87
X	1011000	88
Y	1011001	89
Z	1011010	90

Binary Arithmetic Review

It's probably not a bad idea to review binary arithmetic for just a moment, since it seems to be one of the least understood details of data communications. I promise, this will not be painful; I just want to offer a quick explanation of the numbering scheme and the various codesets that result.

Modern computers are often referred to as digital computers because the values they use to perform their function have discrete values

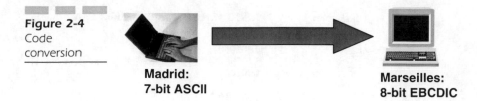

Figure 2-4
Code
conversion

Madrid:
7-bit ASCII

Marseilles:
8-bit EBCDIC

(remember, the word "digital" means "discrete"). Those values are nominally zero and one: In other words, a value can either be one or zero, on or off, positive or negative, presence of voltage or absence of voltage, presence of light or absence of light. There are two possible values for any given situation, and this type of system is called *binary*. The word means a system that comprises two distinct components or values. Computers operate using base two arithmetic, whereas humans use base ten. Let me take you back to second grade.

When we count, we arrange our numbers in columns that have values based on multiples of the number ten, as shown in Figure 2-5. Here we see the number six thousand, seven hundred eighty-three, written using the decimal numbering scheme. We easily understand the number as it is written because we are taught to count in base ten from an early age.

Computers, however, don't speak in base ten. Instead, they speak in base two. Instead of having columns that are multiples of ten, they use columns that are multiples of two, as shown in Figure 2-6. In base ten, the columns are (reading from the right):

- Ones
- Tens
- Hundreds
- Thousands
- Ten thousands
- Hundred thousands
- Millions
- Etc.

Figure 2-5
Base ten
numbering
scheme

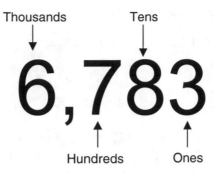

In base two, the columns are

- Ones
- Twos
- Fours
- Eights
- Sixteens
- Thirty-twos
- Sixty-fours
- One hundred twenty-eights
- Two hundred fifty-sixes
- Five hundred twelves
- One thousand twenty-fours
- Etc.

Figure 2-6
Base two
numbering
scheme

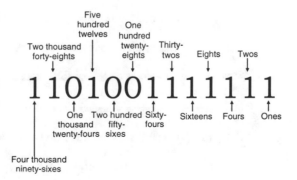

So our number, 6,783, would be written as follows in base two:

1101001111111

From right to left that's one 1, one 2, one 4, one 8, one 16, one 32, one 64, no 128s, no 256s, one 512, no 1,024s, one 1,048s, and one 4,096. Add them all up $(1 + 2 + 4 + 8 + 16 + 32 + 64 + 512 + 2,048 + 4,096)$ and you *should* get 6,783.

That's binary arithmetic. Most PCs today use the seven-bit ASCII character set shown in Table 2-1. The mainframe, however (remember the mainframe?), uses an eight-bit code called EBCDIC. What happens when a seven-bit ASCII PC sends information to an EBCDIC mainframe system that only understands eight-bit characters? Clearly, problems would result. Something, therefore, has to take on the responsibility of translating between the two systems so that they can intelligibly transfer data between each other.

Logical Layering

Another problem that arises has to do with the logical relationship between the applications running in the two systems. While the PC most likely supports the e-mail account of a single user, the mainframe undoubtedly hosts hundreds, perhaps thousands of accounts, and must therefore ensure that each user receives his or her mail and *only* his or her mail. Some kind of user-by-user and process-by-process differentiation is required to maintain the integrity of the system and its applications. This is illustrated graphically in Figure 2-7.

Figure 2-7
Logical session
management

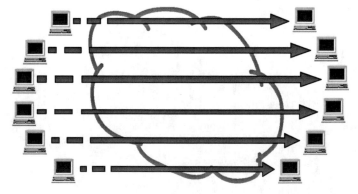

The next major issue has to do with the network over which the information is to be transmitted from Madrid to Marseilles. In the past, information was either transmitted via a dedicated and very expensive point-to-point circuit or over the relatively slow public switched telephone network, or PSTN. Today, however, most modern networks are packet-based, meaning that messages are broken into small, easily routable pieces, called packets, prior to transmission. Of course, this adds an additional layer of complexity to the process: What happens if one of the packets fails to arrive at its destination? Or, what if the packets arrive at the destination out of order? Some process must be in place to manage these challenges and overcome the potentially disastrous results that could occur.

Computer networks have a lot in common with modern freeway systems, including the tendency to become congested. Congestion results in delay, which some applications do not tolerate well. What happens if some or all of the packets are badly delayed, as shown in Figure 2-8? What is the impact on the end-to-end quality of the service?

Another vexing problem that often occurs is errors in the bitstream. Any number of factors including sunspot activity, the presence of electric motors, and the electrical noise from fluorescent lights, can result in ones being changed to zeroes and zeroes being changed to ones, as shown in Figure 2-9. Obviously, this is an undesirable problem, and there must be a technique in place to detect and correct these errors when they occur.

There may also be inherent problems with the physical medium over which the information is being transmitted. There are many different media including twisted copper wire pairs, optical fiber, coaxial cable, and wireless systems, to name a few. None of these are perfect transmission media; they all suffer from the vagaries of backhoes, lightning

Figure 2-8
Problems in the network cause delay and lost data.

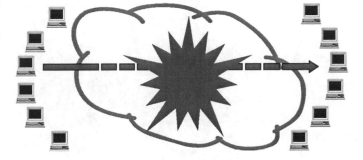

Figure 2-9
Bit errors

01101110111101010111111101

strikes, sunlight, earth movement, squirrels with sharp teeth, kids with BB guns, and other impairments far too numerous to name. When these problems occur, how are they detected? Equally important, how are the transmission impairments that they inevitably cause reported and corrected?

There must also be an agreed upon set of rules that define exactly how the information is to be physically transmitted over the selected medium. For example, if the protocol to be used dictates that information will *always* be transmitted on pin 2 of a data cable, such as that shown in Figure 2-10, then the other end will have a problem, since its received signal will arrive on the same pin that it wants to *transmit* on! Furthermore, there must be agreement on how information is to be physically represented, how and when it is to be transmitted, and how it is to be acknowledged. What happens if a very fast transmitter overwhelms the receive capabilities of a slower receiver? Does the slower receiver have the ability, or even the *right,* to tell it to slow down?

Collectively, all of these problems pose what seem to be insurmountable challenges to the transmission of data from a source to a receiver. And while the process is obviously complex, steps have been taken to simplify it by breaking it into logical pieces. Those pieces, as we described earlier, are protocols. Collections of protocols, carefully

Figure 2-10
Physical
agreements

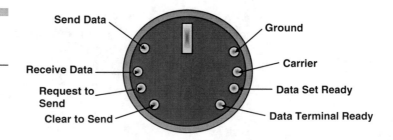

selected to form functional groupings, are what make data communications work properly.

Perhaps the best-known family of protocols is the International Organization for Standardization's *Open Systems Interconnection Reference Model*, usually called the OSI Model for the sake of simplicity. Shown in Figure 2-11 and comprising seven layers, it provides a logical way to study and understand data communications and is based on the following simple rules. First, each of the seven layers must perform a clearly defined set of responsibilities that are unique to that layer and guarantee the requirement of functional modularity. Second, each layer depends upon the services of the layers above and below to do its own job, as we would expect, given the modular nature of the model. Third, the layers have no idea how the layers around them do what they do; they simply know that they do it. This is called transparency. Finally, there is nothing magical about the number seven. If the industry should decide that we need an eighth layer on the model, or that layer five is redundant (and there are those who think it is), then the model will be changed. The key is functionality. There is an ongoing battle within the ranks of OSI Model pundits, for example, over whether there is actually a requirement for *both* layers six and seven, because many believe them to be so similar functionally that one or the other is redundant. Others question whether there is *really* a need for layer five, the functions of which are considered by many to be superfluous and redundant. To these people I recommend the purchase of a dog. Whether the addition or elimination of a layer ever actually happens is not important. The fact that it *can* is what matters.

It is important to understand that the OSI Model is nothing more than a conceptual way of thinking about data communications. It isn't hardware; it isn't software. It merely simplifies and groups the processes of data transmission so that they can be easily understood and manipulated.

Let's look at the OSI Model in a little more detail (see Figure 2-12). I tend to think of it as a seven-drawer chest. In each drawer a collection of standards is stored, and when network implementers set up a network they rummage through each drawer, select the most appropriate standard for their requirements, and set up the network.

As we mentioned earlier, the model is a seven-layer construct within which each layer is tightly dependent upon the layers surrounding it. The Application Layer, at the top of the model, "speaks" to the actual application process that creates the information to be transported by the

Figure 2-11
OSI Model

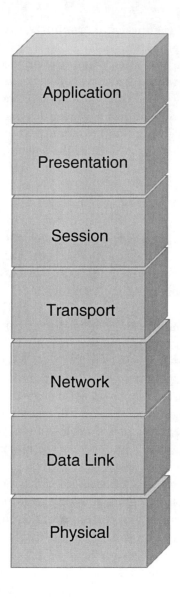

network; it is closest to the customer and the customer's processes, and is therefore the most customizable and manipulable of all the layers. It is highly open to interpretation. On the other end of the spectrum, the Physical Layer dwells within the confines of the actual network, and is totally standards dependent. There is minimal room here for interpretation; a pulse is either a one or a zero—there's nothing in between.

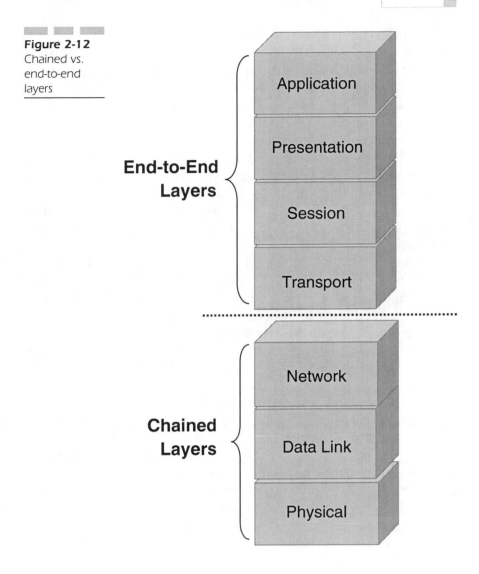

Figure 2-12
Chained vs.
end-to-end
layers

Physical Layer standards, therefore, tend to be highly commoditized, while Application Layer standards tend to be highly specialized. This becomes extremely important as the service provider model shifts from delivering commodity bandwidth to providing customized services—even if they're mass customized—to their customer base. Service providers are clawing their way up the OSI food chain to get as close to the Application Layer end of the model as they can because of the Willie Sutton Rule.

The Willie Sutton Story

Willie Sutton became famous in the 1930s for a series of outrageous robberies during which he managed to outwit the police at every turn. During his career he had two nicknames, "The Actor" and "Slick Willie," because of his ingenious tendency to use a wide array of disguises during his robberies. A sucker for expensive clothes, Sutton was an immaculate dresser. Although he was a bank robber, he had the reputation of being a gentleman; in fact, people who witnessed his robberies stated he was quite polite. One teller remembers him coming into the bank dressed to the nines carrying flowers, which he presented to her in exchange for her money. Another victim said Sutton's robberies were like attending the movies, except that the usher had a gun.

On February 15, 1933, Sutton and an accomplice attempted to rob the Corn Exchange Bank and Trust Company in Philadelphia, Pennsylvania. Sutton, disguised as a mailman, entered the bank early in the morning, but a suspicious passerby caused them to abort the robbery. Roughly a year later, however, on January 15, 1934, he entered the same bank through a ceiling skylight. When the guard arrived, Sutton forced him to admit the employees, whom Sutton handcuffed and locked in a small back room.

Sutton also robbed a Broadway jewelry store in broad daylight, dressed as a telegraph messenger. His other disguises included a policeman, special delivery messenger, and maintenance man.

Sutton was caught in June of 1931 and sentenced to 30 years in prison. He escaped on December 11, 1932 by climbing a prison wall. Two years later he was recaptured and sentenced to serve 25 to 50 years in Eastern State Penitentiary, Philadelphia, for the robbery of the Corn Exchange Bank.

Sutton's career was not over yet, however. On April 3, 1945, Sutton was one of 12 convicts who escaped from Eastern State through a tunnel. He was recaptured the same day by Philadelphia police officers and sentenced to life imprisonment as a fourth-time offender. At that time he was transferred to the Philadelphia County Prison in Homesburg, Pennsylvania, to live out the rest of his days. On February 10, 1947, Sutton tired of prison life. He and several other prisoners, dressed as prison guards, carried two ladders across the prison yard to the wall shortly after dark. When the searchlights froze them in its glare, Sutton yelled, "It's okay," and no one stopped him. They climbed the wall under the watchful eye of the guards and disappeared into the night.

On March 20, 1950, Willie Sutton was added to the FBI's Ten Most Wanted List. Because of his expensive clothing habit, his photograph was given to tailors all over the country in addition to the police. On February 18, 1952, a tailor's 24-year-old son recognized Sutton on the New York subway and followed him to a local gas station. The man reported the incident to the police, who later arrested Sutton.

He did not resist his arrest by New York City Police, but denied any robberies or other crimes since his 1947 escape from the Philadelphia County Prison. When he was arrested, Sutton owed one life sentence plus 105 years to the people of Pennsylvania. Because of his new transgressions (mostly making the police look remarkably incompetent), his sentence was augmented by an additional 30 years to life in New York State Prison.

Shortly after his final incarceration, a young reporter was granted a rare interview with Sutton in his prison cell. When they met, Sutton shook his hand and asked, "What can I do for you, young man?" The reporter, nervous, stammered back, "M-M-Mr. Sutton, why do you rob banks?" Sutton sat back and replied with a smile, "Because, young man, that's where the money is."

That's not the end of the story, however. In 1969, the New York State Prison Authority decided that Sutton did not have to serve his entire sentence of two life sentences plus 105 years because of failing health. So, on Christmas Eve, 1969, Sutton, now 68, was released from Attica State Prison. And in 1970, Sutton did a television commercial to promote the New Britain, Connecticut, Bank and Trust Company's new photo credit card program. You have to love the little ironies in life. Sutton died in 1980 in Spring Hill, Florida, at the age of 79.

So what does Willie Sutton have to do with the OSI Model and service providers? Not much—but his career does. Today's service providers are climbing the food chain because the money is up there with the customers. Yes, of course there is money to be made at the Physical Layer end of the model, but the sustainable, growable revenues are up where services can be customized endlessly to meet the changing needs of customers.

More Now About the Inner Workings of the OSI Model   The functions of the model can be broken into two pieces, as illustrated by the dashed line in Figure 2-12 between layers three and four that divides the model into the *chained layers* and the *end-to-end layers*.

The chained layers are made up of layers one through three: the Physical Layer, the Data Link Layer, and the Network Layer. They are responsible for providing a service called *connectivity*. The end-to-end layers on the other hand comprise the Transport Layer, the Session Layer, the Presentation Layer, and the Application Layer. They provide a service called *interoperability*. The difference between the two services is important.

Connectivity is the process of establishing a physical connection so that electrons can flow correctly from one end of a circuit to the other. There is little intelligence involved in the process; it occurs, after all, pretty far down in the primordial protocol ooze of the OSI Model. Connectivity is critically important to network people—it represents their lifeblood. Customers, on the other hand, are typically only aware of the criticality of connectivity when it isn't there for some reason. No dial tone? Visible connectivity. Can't connect to the ISP? Visible connectivity. Dropped call on a cell phone? Visible connectivity.

Interoperability, however, is something that customers are much more aware of. Interoperability is the process of guaranteeing *logical connectivity* between two communicating processes over a physical network. It's wonderful that the lower three layers give a user the ability to spit bits back and forth across a wide area network. But what do the bits mean? Without interoperability, that question cannot be answered. For example, in our e-mail scenario, the e-mail application running on the PC and the e-mail application running on the mainframe are logically incompatible with each other for any number of reasons, which will be discussed shortly. They can certainly swap bits back and forth, but without some form of protocol intervention, the bits are meaningless. Think about it: If the PC shown on the left side of Figure 2-13 creates an e-mail message that is compressed, encrypted, ASCII encoded, and shipped across logical channel 17, do the intermediate switches that create the path over which the message is transmitted care? Of course not. Only the transmitter and receiver of the message that house the applications that will have to interpret it care about such things. The intermediate switches care that they have electrical connectivity, that they can see the

Figure 2-13
Connectivity

Madrid   **Marseilles**

bits, that they can determine whether they are the *right* bits, and whether they are the intended recipient of those bits—or not. Therefore, the end devices, the sources and sinks of the message, must implement all seven layers of the OSI Model, because they must not only concern themselves with connectivity issues, but also with issues of interoperability. The intermediate devices, however, care only about the functions and responsibilities provided by the lower three layers. Interoperability, because it only has significance in the end devices, is provided by the end-to-end layers—layers four through seven. Connectivity, on the other hand, is provided by the chained layers, layers one through three, because those functions are required in every link of the network chain —hence the name. Let me say this one more time: *The chained layers are the historical domain of the telephone company.*

Layer by Layer

The OSI Model relies on a process called *enveloping*, illustrated in Figure 2-14, to perform its tasks. If we return to our earlier e-mail example, we

Figure 2-14
A schematic
representation
of enveloping

find that each time a layer invokes a particular protocol, it wraps the user's data in an "envelope" of overhead information that tells the receiving device about the protocol used. For example, if a layer uses a particular compression technique to reduce the size of a transmitted file and a specific encryption algorithm to disguise the contents of the file, then it is important that the receiving device be made aware of the techniques employed so that it knows how to decompress and decrypt the file when it receives it. Needless to say, quite a bit of overhead must be transmitted with each piece of user data. The overhead is needed, however, if the transmission is to work properly. So, as the user's data passes down the so-called stack from layer to layer, additional information is added at each step of the way, as illustrated by the series of envelopes. In summary then, the message to be transported is handed to layer seven, which performs Application Layer functions and then attaches a header to the beginning of the message that explains the functions performed by that layer so that the receiver can interpret the message correctly. In our illustration, that header function is represented by information written on the envelope at each layer. When the receiving device is finally handed the message at the Physical Layer, each succeeding layer must open its own envelope until the kernel—the message—is exposed for the receiving application. Thus, OSI protocols really do work like a nested Russian doll. After peeling back layer after layer of the network onion, the core message is exposed.

Let's now go back to our e-mail example, but this time we'll describe it within the detailed context of OSI's layered architecture. We begin with a lesson on linguistics.

Esperanto

There is an old and somewhat comforting cliché, which observes, "Wherever one goes, people speak English." In fact, less than ten percent of the world's population speaks English and, to their credit, many of them speak it as a second language.[1] Many believe there is a real need for a

---

[1]There is an old joke among seasoned international travelers that goes likes this: "What do you call someone who speaks three languages?" *Trilingual.* "OK, what do you call someone who speaks *two* languages?" *Bilingual.* "OK, what do you call someone who speaks *one* language?" *American.*

truly international language. In 1887, Polish physician Ludwig L. Zamenhof published a paper on the need for a universally spoken tongue. He believed that most of the world's international diplomacy disputes resulted from a communication failure between monolingual speakers and the inevitable misunderstandings of nuance that occur when one language is translated into another. Zamenhof set out to solve this "Tower of Babel" problem (origin of the word "babble," by the way), resulting in the creation of the international language called *Esperanto*. In Esperanto, the word "Esperanto" means "one who hopes."

Since its creation, Esperanto has been learned by millions and, believe it or not, is widely spoken—current estimates are approximately two million speakers. And its use is far from being purely academic: Meetings are held in Esperanto, advertising campaigns use it, hotels and restaurants publish literature using it, and professional communities such as health care and scientific research now use Esperanto widely as a way to universally communicate information. Second only to English, it is the lingua franca of the international world. It is most commonly spoken in Central and Eastern Europe, East Asia (particularly mainland China), South America, and Southwest Asia. It is less commonly spoken in North America, Africa, and the Middle East.

Esperanto's success as the language of international communication results from three advantages. It is easy to learn; it is politically neutral; and, there are practical reasons to learn it. The structure of the language is so simple and straightforward that it can typically be learned in less than a quarter of the time it takes to learn a traditional language. For example, all letters have one sound and one sound *only*. There are only 16 grammar rules to learn, compared to the hundreds that pervade English and other Romance or Germanic languages. Furthermore, there are no irregular verb forms (you have to love that!). Even the vocabulary is simple to learn; many words are instantly recognizable (and learnable), such as these:

- Telefono (telephone)
- Biciclo (bicycle)
- Masxino (machine)
- Reto (network)
- Kosmo (outer space)
- Plano (plan)

Speakers of languages other than English will recognize the roots of these words; Reto, for example, is similar to the Spanish word *red* (network). A pretty good Esperanto–English dictionary can be found at http://esperanto-panorama.net/vortaro/eoen.htm.

So what does this have to do with telecommunications and the transmission of e-mail messages? Read on.

Layer Seven: The Application Layer

The network user's application (Eudora, Outlook, Outlook Express, PROFS, etc.) passes data down to the uppermost layer of the OSI Model, called the Application Layer. The Application Layer provides a set of highly specific services to the application above it that have to do with the *meaning* or *semantic content* of the data. These services include, among others, file transfer, remote file access, terminal emulation, network management, mail services, and data interoperability. This interoperability is what allows our PC user and our mainframe-based user to communicate: The Application Layer converts the application-specific information into a common, "canonical" form that can be understood by both systems. A canonical form is a form that can be understood universally. The word comes from "canon," which refers to the body of officially established rules or laws that govern the practices of a church. The word also means an accepted set of principles of behavior that all parties in a social or functional grouping agree to abide by. Hence, the applicability of Esperanto.

Now let's examine a real-world example of a network-oriented canonical form.

When network hardware manufacturers build components—switches, multiplexers, cross-connect systems, modem pools—for sale to their customers, they do so knowing that one of the most important aspects of a successful hardware sale is the inclusion of an element-management system that will allow the customer to manage the device within his or her network. The only problem is that today, most networks are made up of equipment purchased from a variety of vendors. Each vendor develops its own element managers on a device-by-device basis, and they work exceptionally well for each device. This does not become a problem until it comes time to create a management hierarchy for a large network—shown in Figure 2-15—at which time the network management center begins to look a lot like a Macy's television department. Each device or

Figure 2-15
A large
managed
network

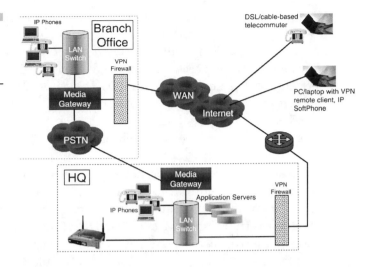

set of devices requires its own display monitor, and when one device in the network fails, causing a waterfall effect, the network manager must reconstruct the entire chain of events to discover what the original causative factor was. This is sometimes called the "Three Mile Island effect." Back in the 1970s when the Three Mile Island nuclear power plant went critical and tried to make Pennsylvania glow in the dark, it became clear to the Monday morning quarterbacks trying to reconstruct the event (and create the "How could this have been prevented" document) that all the information required to turn the critical failure of the reactor into a nonevent was in fact in the control room; that is, buried somewhere in the hundreds of pages of fanfold paper that came spewing out of the high-speed printers scattered all over the control room. There was no procedure in place to receive the output from the many managed devices and processes involved in the complex task of managing a nuclear reactor, analyze the output, and hand a simple, easy-to-respond-to decision to the operator.

The same problem is true in complex networks. Most of them have hundreds of managed devices with simple associated element-management systems that generate primitive data about the health and welfare of each device. The information from these element managers is delivered to the network management center, where it is displayed on one of many monitors that the network managers themselves use to track and respond to the status of the network. What they *really* need is a single map of the network that shows all of the managed devices in green if they are healthy. If a device begins to approach a preestablished threshold of performance, the icon on the map that represents that device turns yellow, and if it fails entirely it turns red, yells loudly, and automatically reports and escalates the trouble. In one of his many books on American management practices, USC Professor Emeritus Warren Bennis observes that "the business of the future will be run by a person and a dog. The person will be there to feed the dog; the dog will be there to make sure the person doesn't touch anything." Clearly that model applies here.

So, how can this ideal model of network management be achieved? Every vendor will tell you that its element-management system is the best element manager ever created. None of them are willing to change the user interface that they have created so carefully. Using a canonical form, however, there is no reason to. All that has to be done is to exact an agreement from all vendors that stipulates that while they do not have to change their user interface, they must agree to speak some form of technological Esperanto on the back side of their device. That way, the users still get to use the interface they have grown accustomed to, but on the network side, every management system will talk to every other management system using a common and widely accepted form. Again, it's just like a canonical language: If people from five different language groups need to communicate, they have a choice. They can each learn everybody else's language (four additional languages per person) or they can agree on a canonical language (Esperanto), which reduces the requirement to a single language each.

In network management, there are several canonical forms. The most common are ISO's Common Management Information Protocol (CMIP), the IETF's Simple Network Management Protocol (SNMP), and the Object Management Group's Common Object Request Brokered Architecture (CORBA). As long as every manager device agrees to use one of these on the network side, every system can talk to every other system.

Other canonical forms found at the Application Layer include ISO's X.400/X.500 Message Handling Service (Quiz: Where would you find these?) and the IETF's Simple Mail Transfer Protocol (SMTP) for e-mail applications; ISO's File Transfer, Access, and Management (FTAM), the IETF's File Transfer Protocol (FTP), and the Hypertext Transfer Protocol (HTTP) for file transfer; and a host of others. Note that the services provided at this layer are highly specific in nature: They perform a limited subset of tasks.

Layer Six: The Presentation Layer

For our e-mail example, let's assume that the Application Layer converts the PC-specific information to X.400 format and adds a header that will tell the receiving device to look for X.400-formatted content. This is not a difficult concept; think about the nature of the information that must be included in any e-mail encoding scheme. Every system must have a field for the following:

- Sender (From)
- Recipient (To)
- Date
- Subject
- Cc
- Bcc
- Attachment
- Copy
- Message body
- Signature (optional)
- Priority
- Various other miscellaneous fields

The point is that the number of defined fields is relatively small, and as long as each mail system knows what the fields are and where they exist in the coding scheme of the canonical standard, it will be able to map its own content to and from X.400 or SMTP. Problem solved.

Once the message has been encoded properly as an e-mail message, the Application Layer passes the now slightly larger message down to the Presentation Layer. It does this across a layer-to-layer interface using a simple set of commands called *service primitives*.

The Presentation Layer provides a more general set of services than does the Application Layer, which have to do with the structural *form* or *syntax* of the data. These services include code conversion, such as seven-bit ASCII to eight-bit EBCDIC translation; compression, using such services as PKZIP, British Telecom Lempel-Ziv, the various releases of the Moving Picture Experts Group (MPEG), the Joint Photographic Experts

Group (JPEG), and a variety of others; and encryption, including Pretty Good Privacy (PGP), Public Key Cryptography (PKC), and so on. Note that these services can be used on any form of data: Spreadsheets, word processing documents, and rock music can all be compressed and encrypted. Compression is typically used to reduce the number of bits required to represent a file through a complex manipulative mathematical process that identifies redundant information in the image, removes it, and sends the resulting smaller file off to be transmitted or archived. To explain how compression works, let's examine JPEG.

JPEG was developed jointly by ISO and the ITU-T as a technique for the compression of still images while still retaining varying degrees of quality as required by the user's application. Here's how it works. Please refer to Figures 2-16a and b, which are photographs of my daughter, Cristina.

Figure 2-16a shows the original photograph, a reasonably good quality picture that has, in fact, been substantially compressed using JPEG. Figure 2-16b is a small portion of the first image, specifically Cristina's right eye. Notice the small boxes that make up the image. Those boxes are called *picture elements,* or pixels. Each pixel requires substantial computer memory and processing resources: eight bits for storage of the red components of the image, eight bits for green, and eight bits for blue —the three primary colors (and the basis for the well-known RGB color scheme). That's 24-bit color, and every pixel on a computer screen requires them. Furthermore, there are a lot of pixels on a screen: Even a relatively low-resolution monitor that operates at 640x480 has 307,200 pixels, with 24 bits allocated per pixel. That equates to 921,600 bytes of information, or roughly one megabyte. And that's just color information. Just for fun, let's see what happens when we make the image move, as we will do if we're transporting video. Since typical video generates 30 frames per second, that's 221,184,000 bits that have to be allocated per second—a 222 Mbps signal. That's faster than a 155 Mbps SONET OC-3c signal! The message is that we'd better be doing *some* kind of compression!

JPEG uses an ingenious technique to reduce the bit count in still images. First, it clusters the pixels in the image (look at Figure 2-16b) into 16-pixel-by-16-pixel groups, which it then reduces to eight-by-eight groups by eliminating every other pixel. The JPEG software then calculates an average color, hue, and brightness value for each eight-by-eight block, which it encodes and transmits to the receiver. In some cases the

Figure 2-16a
My daughter,
Cristina

Figure 2-16b
Cristina's eye
showing pixels

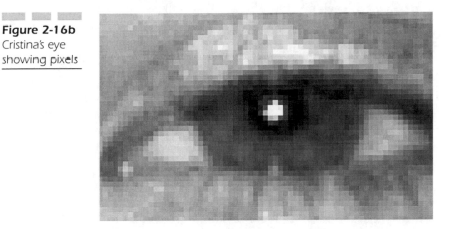

image can be further compressed, but the point is that the number of bits required to reconstruct a high-quality image is dramatically reduced by using JPEG.

More About Compression

Compression schemes do a good job of compressing and faithfully reconstituting images, particularly when the image being compressed is a photograph or video clip. To understand the dynamics of this relationship, let's take a moment to consider what it takes to create a digital photograph displayed on a computer screen.

A laptop computer display is often referred to as being "640 by 480," "800 by 600," or "1024 by 768." These numbers refer to the number of *picture elements,* more commonly called *pixels,* that make up the display. Look closely at the screen of your computer and you will find that it is made up of thousands of tiny spots of light (the pixels), each of which can take on whatever characteristics are required to correctly and faithfully paint the image on the screen. These characteristics include color components (sometimes called chrominance), black and white components (sometimes called luminance), brightness (the intensity of the signal), and hue (the actual wavelength of the color). These characteristics are important in video and digital imaging systems because they determine the quality of the final image. The image, then, is a mosaic of light; the "tiles" that make up the mosaic are light-emitting diodes that create the proper light at each pixel location.

Each pixel has a red, green, and blue "light generator," as shown in Figure 2-17. Red, green, and blue are called the *primary colors*, because as colors they form the basis for the creation of all other colors. It is a well known fact that if three white lights are covered with red, green, and blue color gels respectively, and the lights are shined at roughly the same spot as shown in Figure 2-18, the result will be a light spot for each color, but where the three colors intersect the result will be white light. *The combination of the three primary colors creates white.*

Each primary color also has a *complimentary color* in the overall spectrum. As Figure 2-19 shows, the complementary color for red is cyan, while the complimentary colors for green and blue are magenta and yellow, respectively. Table 2-2 shows the relationships that exist between the primary and complementary colors.

For full-color, uncompressed images, each of the red, green, and blue elements requires eight bits, or 24 bits per pixel. This yields what is known as "256-bit color" (2^8). Now consider the storage requirements for an image that is 640 by 480 pixels in size.

$$640 \times 480 = 307{,}200 \text{ pixels}$$

$$307{,}200 \text{ pixels} \times 24 \text{ bits/pixel} = 7{,}372{,}800 \text{ bits}$$

$$7{,}372{,}800 \text{ bits} \div 8 \text{ bits per byte} = 921{,}600 \text{ bytes per image}$$

In other words, an uncompressed, relatively low quality image requires one megabyte of storage. A larger 1,024 by 768 pixel image requires 6.3 MB of storage capacity in its uncompressed form. Given

Figure 2-17
The red, green, and blue (RGB) components of a pixel

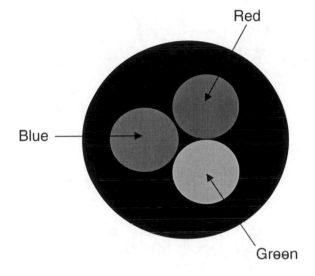

Figure 2-18
Red, green, and blue light, shining on the same place, create white light.

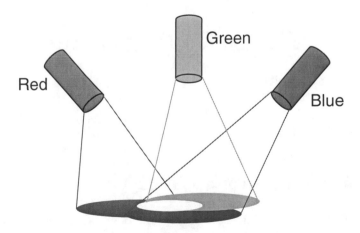

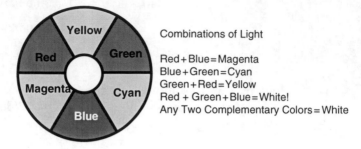

Figure 2-19
Complementary
colors

Combinations of Light

Red+Blue=Magenta
Blue+Green=Cyan
Green+Red=Yellow
Red + Green+Blue=White!
Any Two Complementary Colors=White

Table 2-2

Primary and
complemen-
tary color
combinations

If you combine:	The result is:
Red + Blue	Magenta
Green + Red	Yellow
Blue + Green	Cyan
Red + Green + Blue	White
Any Two Complementary Colors	White

today's relatively low bandwidth access solutions (ISDN, DSL, cable modems, wireless), we should be thankful that compression technologies exist to reduce the bandwidth required to move them through a network!

Image Coding Schemes

Images are encoded using a variety of techniques. These include Windows Bitmap (BMP), JPEG, the Graphical Interchange Format (GIF), and the Tag Image File Format (TIFF). They are described in detail in the following sections.

Windows Bitmap (BMP)

Windows bitmap files are stored in a device-independent bitmap format that allows the Windows operating system to display the image on any type of display device. The phrase "device independent" means that the bitmap specifies pixel color in a format that is independent of the method used by a display device to represent color. The default filename extension of a Windows file is .BMP.

The Joint Photographic Experts Group (JPEG)

As we mentioned earlier, JPEG is a standard designed to control image compression. The acronym stands for Joint Photographic Experts Group, the original name of the international body that created the standard. Made up of both technologists and artists, the JPEG committee created a highly complete and flexible standard that compresses both full-color and gray-scale images. It is effective when used with photographs, artwork, and medical images, and works less effectively on text and line drawings. JPEG is designed for the compression of still images, although there is a related standard (still in draft at the time of this writing) called Motion JPEG 2000 (MJP2). MJP2 is not really a formal standard, although various vendors and developers have tried to formalize it; MPEG, discussed later, is designed for the compression of moving images such as multimedia, movies, and real-time diagnostic images. However, according to the MJP2 committee, MJP2 will be the compression technology of choice for medical imaging, security systems, and digital cameras.

JPEG is considered to be a "lossy" solution, meaning that once the original image has been compressed, the decompressed image loses a slight degree of integrity when it is viewed. There are, of course, lossless compression algorithms; JPEG, however, achieves more efficient compression than is possible with competing lossless techniques. On first blush this would appear to be a problem for many applications, since the compression process actually "squeezes" information out of the original image, leaving a slightly inferior artifact. For example, a diagnostician examining a medical image might be concerned with the fact that the compressed image is not as good as the original. Luckily, this is not a problem for one very simple reason: The human eye is an imperfect viewing device. JPEG is designed to take advantage of well-understood limitations in the human eye, particularly the fact that small color changes are not perceived as discretely as small brightness changes. Clearly, JPEG is designed to compress images that will be viewed by people and, therefore, do not have to be absolutely faithful reproductions of the original image from which they were created. A machine analysis of a JPEG image would certainly yield inferior results, but to the human eye it is perfectly acceptable.

Of course, the degree of loss that occurs during a JPEG transformation can be controlled by adjusting a variety of compression parameters relative to each other. For example, file size and image quality can be adjusted relative to one another. A medical image, which requires extremely high quality on the image output side, would require a large file size, while a compressed text document could easily suffer significant loss without losing readability, resulting in a very small file.

The hardware or software coders that create (compress) and expand (decompress) JPEG images are called CODECs. If high image quality is not critically important, a low cost, higher loss CODEC can be used, thus reducing the overall cost of the deployed hardware or software solution.

JPEG is used as a compression tool for two primary reasons: to reduce file sizes for transmission or archival storage and to archive 24-bit color images instead of eight-bit color images. Clearly, the ability to reduce the number of bytes required to store an image is an advantage. It reduces the cost of networking by reducing the amount of connect time required to transmit an image across a network and lowers IT costs by reducing the amount of disk space required to store the image. JPEG can easily achieve compression ratios in excess of 20:1, meaning that a 2 MB file becomes a 100 KB entity following JPEG compression.

The second fundamental advantage of JPEG is that it stores full, 24-bit color information. GIF, the other image format that is widely used on the Web (discussed later), stores eight bits/pixel—256-bit color. GIF is designed for inexpensive computer displays. However, high-end display units are becoming quite cost-effective, and JPEG photos are far richer than GIF images when displayed on lower-cost displays. For this reason, GIF is seen by many as becoming obsolete.

The truth is, however, that JPEG will not completely replace GIF because for certain forms of images GIG continues to be a superior solution. For the most part, JPEG is better than GIF for storing full-color or gray-scale images of natural scenes such as scanned photographs and continuous-tone artwork. Any smooth variation in color, such as that which occurs in the highlighted or shaded areas of a subtle image, will be represented far more faithfully and in less space by JPEG.

On the other hand, GIF does a better job on images with only a few distinct colors such as sketches, maps, line drawings, and cartoons. Not only is GIF considered lossless for these images, it often achieves compression ratios far higher than JPEG can achieve. For example, large numbers of clustered pixels that are the same color are compressed *very* efficiently by GIF. However, JPEG has a hard time compressing such data without introducing visible defects.

Computer images, such as vector drawings or ray-traced graphics, are typically somewhere between photographs and line drawings in terms of their complexity in the eyes of the compression algorithm. The more complex the image, the more likely JPEG will be able to achieve significant levels of compression of the image. This is equally true with natural artwork. On the other hand, icons made up of only a few colors are better handled by GIF.

JPEG has difficulty achieving satisfactory compression with long, sharp edges. For example, if the image to be compressed has a row of black pixels immediately adjacent to a row of white pixels, the edges will often appear blurred unless a very high-quality setting is used—which, of course, reduces the degree of compression that JPEG will attempt to achieve. The good news is that such long, sharp edges are relatively uncommon in scanned photographs but are common in GIF files. Remember, straight lines are rare in nature, the subject of most photographs. They are quite common, however, in line drawings and illustrations. As a result, GIF is typically a better choice for compression of these image types.

As a general rule, two-level black-and-white images should not be converted to JPEG because they (by their very nature) violate the conditions listed previously. Gray-scale images that have 16 gray levels are far more acceptable to JPEG. However, GIF is considered to be a lossless encoding scheme for images of up to 256 levels, while JPEG continues to be lossy.

How JPEG Works The actual mechanical and mathematical processes that govern JPEG's inner workings are quite complex and will not be covered in detail here. Readers interested in more detail are directed to read the tutorial information that can be found at www.dcs.ed.ac.uk/home/mxr/gfx/2d/JPEG.txt. However, a high-level description of the process follows.

Earlier, we discussed the fact that a 640x480–pixel image requires nearly a full megabyte of disk space for uncompressed storage. To reduce that requirement, JPEG performs a mathematical permutation of the image that removes redundant information contained in the image, thus reducing storage requirements for the image. This results in loss of information and, while this appears to be a bad thing, it really isn't because of the limitations of the human eye described earlier.

Consider the image of my son Steve shown in Figure 2-20. Using a high-quality digital still camera, I took this portrait of him. I then enlarged a small section of the image to illustrate how the JPEG compression algorithm actually works. Keep in mind that a digital image comprises a large number of individual pixels, each of which has individual color characteristics associated with it. In the original photograph, Steve's skin tones are relatively uniform across his face except for the areas that are in shadow, such as his eyes and parts of his chin. However, when we enlarge a small area of the image, the pixels vary widely in terms of their color components. This disparity is what JPEG exploits as a way of getting the job done. It knows that the human eye cannot see the difference between these pixels until they are dramatically enlarged, as I have done in Figure 2-21.

What JPEG does is cluster the pixels in 16-by-16–pixel groups. It then mathematically removes every other pixel, resulting in eight-by-eight–pixel arrays. It then performs a mathematical transformation of the 64-pixel arrays called a *discrete cosine transform* that calculates an average

Figure 2-20
My son, Steve

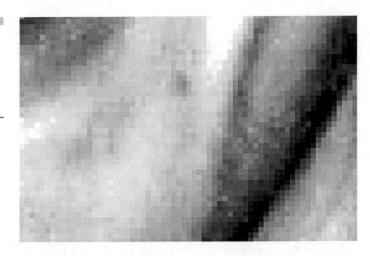

Figure 2-21
Steve's chin, enlarged to show color ranges among pixels

value for each group of pixels. Additional steps in the process further reduce the pixel count, resulting in a dramatic reduction in the total number of pixels required to represent the original image—hence the term "lossy." To restore the image, the reverse process is invoked.

JPEG is a 24-bit color scheme that yields high degrees of color fidelity. Unfortunately, many users do not have access to 24-bit color monitors, which means that a JPEG-encoded image will not display the image correctly. To display a full-color image, the host computer must analyze the file, choose an appropriate set of representative colors from a color "chart," and map the image into the new color domain. This process is known as *color quantization*.

The speed and image quality of a JPEG viewer used on a lower quality machine is largely determined by its quantization algorithm. If the algorithm is complex and well designed, the user will see minimal impact on the image from the quantization process. If the encoder is inferior, however, the result will be equally inferior.

Of course, there are ways around this. GIF image are already quantized to 256 or fewer colors (GIF doesn't allow more than 256 color palette entries.) GIF has the advantage that the creator of the image calculates the degree of color quantization so that viewers do not have to. For this reason, GIF viewers are typically faster than JPEG viewers. Of course, this also limits the user to whatever constraints the GIF viewer places on the image. If the creator of the image quantized the original image to a level that cannot be achieved by a receiving device, the image will have to be further quantized, resulting in poorer image quality.

GIF clearly offers some advantages over JPEG, but in the long term JPEG offers better image quality than GIF. Of course, eight-bit displays are rapidly disappearing from the landscape in favor of higher resolution monitors. For them, GIF has already become an academic argument because JPEG is a far better solution for image display.

The Graphics Interchange Format (GIF)

GIF defines a protocol designed to transmit raster-based graphic data that is independent of the hardware used to create or display them. GIF is defined in terms of blocks and subblocks that contain relevant information used to reproduce a graphic.

CompuServe released GIF as a free and open specification in 1987. It soon became a global standard and also played an important role within the Internet community as users started to share graphics files. It was well supported by CompuServe's Information Service but did not require a subscription to CompuServe. The format was relatively simple and well documented.

Like most graphic management tools, GIF compresses images to reduce the file size, using a technique called Lempel-Ziv-Welch (LZW) compression. Unisys holds a patent on the LZW procedure, which soon became a popular technique for data compression. GIF is not the only tool that relies on LZW; TIFF, described later, also includes LZW compression among its favored compression techniques.

JPEG vs. GIF

JPEG can typically achieve compression ratios of 10:1 to 20:1 without perceptible loss; 30:1 to 50:1 compression is possible with small to moderate visible artifacts, while 100:1 compression can be achieved— although quality suffers dramatically.

By comparison, a GIF image loses most of the color information in the process of reducing the 24-bit image to the 256-color palette, providing a 3:1 compression ratio. GIF's LZW compression scheme doesn't work well on photographs, yielding maximum compression levels of 5:1—and sometimes far less.

Because the human eye is more sensitive to luminance variations than it is to variations in color, JPEG compresses color data more than it does brightness data. Generally speaking, a gray-scale image that is

JPEG encoded only achieves a 10 to 25 percent reduction of a full-color JPEG file of similar quality. The uncompressed gray-scale image comprises eight bits per pixel, or roughly a third of the color data. As a result, the compression ratio is much lower.

Tag Image File Format (TIFF)

TIFF is a tag-based image format designed to promote the interoperability of digital images. The format came into being in 1986 when Aldus Corporation, working with leading scanner vendors, created a standard file format for images to be used in desktop publishing applications. The first version of the specification was published in July 1986; more current versions are released regularly by Adobe and are available at their Web sites.

The format that defines a file specifies both the structure of the file and its content. TIFF content consists of a series of definitions of individual fields. The structure, on the other hand, describes how to actually find the fields. These "pointers" are called tags.

TIFF provides a general-purpose format that is compatible with a variety of scanners and image-processing applications. It is device-independent and is acceptable to most operating systems, including Windows, Macintosh, and UNIX. The standard has been integrated into most scanner manufacturers' software and desktop publishing applications.

Adobe continues to enhance TIFF within publishing applications and maintains backward compatibility whenever possible.

Compressing Moving Images

Growth in videoconferencing, on-demand training, and gaming is fueling the growth in digital video technology but the problems mentioned previously still loom large. Recent advances have had an impact; for example, storage and transport limitations can often be overcome with compression.

The most widely used compression standard for video is MPEG, created by the Moving Pictures Expert Group, the joint ISO/IEC/ITU-T organization that oversees standards development for video. MPEG is relatively straightforward. There are three types of MPEG frames created during the compression sequence. They are intra (I) frames, predicted (P) frames, and bidirectional (B) frames. An intraframe or I frame is nothing more than a frame that is coded as a still image and used as a reference frame. Predicted frames, on the other hand, are predicted from the most recently reconstructed I or P frame. B frames are predicted from the closest two I or P frames, one from the past and one in the future. Imagine the following scenario: You are converting a piece of video that you shot at the beach to MPEG. The scene, shown in Figure 2-22, lasts 6 seconds, and is nothing more than footage of the fishing boat moving slowly in front of the camera, which is locked down on a tripod. Remember that video captures a series of still frames, one every $1/30$ of a second (30 frames per second). What MPEG does, in effect, is an analysis of the video based on the reference (I) frames, the predicted frames, and the bidirectional frames. From the image shown in the illustration, it should be clear that very little changes from one frame to another in a $1/30$-second interval. The boat may move slightly (but very slightly), and the foam that the propeller is churning up will change. Other than that, very little in the scene changes. Without going into too much technical detail, what MPEG does is reuse those elements of the I frame that don't change—or that change infrequently—so that it does not have to recreate them, thus reducing overall compression time. In our fishing boat scene it should be fairly obvious that the background certainly won't change much (unless a bird flies into it), the immediate foreground won't change, and the color and shape of the boat are constant. This is a fairly predictable scene. As a result, the number of I frames that will be interspersed among the P and B frames is relatively small. If the scene were different—a fishing boat being tossed about on rough seas, for example —then the number of minimally uncompressed reference frames would be greater because of the constantly changing point of reference. MPEG looks backward to establish patterns of behavior of past frames, then looks at the reference frame, and finally predicts what future frames will probably look like based on past history. Ultimately, the sequence of frames is as follows:

IBBPBBPBBPBBIBBPBBPB . . .

There are 12 frames from I to I, and the ratio of P frames to B frames is based on experience. Sequentially, the frames would look like the sequence shown in Figure 2-23.

Figure 2-22
A fishing boat —how much of the image must actually change as the result of MPEG compression?

Figure 2-23
The sequence of I, P, and B frames in an MPEG-compressed sequence

IBBPBBPBBPBBI

The MPEG Standards

There are several different versions of MPEG compression. They are described in the following sections.

MPEG-1

MPEG-1 was created to solve the transmission and storage challenges associated with relatively low-bandwidth situations, such as PC to CD-ROM or low bit rate data circuits. MPEG-2, on the other hand, was designed to address much more sophisticated transmission schemes in the 6 to 40 megabit per second range. This positions it to handle such applications as broadcast television and HDTV, as well as variable bit rate video delivered over packet, Ethernet, and Token Ring networks.

MPEG-2

The original MPEG-1 standard was optimized for CD-ROM applications at 1.5 Mbps. Video, one of the applications that users wanted to place on CD, was strictly noninterlaced format (progressive scan). Unfortunately, today's television pattern is interlaced. This created a problem; luckily MPEG-2 handles the conversion well.

MPEG-2 initially gained the attention of both the telephone and the cable television industries as the near-term facilitator of video-on-demand services. Today, it is widely used in DVD production because it has the ability to achieve high compression levels without loss of quality, as well as in cable television, DBS, and HDTV. MPEG-2 has also found a home in corporate videos and training products because of its ability to achieve CD-compatible compression levels.

MPEG-3

MPEG-3, also known as MP3, has become both famous and infamous of late and has made Napster and Metallica household names. MPEG layer 3 is a type of audio CODEC that delivers significant compression of as much as 12:1 of an audio signal with very little degradation of sound quality. Higher compression levels can be achieved, but sound quality suffers.

The standard bit rates used in MP3 CODECs are 128 and 112 Kbps. One advantage of MP3 is that a sound file can be broken into pieces and each piece remains independently playable. The feature that makes this possible means that MP3 files can be streamed across the net in real time, although network latency can limit the quality of the received signal.

The only real disadvantage of MP3 compression is that significant processor power is required to encode and play files. Dedicated MP3 players have recently emerged on the market and are proving to be quite popular, led by Apple's iPod. I received one as a gift from my wife this year; it has a 40 GB hard drive, on which I have now loaded nearly 4,000 songs —and used precisely 8 GB of space. Amazing.

MPEG-4

MPEG-4 is a specially designed standard for lower bit rate transmissions such as dial-up video. MPEG-4 is the result of an international effort that involves hundreds of researchers from all over the world. Its development was finalized in October 1998 and became an International Standard in the first months of 1999.

MPEG-4 builds on the proven success of three fields: digital television, interactive graphics applications, and interactive multimedia such as the distribution of HTML-encoded Web pages.

MPEG-7

MPEG-7, officially known as the Multimedia Content Description Interface, is an evolving standard used to describe multimedia content data that allows some degree of interpretation of the information's actual meaning, which can be accessed by a computer. MPEG-7 does not target any particular application; instead it attempts to standardize the interpretation of image elements so that it can support a broad range of applications.

Consider the following scenario. A video editor, attempting to assemble a final cut of a video, finds that she needs a specific piece of B-roll (fill-in footage) to serve as a segue and to cover a voice-over during one particular scene in the video. She knows that the footage she needs exists in the company's film vault; she just can't remember what project it is filed under. Using MPEG-7, she requests a 7-second clip that has the following characteristics:

- Bright blue sky, midday
- White beach in foreground, light surf
- Red and green beach umbrella in lower left corner of shot
- Sea gull enters from upper left, exits upper right
- Ship on horizon moving slowly from right to left

With MPEG-7, the clip can be found using these descriptors.

The elements that MPEG-7 attempts to standardize support a broad range of applications such as digital libraries, the selection of broadcast media, postproduction and multimedia editing, and home entertainment devices. MPEG-7 will also add a new level of search capability to the Web, making it possible to search for specific multimedia content as easily as text can be searched for today. This capability will be particularly valuable to managers and users of large content archives and multimedia catalogues that allow people to select content for purchase. The same information used for successful content retrieval will also be used to select and filter *push* content for highly targeted advertising applications.

Any domain that relies on the use of multimedia content will benefit from the deployment of MPEG-7. Consider the following list of examples:

- Architecture, real estate, and interior design (e.g., searching for ideas)
- Broadcast media selection (e.g., radio channels, TV channels)
- Cultural services (e.g., history museums, art galleries)
- Digital libraries (e.g., image catalogues; music dictionaries; biomedical imaging catalogues; film, video, and radio archives)
- E-Commerce (e.g., personalized advertising, online catalogues, and directories of e-shops)
- Education (e.g., repositories of multimedia courses, multimedia searches for support material)
- Home entertainment (e.g., systems for the management of personal multimedia collections including manipulation of content, home video editing, searching a game, karaoke)

- Investigation services (e.g., human characteristics recognition, forensics)
- Journalism (e.g., searching speeches of a certain politician using his or her name, voice, or face)
- Multimedia directory services (e.g., yellow pages, tourist information, geographical information systems)
- Multimedia editing (e.g., personalized electronic news service, media authoring)
- Remote sensing (e.g., cartography, ecology, natural resources management)
- Shopping (e.g., searching for clothes)
- Social (e.g., dating services)
- Surveillance (e.g., traffic control, surface transportation, nondestructive testing in hostile environments)

The standard also lists the following examples of how the capabilities of MPEG-7 might be used:

- A musician plays a few notes on a keyboard and retrieves a list of musical pieces similar to the required tune or images that match the notes in a certain way, as described by the artist.
- An artist draws a few lines on a screen and finds a set of images containing similar graphics, logos, ideograms.
- A clothing designer selects objects, including color patches or textures, and retrieves examples from which he or she selects the interesting objects with which to compose a design.
- Using an excerpt of Pavarotti's voice, an opera fan retrieves a list of Pavarotti's records and video clips in which Pavarotti sings a particular piece, as well as photographs of Pavarotti.

This is a remarkable (and remarkably complex!) standard that holds *enormous* promise.

The previous section focused on the layer six function known as compression. Encryption, on the other hand, is used when the information contained in a file is deemed sensitive enough to require that it be hidden from all but those eyes with permission to view it.

Encryption is one aspect of a very old science called cryptography. Cryptography is the science of writing in code; its first known use dates to 1900 BC when an Egyptian scribe used nonstandard hieroglyphs to capture the private thoughts of a customer. Some historians feel that cryptography first appeared as the natural result of the invention of the written word; its use in diplomatic messages, business strategy, and battle plans certainly supports the theory.

In data communications and telecommunications, encryption is required any time the information to be conveyed is sensitive and the possibility exists that the transmission medium is insecure. This can occur over any network, although the Internet is most commonly cited as being the most insecure of all networks.

All secure networks require a set of specific characteristics if they are to be truly secure. The most important of them are listed here:

- *Privacy/confidentiality:* The ability to guarantee that no one can read the message except the intended recipient

- *Authentication:* The guarantee that the identity of the recipient can be verified with full confidence

- *Message integrity:* Assurance that the receiver can confirm that the message has not been changed in any way during its transit across the network

- *Nonrepudiation:* A mechanism to prove that the sender really sent this message and that it was not sent by someone pretending to be the sender

Cryptographic techniques, including encryption, have two responsibilities: They ensure that the transmitted information is free from theft or any form of alteration and provide authentication for both senders and receivers. Today, three forms of encryption are most commonly employed: secret-key (or symmetric) cryptography, public-key (or asymmetric) cryptography, and hash functions. How they work is beyond the scope of this book, but there are numerous resources available on the topic. One of the best resources is *An Overview of Cryptography* by my good friend Gary Kessler. The paper, which Kessler updates routinely, is available at www.garykessler.net/library/crypto.html.

Another interesting encryption technology that has taken on an air of some importance in the last few years is a technique called *steganography*. Steganography is a technique used to surreptitiously hide one message inside another. The name derives from Johannes Trithemius's *Steganographia*, published in 1621, a work on cryptography and steganography that was, in fact, disguised as a book on black magic.

Steganographic messages are typically encrypted before being embedded in another message, after which a *covertext* is created to contain the encrypted message. This is called *stegotext*. For example, consider a

JPEG image, which is made up of groups of eight-bit bytes that ultimately encode the displayed image. Because JPEG is a compression technique, some quality of the original image is lost in the process, although with modern compression algorithms the loss is kept to an absolute minimum. Like all bytes, those found in a JPEG image have both a least significant bit and a most significant bit. In fact, changing the least significant bit from a one to a zero or vice versa results in no appreciable change to the output image. It makes sense, therefore, that someone could write a relatively simple routine to selectively encode the least signficant bit of each byte, collect all of the encoded bits, and group them into an embedded message. Make sense? Naturally, the larger the image, the more data can be hidden within it.

Consider, for example: A 24-bit bitmap that has eight bits representing each of the three primary colors (red, green, and blue—RGB) for each pixel. The red color alone, represented by eight bits, has 256 possible values (2^8). The difference between a red value of 01111111 and a red value of 01111110 will be undetectable by the human eye. Therefore, the least significant bit can be used for something other than color information. If we then encode the green and the blue as well, we will have three encodable bits, which means that three pixels will yield nine bits, one more than is required to encode an eight-bit character.

So why do we care about this? In October 2001, the *New York Times* claimed that terrorists were using steganographic techniques to encode messages into images (by some acounts, pornographic messages at adult Web sites). While these claims were largely dismissed by security experts, it is certainly a possibility. Needless to say, security analysts have created steganalysis tools that can be used to detect and read messages embedded steganographically in images and other file types.

From a more practical perspective, steganographic techniques can be used to embed important information such as digital watermarks, copyright information, and the like.

Let's turn our attention now back to our e-mail message.

The Session Layer

We have now left the Presentation Layer. Our e-mail message is encrypted, compressed, and may have gone through an ASCII-to-EBCDIC code conversion before descending into the complexity of the

Session Layer. As before, the Presentation Layer added a header containing information about the services it employed.

For being such an innocuous layer, the Session Layer certainly engenders a lot of attention. Some believe that the Session Layer could be eliminated by incorporating its functions into the layer above or the layer below, thus simplifying the OSI model. Whatever, the bottom line is that it *does* perform a set of critical functions that cannot be ignored.

First of all, the Session Layer ensures that a logical relationship is created between the transmitting and receiving applications. It guarantees, for example, that our PC user in Madrid receives his or her mail and *only* his or her mail from the mainframe, which is undoubtedly hosting large numbers of other e-mail users. This requires the creation and assignment of a logical session identifier.

Many years ago, I recall an instance when I logged into my e-mail account and found, to my horror, that I was actually logged into my vice president's account. Needless to say, I back-pedaled out of there as fast as I could. Today I know that this occurred because of an execution glitch in the Session Layer.

Layer five also shares responsibility for security with the Presentation Layer. You may have noticed that when you log in to your e-mail application, the first thing the system does is ask for a login ID, which you dutifully enter. The ID appears in the appropriate field on the screen. When the system asks for your password, however, the password does not appear on the screen—the field remains blank or is filled with stars, shown graphically in Figure 2-24. This is because the Session Layer knows that the information should not be displayed. When it receives the correct login ID, it sends a command to the terminal (your PC) asking you to enter your password. It then immediately sends a second message to the terminal telling it to turn off "local echo" so that your keystrokes are not echoed back onto the screen. As soon as the password has been transmitted, the Session Layer issues a command to turn local echo back on again, allowing you to once again see what you type.

Figure 2-24
Session Layer
turns off echo
to protect user.

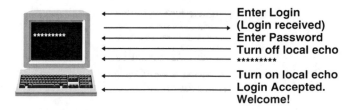

Enter Login
(Login received)
Enter Password
Turn off local echo
****

Turn on local echo
Login Accepted.
Welcome!

Another responsibility of the Session Layer that is particularly important in mainframe environments is a process called *checkpoint restart*.
This is a process that is analogous to the autosave function that is available on many PC-based applications today. I call it the Hansel and Gre-
tel function: As the mainframe performs its many tasks during the online
day, the Session Layer keeps track of everything that has been done,
scattering a trail of digital bread crumbs along the way as processing is
performed. Should the mainframe fail for some reason (the dreaded
ABEND), the Session Layer will provide a series of recovery checkpoints.
As soon as the machine has been rebooted, the Session Layer performs
the digital equivalent of walking back along its trail of bread crumbs. It
finds the most recent checkpoint and the machine uses that point as its
most recent recovery data, thus eliminating the possibility of losing huge
amounts of recently processed information.

So, the Session Layer may not be the most glamorous of the seven layers, but its functions are clearly important. As far as standards go, the
list is fairly sparse; see the ITU-T's X.225 standard for the most compre-
hensive document on the subject.

After adding a header, layer five hands the steadily growing Protocol
Data Unit, or PDU, down to the Transport Layer. This is the point where
we first enter the network. Until now, all functions have been software-
based and, in many cases, a function of the operating system.

The Transport Layer's job is simple: to guarantee end-to-end, error-
free delivery of the entire transmitted message—not bits, not frames or
cells, not packets, but the entire message. It does this by taking into
account the nature and robustness of the underlying physical network
over which the message is being transmitted, including the following
characteristics:

- Class of service required
- Data transfer requirements
- User interface characteristics
- Connection management requirements
- Specific security concerns
- Network management and reporting status data

There are two basic network types: dedicated and switched. We will
examine each in turn before discussing Transport Layer protocols.

Dedicated networks are exactly what the name implies: an always-on
network resource, often dedicated to a specific customer, which provides

very high-quality transport service. That's the good news. The bad news is that dedicated facilities tend to be expensive, particularly because the customer pays for them whether they use the facility or not. Unless they are literally using it 100 percent of the time, the network is costing them money. The other downside of a dedicated facility is susceptibility to failure: Should a terrorist backhoe driver decide to take the cable out, there is no alternative route for the traffic to take. It requires some sort of intervention on the part of the service provider that is largely manual. Furthermore, dedicated circuits tend to be inflexible, again, because they are dedicated.

Switched resources, on the other hand, work in a different fashion and have their own set of advantages and disadvantages to consider. First and foremost, they require an understanding of the word "virtual."

When a customer purchases a dedicated facility, he or she literally owns the resources between the two communicating entities as shown in Figure 2-25. Either the circuit itself is physically dedicated to them (common in the 1980s), or a timeslot on a shared physical resource such as a T-Carrier is dedicated to them. Data is placed on the timeslot or the circuit and it travels to the other end of the established facility—very simple and straightforward. There is no possibility of a misdelivered message, because the message only has a single possible destination. Imagine turning on the spigot in your front yard to water the plants and having water pour out of your neighbor's hose—it would be about that ridiculous.

In a switched environment, things work quite differently. In switched networks the only thing that is actually dedicated is a timeslot, because everything in the network that is physical is shared among many different users. Imagine what a wonderful boon to the service providers this technology is: It gives them the ability to properly support the transport requirements of large numbers of customers while selling the same physical resources to them, over and over and over again. Imagine!

To understand how this technology works, please refer to Figure 2-26. In this example, device D on the left needs to transmit data to device K on the right. Notice that access to the network resources (the switches in

Figure 2-25
Point-to-point
circuit

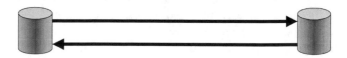

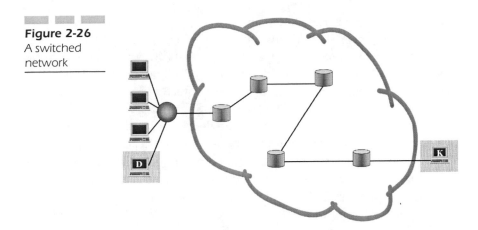

Figure 2-26
A switched
network

the cloud) is shared with three other machines. In order for this to work, each device must have a unique identifier so that the switches in the network can distinguish among all the traffic streams that are flowing through them. This identifier, often called a *virtual circuit identifier,* is assigned by the Transport Layer as one of its many responsibilities. As examples, X.25, frame relay, and Asynchronous Transfer Mode (ATM) are all switched network technologies that rely on this technique. In X.25 the identifier is called a virtual circuit identifier; in frame relay, a data link connection identifier (DLCI, pronounced "Delsey"); and in ATM it is called a virtual circuit identifier as well. Each of these will be described in greater detail later in the book.

When device D generates its message to device K for transport across the network, the Transport Layer "packages" the data for transmission. Among other things, it assigns a logical channel that the ingress switch uses to uniquely identify the incoming data. It does this by creating a unique combination of the unique logical address with the shared physical port to create an entirely unique virtual circuit identifier.

When the message arrives at the first switch, the switch enters the logical channel information in a routing table that it uses to manage incoming and outgoing data. There can be other information in the routing table as well, such as quality of service indicators; these details will be covered later, when we discuss the Network Layer (layer three).

The technology that a customer uses in a switched network is clearly not dedicated, but it gives the appearance that it is. This is called *virtual circuit service,* because it gives the appearance of being there when, in fact, it isn't. Virtual private networks (VPNs), for example, give a customer the appearance that they are buying private network service. In a sense they are: They do have a dedicated logical facility. The difference is that they share the physical facilities with many other users, which allows the service provider to offer the transport service for a lower cost. Furthermore, secure protocols protect each customer's traffic from interception. VPNs are illustrated in Figure 2-27.

As you may have intuited by now, the degree of involvement that the Transport Layer has varies with the type of network. For example, if the network consists of a single, dedicated, point-to-point circuit, then there is very little that could happen to the data during the transmission, since the data would consist of an uninterrupted, "single-hop" stream—there are no switches along the way that could cause pieces of the message to go awry. The Transport Layer, therefore, would have little to do to guarantee the delivery of the message.

However, what if the architecture of the network is not as robust as a private line circuit? What if this is a packet network, in which case the message is broken into segments by the Transport Layer and these segments are independently routed through the fabric of the network? Furthermore, what if there is no guarantee that all of the packets will take the same route through the network wilderness? In that case, the "route" actually consists of a *series* of routes between the switches, like a string of sausage links. In this situation, there is no guarantee that the components of the message will arrive in sequence; in fact, there is no guarantee that they will arrive at all! The Transport Layer, therefore, has a major responsibility to ensure that all of the message components arrive and that they carry enough additional information in the form of yet another header—this time on each packet—to allow them to be properly resequenced at the destination. The header, for example, contains

Figure 2-27
Virtual private
network (VPN)

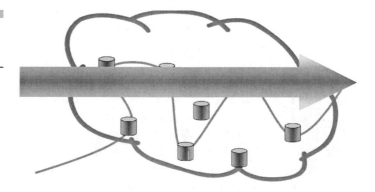

sequence numbers that the receiving Transport Layer can use to reassemble the original message from the stream of random packets.

Consider the following scenario: A transmitter fires a message into the network, where it passes through each of the upper layers until it reaches the originating Transport Layer, which segments the message into a series of five packets labeled one of five, two of five, three of five, and so on. The packets enter the network and proceed to make their way across the wilderness of the network fabric. Packets one, two, three, and five arrive without incident, although they do arrive out of order. Packet four, unfortunately, gets caught in a routing loop in New Mexico. The receive Transport Layer, tasked with delivering a complete, correct message to the layers above, puts everything on hold while it awaits the arrival of the errant packet. The layer, however, will wait only so long. It has no idea where the packet is; it does, however, know where it is *not*. After some predetermined period of time, the receive Transport Layer assumes that the packet isn't going to make it and initiates recovery procedures that result in the retransmission of the missing packet.

Meanwhile, the lost packet has finally stopped and asked for directions, extricated itself from the traffic jams of Albuquerque, and made its way to the destination. It arrives, covered with dust, an "I've Seen Crystal Caverns" bumper sticker on its trailer, expecting to be incorporated into the original message. By this time, however, the packet has been replaced with the resent packet. Clearly, some kind of process must be in place to handle duplicate packet situations, which happen rather frequently.

The Transport Layer then becomes the center point of message integrity. Transport Layer standards are diverse and numerous. ISO, the ITU-T, and the IETF publish recommendations for layer four. The ITU-T publishes X.224 and X.234, which detail the functions of both connection-oriented and connectionless networks. ISO publishes ISO 8073, which defines a transport protocol with five layers of functionality ranging from TP0 through TP4, as shown.

- Class 0 (TP0): Simple class
- Class 1 (TP1): Basic error recovery class
- Class 2 (TP2): Multiplexing class
- Class 3 (TP3): Error recovery and multiplexing class
- Class 4 (TP4): Error detection and recovery class

TP0 has the least capability; it is roughly equivalent to the IETF's User Datagram Protocol (UDP), which will be discussed a bit later. TP4 is the most common ISO Transport Layer protocol and is equivalent in capability to the IETF's Transmission Control Protocol (TCP). It provides an ironclad transport function and operates under the assumption that the network is wholly unreliable and must therefore take extraordinary steps to safeguard the user's data.

Before we descend into the wilds of the Network Layer, let's introduce the concepts of switching and routing.

Modern networks are often represented as a cloud filled with boxes representing switches or routers. Depending upon such factors as congestion, cost, number of hops between routers, and other considerations, the network selects the optimal end-to-end path for the stream of packets created by the Transport Layer. Depending upon the nature of the Network Layer protocol that is in use, the network will take one of two actions. It will either establish a single path over which all the packets will travel, in sequence, or, the network will simply be handed the packets and told to deliver them as it sees fit. The first technique, which establishes a seemingly dedicated path, is called *connection-oriented service*; the other technique, which does *not* dedicate a path, is called *connectionless service*. We will discuss each of these in turn. Before we do, however, let's discuss the evolution of switched networks.

Switched Networks

Modern switched networks typically fall into one of two major categories: *circuit switched*, in which the network preestablishes a path for transport of traffic from a source to a destination, as is done in the traditional telephone network; and *store-and-forward networks*, in which the traffic is handed from one switch to the next as it makes its way across the network fabric. When traffic arrives at a switch in a store-and-forward network, it is stored, examined for errors and destination information, and forwarded to the next hop along the path—hence the name, store and forward. Packet switching is one form of store-and-forward technology.

Store-and-Forward Switching

Thus, from some far-away and beleaguered island, where all day long the men have fought a desperate battle from their city walls, the smoke goes up to heaven; but no sooner has the sun gone down than the light from the line of beacons blazes up and shoots into the sky to warn the neighboring islanders and bring them to the rescue in their ships.

The Iliad by Homer, circa 700 BC

The first store-and-forward networks were invented and used . . . by the early Greeks and Romans. Indeed, Mycenae learned of the fall of Troy because of a line of signal towers between the two cities that used fire in each tower to transmit information from tower to tower. An opening on the side of each tower could be alternately opened and blocked, and using a rudimentary signaling code, short messages could be sent between them in a short period of time. A message could be conveyed across a large country, such as France, in a matter of hours, as Napoleon discovered and used to his great advantage.

The earliest *modern* store-and-forward networks were the telegraph networks. When a customer handed over a message in the form of a flimsy yellow paper that was to be transmitted, the operator would transmit the message in code over the open wire telegraph lines to the next office, where the message printed out on a streaming paper tape. On the tape would appear a sequence of alternating pencil marks and gaps, or spaces, combinations of which represented characters—a mark represented a one, while a space represented a zero. A point of historical interest is that the terminology "mark" and "space" is common parlance in modern networks: In T-Carrier, the encoding scheme is called Alternate Mark Inversion (AMI), because every other one alternates in polarity from the ones that surround it. Similarly, Alternate Space Inversion is used in signaling schemes such as on the ISDN D-Channel.

At any rate, the entire message would be delivered in this fashion, from office to office to office, ultimately arriving at its final destination, a technique called *message switching*. Over time, of course, the process became fully mechanized and the telegraph operators disappeared.

There was one major problem with this technique. What happened if the message, upon arrival, was found to be corrupt, or if it simply did not arrive for some odd reason? In that case, the entire message would have to be resent at the request of the receiver. This added overall delay in the system and was awfully inefficient because, in most cases, only a few characters were corrupted. Nevertheless, the entire message was retransmitted. Once the system was fully mechanized, it meant that the switches had to have hard drives on which to store the incoming messages, which added yet more delay, since hard drives are mechanical devices and by their very nature relatively slow. Improvements didn't come along until the advent of *packet switching*.

Packet Switching

With the arrival of low-cost, high-speed solid-state microelectronics in the 1970s, it became possible to take significant steps forward in switching technology. One of the first innovative steps was *packet switching.* In packet switching, the message that was transmitted in its entirety over the earlier message-switched store-and-forward networks is now broken into smaller, more manageable pieces that are numbered by the Transport Layer before being passed into the network for routing and delivery. This innovation offers several advantages. First, it eliminates the need for the mechanical, switch-based hard drive, because the small packets can now be handled blindingly fast by solid-state memory. Second, should a packet arrive with errors, it and it alone can be discarded and replaced. In message-switched environments, an unrecoverable bit error resulted in the inevitable retransmission of the entire message—not a particularly elegant solution. Packet switching, then, offers a number of distinct advantages.

As before, of course, there are also disadvantages. There is no longer (necessarily) a physically or logically dedicated path from the source to the destination, which means that the ability to guarantee quality of service on an end-to-end basis is severely restricted. There are ways around this, as you will see in the section that follows, but they are often costly and *always* complex. This is one of the reasons that IP telephony had a difficult time achieving widespread deployment. It worked fine in controlled, relatively small corporate environments where traffic patterns could be scrutinized and throttled as required to maintain quality of service (QoS). In the public IP environment, however (read *the Internet*), there was no way to ensure that degree of control. Today, however, such concerns have been largely eliminated; we will discuss VoIP in greater detail in a later chapter.

Packet switching can be implemented in two very different ways. We'll discuss them now.

Connection-Oriented Networks

Capt. Lewis is brave, prudent, habituated to the woods, & familiar with Indian manners & character. He is not regularly educated, but he possesses a great mass of accurate observation on all the subjects of nature which present themselves here, & will therefore readily select those only in his new route which shall be new. He has qualified himself for those observations of longitude & latitude necessary to fix the line he will go over.

> Thomas Jefferson to Dr. Benjamin Rush of Philadelphia on why he picked Meriwether Lewis for the Corps of Discovery

Six papers of ink powder; sets of pencils; "Creyons," two hundred pounds of "best rifle powder"; four hundred pounds of lead; 4 Groce fishing Hooks assorted; twenty-five axes; woolen overalls ad other clothing items, including 30 yds. Common flannel; one hundred flints; 30 Steels for striking or making fire; six large needles and six dozen large awls; three bushels of salt.

> Partial list of items purchased by Lewis for the trip

When Meriwether Lewis and William Clark left St. Louis with the Corps of Discovery in 1803 to travel up the Missouri and Columbia Rivers to the Pacific Ocean, they had no idea how to get where they were going. They traveled with and relied on a massive collection of maps, instruments, transcripts of interviews with trappers and Native American guides, an awful lot of courage, and the knowledge of Sacagawea—the wife of independent French-Canadian trader Toussaint Charbonneau, who accompanied them on their journey. As they made their way across the wilderness of the Northwest, they marked trees every few hundred feet by cutting away a large and highly visible swath of bark, a process known as "blazing." By blazing their trail, others could easily follow them without the need for maps, trapper lore, or guides. They did not need to bring compasses, sextants, chronometers, or to hire local guides; they simply followed the well-marked trail.

If you understand this concept, then you also understand the concept of connection-oriented switching, sometimes called *virtual circuit switching,* one of the two principal forms of switching technologies. When a device sends packets into a connection-oriented network, the first packet, often called a *call setup packet* or *discovery packet*, carries embedded in it the final destination address that it is searching for. Upon arrival at the first switch in the network, the switch examines the packet, looks at the destination address, and selects an outgoing port that will get the packet closer to its destination. It has the ability to do this because presumably, somewhere in the recent past, it has recorded the "port of arrival" of a packet from the destination machine, and concludes that if a packet arrived on that port from the destination host, then a good way to get closer to the destination is to go out the same port that the arriving packet came in on. The switch then records in its routing tables an entry that dictates that all packets originating from the same source (the source being a virtual circuit address/physical port address combination that identifies the logical source of the packets) should be transmitted out the same switch port. This process is then followed by every switch in the path, from the source to the destination. Each switch makes table entries, similar to the blazes left by the Corps of Discovery.[2]

With this technique, the only packet that requires a complete address is the initial one that blazes the trail through the network wilderness. All subsequent packets carry nothing more than a short identifier—a virtual circuit address—that instructs each switch they pass through how to handle them. Thus, all the packets with the same origin will follow the same path through the network. Consequently, they will arrive in order and will all be delayed the same amount of time as they traverse the network. The service provided by connection-oriented networks is called *virtual circuit service*, because it simulates the service provided by a dedicated network. The technique is called connection-oriented because the switches perceive that there is a relationship, or connection, between all of the packets that derive from the same source.

As with most technologies, there is a downside to connection-oriented transmission: In the event of a network failure or heavy congestion somewhere along the predetermined path, the circuit is interrupted and will require some form of intervention to correct the problem, because the network is not self-healing from a protocol point of view. There are certainly network management schemes and stopgap measures in place to reduce the possibility that a network failure might cause a service interruption, but in a connection-oriented network these measures are external. Nevertheless, because of its ability to emulate the service provided by a dedicated network, connection-oriented services are widely deployed and very successful. Examples include frame relay, X.25 packet-based service, and ATM. All will be discussed in detail later in the book.

Connectionless Networks

The alternative to connection-oriented switching is *connectionless switching*, sometimes called *datagram service*. In connectionless net-

---

[2]The alternative is to have a network administrator manually preconfigure the routes from source to destination. This guarantees a great deal of control, but also obviates the need for an intelligent network.

works there is no predetermined path from the source to the destination. There is no call setup packet; all data packets are treated independently, and the switches perceive no relationship between them as they arrive— hence the name, "connectionless." Every packet carries a complete destination address, since it cannot rely on the existence of a preestablished path created by a call setup packet.

When a packet arrives at the ingress switch of a connectionless network, the switch examines the packet's destination address. Based on what it knows about the topology of the network, congestion, cost of individual routes, distance (sometimes called *hop count*), and other factors that affect routing decisions, the switch will select an outbound route that optimizes whatever parameters the switch has been instructed to concern itself with. Each switch along the path does the same thing.

For example, let's assume that the first packet of a message, upon arrival at the ingress switch, would normally be directed out physical port number 7 because, based upon current known network conditions, that port provides the shortest path (lowest hop count) to the destination. However, upon closer examination, the switch realizes that while port 7 provides the shortest hop count, the route beyond the port is severely congested. As a result, the packet is routed out port 13, which results in a longer path but avoids the congestion. And since there is no preordained rout through the network, the packet will simply have to get directions when it arrives at the next switch.

Now, the second packet of the message arrives. Because this is a connectionless environment, however, the switch does not realize that the packet is related to the packet that preceded it. The switch examines the destination address on the second packet, then proceeds to route the packet as it did with the preceding one. This time, however, upon examination of the network, the switch finds that port 7, the shortest path from the source to the destination, is no longer congested. It therefore transmits the packet on port 7, thus ensuring that packet two will, in all likelihood, arrive before packet one! Clearly, this poses a problem for message integrity and illustrates the criticality of the Transport Layer, which, you will recall, provides end-to-end message integrity by reassembling the message from a collection of out-of-order packets that arrive with varying degrees of delay because of the vagaries of connectionless networks.

Connectionless service is often called unreliable because it fails to guarantee delay minimums, sequential delivery, or for that matter, any kind of delivery. This causes many people to question why network designers would rely on a technology that guarantees so little. The answer lies within the layered protocol model. While connectionless networks do not guarantee sequential delivery or limits on delay, they *will* ultimately deliver the packets. Because they are not required to transmit along a fixed path, the switches in a connectionless network have the freedom to "route around" trouble spots by dynamically selecting alternate pathways, thus ensuring delivery, albeit somewhat unpredictable. If this process results in out-of-order delivery, no problem—that's what the Transport Layer is for. Data communications is a team effort and requires the capabilities of many different layers to ensure the integrity and delivery of a message from the transmitter to the receiver. Thus, even an "unreliable" protocol has distinct advantages.

An example of a well-known connectionless protocol is the Internet Protocol, or IP. It relies on the TCP, a Transport Layer protocol, to guarantee end-to-end correctness of the delivered message. There are times, however, when the foolproof capabilities of TCP and TCP-like protocols are considered overkill. For example, network management systems tend to generate large volumes of small messages on a regularly scheduled basis. These messages carry information about the health and welfare of the network and about topological changes that routing protocols need to know about if they are to maintain accurate routing tables in the switches. The problem with these messages is that they (1) are numerous, and (2) often carry information that hasn't changed since the *last* time the message was generated, 30 seconds ago.

TCP and TP4 protocols are extremely overhead-heavy compared to their lighter-weight cousins UDP and TP1. Otherwise, they would not be able to absolutely, positively guarantee the delivery of the message. In some cases, however, there may not be a need to absolutely, positively guarantee delivery. After all, if I lose one of those status messages, no problem: It will be generated again in 30 seconds anyway. The result of this is that some networks choose not to employ the robust and capable protocols available to them, simply because the marginal advantage they provide doesn't merit the transport and processing overhead they create in the network. Thus, connectionless networks are extremely widely deployed. After all, those billion (or so) Internet users must be *reasonably* happy with the technology.

Let's now examine the Network Layer. Please note that we are now entering the realm of the chained layers, which you will recall are used by all devices in the path—end-used devices as well as the switches or routers themselves.

The Network Layer

The Network Layer, which is the uppermost of the three chained layers, has two key responsibilities: routing and congestion control. We will also briefly discuss switching at this layer, even though many books consider it to be a layer two process. So for the purists in the audience, please bear with me—there's a method to my madness.

When the telephone network first started its remarkable growth path at the sunrise of the twentieth century, there was no concept of switching. If a customer wanted to be able to speak with another customer, he or she had to have a phone in his or her house with a dedicated path to that person's home. Another person required another phone and phone line—and you quickly begin to see where this is leading. Figure 2-28 illustrates the problem: The telephone network's success would bring on the next ice age, blocking the sun with all the aerial wire the telephone network would be required to deploy. Consider this simple mathematical model: In order to fully interconnect, or mesh, five customers as shown in Figure 2-29 so that any one of them can call any other, the telephone company would have to install ten circuits, according to the equation $n(n-1)/2$, where n is the number of devices that wish to communicate. Extrapolate that out now to the population of even a small city—say, 2,000 people. That boils down to 3,997,999 circuits that would have to be

Figure 2-28
Aerial
telephone wire
(Courtesy
Lucent
Technologies)

Figure 2-29
Meshed
network; five
users, ten
circuits

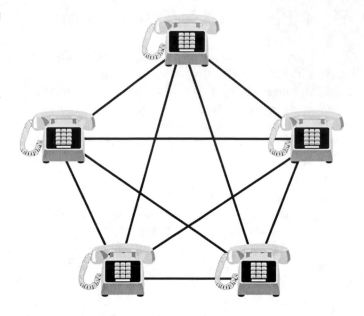

installed—all to allow 2,000 people to call each other. Obviously, some alternative solution was greatly needed. That solution was the switch.

The first "switches" did not arrive until 1878 with the near-disastrous hiring of young boys to work the cord boards in the first central offices. John Brooks, author of *Telephone: The First 100 Years,* offers the following:

> The year of 1878 was the year of male operators, who seem to have been an instant and memorable disaster. The lads, most of them in their late teens, were simply too impatient and high spirited for the job, which, in view of the imperfections of the equipment and the inexperience of the subscribers, was one demanding above all patience and calm. According to the late reminiscences of some of them, they were given to lightening the tedium of their work by roughhousing, shouting constantly at each other, and swearing frequently at the customers. An early visitor to the Chicago exchange said of it, "The racket is almost deafening. Boys are rushing madly hither and thither, while others are putting in or taking out pegs from a central framework as if they were lunatics engaged in a game of fox and cheese. It was a perfect bedlam."

Later in 1878 the boys were grabbed by their ears and removed from their operator positions, replaced quickly—and, according to a multitude of accounts, to the enormous satisfaction of the customers—by women, shown in Figure 2-30.

These operators were, in fact, the first switches. Now, instead of needing a dedicated circuit running from every customer to every other customer, each subscriber needed a single circuit that ran into the central exchange, where it appeared on the jack field in front of an operator (each operator managed approximately 100 positions, the optimum number according to Bell System studies). When a customer wanted to make a call, he or she would crank the handle on the phone, generating current which would cause a flag to drop, a light to light or a bell to ring in front of the operator. Seeing the signal, the operator would plug a headset into the customer's jack appearance and announce that she was ready to receive the number to be dialed. The operator would then simply "cross-connect" the caller to the called party, then wait for the receiver to be picked up on the other end. The operator would periodically monitor the call and pull the patch cord down when the call was complete. Incidentally, note the roller skates on the feet of the supervisors. The cord boards were so long that the process of collecting tickets took far too long, so they added mobility!

This model meant that instead of needing 3,997,999 circuits to provide universal connectivity for a town of 2,000 people, 2,000 were needed—a rather *significant* reduction in capital outlay for the telephone company, wouldn't you say? Instead of looking like Figure 2-29, the network now looked like Figure 2-31.

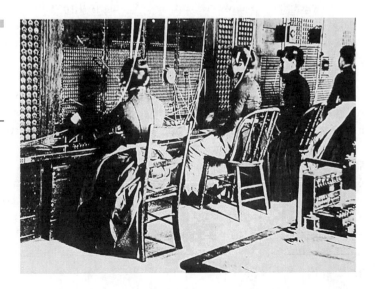

Figure 2-30
First women operators (Courtesy Lucent Technologies)

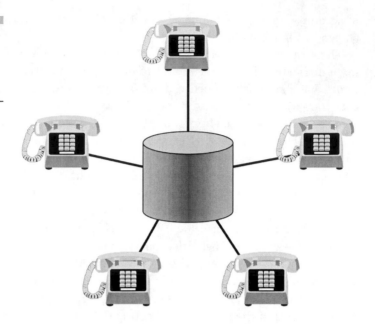

Figure 2-31
Switched
network; five
users, five
circuits

Over time, manual switching slowly disappeared, replaced by mechanical switches initially followed by more modern all-electronic switches. The first true mechanical switches didn't arrive until 1892, when Almon Strowger's Step-by-Step switch was first installed by his company, Automatic Electric.

Strowger's story is worth telling because it illustrates the serendipity that characterized so much of this industry's development. It seems that Almon Strowger was not an inventor, nor was he a telephone person. He was, in fact, an undertaker in a small town in Missouri. One day, he came to the realization that his business was (OK, I won't say dying) declining, and upon closer investigation determined that the town's operator was married to his competitor! As a result, any calls that came in for the undertaker naturally went to her husband—and *not* to Strowger.

To equalize the playing field, Strowger called upon his considerable talents as a tinkerer and designed a mechanical switch and dial telephone, which is still in use today in a number of developing countries.

The bottom line to all this is that switches create temporary end-to-end paths between two or more devices that wish to communicate with each other. They do it in a variety of ways; circuit-switched networks create a "virtually dedicated path" between the two end points and offer constant end-to-end delay, making them acceptable for delay-sensitive applications or connections with long hold times. Store-and-forward net-works, particularly packet networks, work well for short, bursty mes-sages with minimal delay sensitivity.

To manage all this, however, is more complicated than it would seem at first blush. First of all, the switches must have the ability to select not only a path, but also the *best* path, based on QoS parameters. This con-stitutes intelligent routing. Second, they should have some way of moni-toring the network so that they always know its current operational conditions. Finally, should they encounter unavoidable congestion, the switches should have one or more ways to deal with it.

Routing Responsibilities

So, how are routing decisions made in a typical network? Whether con-nectionless or connection-oriented, the routers and switches in the net-work must take into account a variety of factors to determine the best path for the traffic they manage. These factors fall into a broad category of rule sets called *routing protocols*. For reference purposes, please refer to the "tree" shown in Figure 2-32.

Once the Transport Layer has taken whatever steps are necessary to prepare the packets for their transmission across the network, they are passed to the Network Layer.

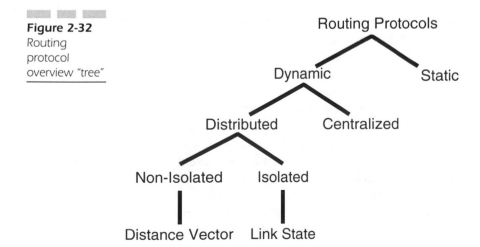

Figure 2-32
Routing
protocol
overview "tree"

The Network Layer has two primary responsibilities in the name of network integrity: *routing* and *congestion control*. Routing is the process of intelligently selecting the most appropriate route through the network for the packets; congestion control is the process that ensures that the packets are minimally delayed (or at least equally delayed) as they make their way from the source to the destination. We will begin with a discussion of routing.

Routing Protocols

Routing protocols are divided into two main categories—*static routing protocols* and *dynamic routing protocols*. Static routing protocols are those that require a network administrator to establish and maintain them. If a routing table change is required, the network administrator must manually make the change. This ensures absolute security, but is labor intensive and therefore less frequently used other than in highly secure environments (e.g. military, health care) or network architectures that are designed around static routing because the routes are relatively stable anyway (IBM's Systems Network Architecture, SNA, for example). More common are dynamic routing protocols, in which network devices make their own decisions about optimum route selection. They do this in the following general way: They pay attention to the network around them; collect information from their neighbors about best routes to particular destinations based on such parameters as least number of hops, least delay, lowest cost, or highest bandwidth; archive those bits of information in tables; and then selectively flush the tables periodically to ensure that the information contained in them is always as current as possible. Because dynamic routing protocols assume intelligence in the switch and can therefore reduce the amount of human intervention required, they are commonly used; in fact, they are the most widely deployed routing protocols.

Dynamic routing protocols are further divided into two subcategories, *centralized* and *distributed*. Centralized routing protocols concentrate the route decision-making processes in a single node, thus ensuring that all nodes in the network receive the same and most current information possible. When a switch or router needs routing information that is not contained in its own table, it sends a request to the root node asking for direction. There are significant downsides to this technique. First, by concentrating the decision-making capability in a single node, the likelihood of a catastrophic failure is dramatically increased. If that node fails, the entire network's ability to seek optimal routing decisions fails. Second, because all nodes in the network must go to that central device for routing instructions, a significant choke point can result.

There are several options that can reduce the vulnerability of a single point of failure. The first, of course, is to distribute the routing function. This conflicts with the concept of centralized routing, but only somewhat. Consider the Internet, for example. It uses a sort of hybrid of centralized and distributed routing protocols in its Domain Name Server (DNS) function. A limited number of devices are tasked with the responsibility of maintaining knowledge of the topology of the network and the location of domains, providing something of a AAA trip-planning service for data packets.

Another option is to designate a backup machine tasked with the responsibility of taking over in the event of a failure of the primary routing machine. This technique, used in the early Tymnet packet networks, relied on the ability of the primary machine to send a "sleeping pill packet" to the backup machine, with these instructions: "Pay attention to what I do, memorize everything I learn, but just sit in the corner and be a potted plant. Take no action, make no routing decisions—just learn. Let the sleeping pill keep you in a semicomatose state. If for some reason I fail, I will stop sending the sleeping pills, at which time you will wake up and take over until I recover." An ingenious technique, but overly complex and far too failure prone for modern network administrators. Distributed routing protocols are far more common today.

In distributed routing protocol environments, each device collects information about the topology of the network and makes independent routing decisions based upon the information it accumulates. For example, if a router sees a packet from source X arrive on port 12, it knows that somewhere out port 12, it will find destination X. It doesn't know how far out there necessarily, just that the destination is somewhere out there over the digital horizon. Thus, if a packet arrives on another port looking to be transmitted to X, the router knows that by sending the packet out port 12 it will at least get closer to its destination. It therefore makes an entry in its routing tables to that effect, so that the next time a packet arrives with the same destination, the switch can consult its table and route the packet quickly.

These routing protocols are analogous to the process of stopping and asking for directions on a road trip (or not), reputedly one of the great male-female differentiators—right after who controls the TV remote. Anthropologists must have a field day with this kind of stuff. According to apocryphal lore, women have no problem whatsoever stopping and asking for directions, while men are loathe to do it—one of those silly threats to the manhood things. Anyway, back to telecomm. If you were planning a road trip across the country, you could do so using one of two philosophies. You could go to AAA or a travel agent and have them plan out the entire route for you, or you could do the Jack Kerouac thing and simply get in the car and drive. Going to AAA seems to be the simplest option because, once the route is planned, all you have to do is follow the directions—Lewis's blazed trail, as it were. The downside is that if you make it as far as Scratch Ankle, Alabama (yes, it's a real place), and the road over which you are supposed to travel is closed, you are stuck—you have to stop and ask for directions anyway.

The alternative is to simply get in the car, drive to the nearest gas station, and tell them that you are trying to get to Dime Box, Texas. The attendant will no doubt tell you the following: "I don't know where Dime Box is, but I know that Texas is down in the Southwest. So if you take this highway here to Kansas City, it'll get you closer. But you'll have to ask for better directions when you get there." The next gas station attendant may tell you, "Well, the quickest way to central Texas is along this highway, but it's rush hour and you'll be stuck for days if you go that way. I'd take the surface street. It's a little less comfortable, but there's no congestion." By stopping at a series of gas stations as you traverse the country and asking for help, you will eventually reach your destination, but the route may not be the most efficient. That's okay, though, because you will never have to worry about getting stuck with bad directions. Clearly the first example of these is connection oriented; the second is connectionless. Connection-oriented travel is a far more comfortable, secure way to go; connectionless is riskier, less sure, more flexible, and much more fun. Obviously, distributed routing protocols are centrally important to the traveler as well as to the gas station attendant who must give them reliable directions, and equally important to routers and switches in connectionless data networks.

Distributed routing protocols fall into two categories: *distance vector* and *link state*. Distance vector protocols rely on a technique called *table swapping* to exchange information about network topology with each other. This information includes destination/cost pairs that allow each device to select the least cost route from one place to another. On a scheduled basis, routers transmit their entire routing tables on all ports to all adjacent devices. Each device then adds any newly arrived information to its own tables, thus ensuring currency. The only problem with this technique is that it results in a tremendous amount of management traffic (the tables) being sent between network devices, and if the network is relatively static—that is, changes in topology don't happen all that often —then much of the information is unnecessary and can cause serious congestion. In fact, it is not uncommon to encounter networks that have more management traffic traversing their circuits than actual user traffic. What's wrong with this picture? Distance vector works well in small networks where the number of multihop traverses is relatively low, thus minimizing the impact of its bandwidth-intensive nature.

Distance vector protocols have that name because of the way they work. Recovering physicists will remember that a vector is a measure of something that has both direction and magnitude associated with it. The name is appropriate in this case, because the routing protocol optimizes on a direction (port number) and a magnitude (hop count).

Since networks are growing larger, traffic routinely encounters route solutions with large hop counts. This reduces the effectiveness of distance vector solutions. A better solution is the link state protocol. Instead of transmitting entire routing tables on a scheduled basis, link state protocols use a technique called *flooding* to only transmit changes that occur to adjacent devices *as they occur*. This results in less congestion and more efficient use of network resources and reduces the impact of multiple hops in large scale networks.

Both distance vector and link state protocols are in widespread use today. The most common distance vector protocols are the Routing Information Protocol (RIP), and Cisco's Interior Gateway Routing Protocol (IGRP) and Border Gateway Protocol (BGP). Link state protocols include Open Shortest Path First (OSPF), commonly used on the Internet, as well as the Netware Link Services Protocol (NLSP), used to route IPX traffic.

Clearly, both connection-oriented and connectionless transport techniques, as well as their related routing protocols, have a place in the modern telecommunications arena. As QoS becomes such a critical component of the service offered by network providers, the importance of both routing and congestion control becomes apparent. We now turn our attention to the second area of responsibility at the Network Layer, *congestion control*.

Congestion Control

At its most fundamental level, congestion control is a mechanism for reducing the volume of traffic on a particular route through some form of load balancing. No matter how large, diverse, or capable a network is, some degree of congestion is inevitable. It can result from sudden, unexpectedly high utilization levels in one area of the network, from failures of network components, or from poor engineering. In the telephone network, for example, the busiest calling day of the year in the United States is Mother's Day. To reduce the probability that a caller will not be able to complete a call to Mom, network traffic engineers take extraordinary steps to load balance the network. For example, when subscribers on the east coast are making long-distance calls at 9:00 in the morning, west coast subscribers haven't even turned on their latte machines yet. Network resources in the west are underutilized during that period, so engineers route east coast traffic westward and then hairpin it back to its destination, to spread the load across the entire network. As the day gets later, they reduce the volume of westward-bound traffic to ensure that California has adequate network resources for its own calls.

There are two terms that are important in this discussion. One is congestion; the other is delay. The terms are often used interchangeably, but they are not the same thing, nor are they always related to one another.

Years ago I lived in the San Francisco Bay area, where traffic congestion is a way of life. I often had to drive across the many bridges that crisscross San Francisco Bay, the Suisun Straits, or the Sacramento River. Many of those bridges require drivers to stop and pay a toll, resulting in localized delay. The time it takes to stop and pay the toll is mere seconds, yet traffic often backs up for miles as a result of this local phenomenon, causing a global effect.

This is the relationship between the two: Local delay often results in widespread congestion. And congestion is usually caused by inadequate buffer or memory space. Increase the number of buffers—the lanes of traffic on the bridge, if you will—and congestion drops off. Open another line or two at Home Depot ("No waiting on line seven!")—and congestion drops off.

The various players in the fast-food industry manage congestion in different ways—and with dramatically different results. Without naming them, some use a single queue with a single server to take orders, a technique that works well until the lunch rush begins. Then things back up dramatically. Others use multiple queues with multiple servers, a technique that is better, except that one queue can experience serious delays should someone place an order for a nonstandard item or try to pay with a credit card. That line then experiences serious delay. The most effective restaurants stole an idea from the airlines, and use a single queue with multiple servers. This keeps things moving because the instant a server is available, the next person in line is served.

Remember when Jeff Goldblum, the chaos theoretician in *Jurassic Park,* talked about the Butterfly Effect? How a butterfly flapping its wings in the Amazon Basin can kick off a chain of events that affect weather patterns in New York City? That aspect of Chaos Theory contributes greatly to the manner in which networks behave—and the degree to which their behavior is immensely difficult to predict under load.

So how is congestion handled in data networks? The simplest technique, used by both frame relay and ATM, is called packet discard, shown in Figure 2-33. In the event that traffic gets too heavy based on preestablished management parameters, the switches simply discard excess packets. They can do this because of two facts: First, networks are highly capable and the switches rarely have to resort to these measures; second, the devices on the ends of the network are intelligent and will detect the loss of information and take whatever steps are required to have the discarded data resent. As drastic as this technique seems, it is not all that catastrophic. Modern networks are heavily dependent on optical fiber and highly capable digital switches; as a result, packet discard, while serious, does not pose a major problem for the network. Even when it is required, recovery techniques are fast and accurate.

Because congestion occurs primarily as the result of inadequate buffer capacity, one solution is to preallocate buffers to high-priority traffic. Another is to create multiple queues with varying priority levels. If a voice or video packet arrives that requires high-priority, low-delay

Figure 2-33
Congestion
control with
packet discard

treatment, it will be deposited into the highest priority queue for instan-taneous transmission. This technique holds great promise for the evolv-ing all-services Internet, because its greatest failing is its inability to deliver multiple, dependable, sustainable grades of service quality.

Other techniques are available that are somewhat more complex than packet discard but do not result in loss of data. For example, some devices, when informed of congestion within the network, will delay transmission to give the network switches time to process their overload before receiving more. Others will divert traffic to alternate, less-congested routes, or trickle packets into the network, a process known as "choking." This is shown in Figure 2-34. Frame relay, for example, has the ability to send what is called a *choke packet* into the network, implicitly telling the receiving device that it should throttle back to reduce conges-tion in the network. Whether it does or not is another story, but the point is that the network has the intelligence to invoke such a command when required. This technique is now used on freeways in large cities: Traffic lights are installed on major onramps that meter the traffic onto the roadway. This results in a dramatic reduction in congestion. Other net-works have the intelligence to diversely route traffic, thus reducing the congestion that occurs in certain areas.

Clearly, the Network Layer provides a set of centrally important capa-bilities to the network itself. Through a combination of network proto-cols, routing protocols, and congestion control protocols, routers and switches provide granular control over the integrity of the network.

Figure 2-34
Congestion
control using
throttling

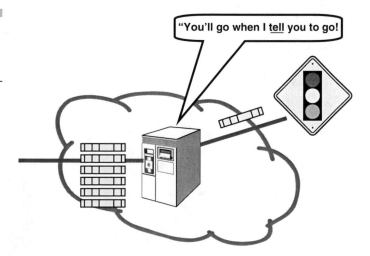

Back to the E-Mail Message

Let us return now to our e-mail example. The message has been divided into packets by the Transport Layer and delivered in pieces to the Network Layer, which now takes whatever steps are necessary to ensure that the packets are properly addressed for efficient delivery to the destination. Each packet now has a header attached to it that contains routing information and a destination address. The next step is to get the packet correctly to the next link in the network chain—in this case, the next switch or router along the way. This is the responsibility of the Data Link Layer.

The Data Link Layer

The Data Link Layer is responsible for ensuring bit-level integrity of the data being transmitted. In short, its job is to make the layers above believe that the world is an error-free and perfect place. When a packet is handed down to the Data Link Layer from the Network Layer, it wraps the packet in a *frame*. In fact, the Data Link Layer is sometimes called the *frame layer*. The frame built by the Data Link Layer is made up of several fields, shown graphically in Figure 2-35, that give the network devices the ability to ensure bit-level integrity and proper delivery of the packet, which is now encased in a frame, *from switch to switch*. Please note that this is different from the Network Layer, which concerns itself with routing packets *to the final destination*. Even the addressing is unique: Packets contain the address of the ultimate destination, used by the network to route the packet properly; frames contain the address of the next link in the network chain (the next switch), used by the network to move the packet along, switch by switch.

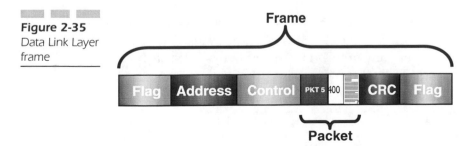

Figure 2-35
Data Link Layer
frame

As the diagram illustrates, the beginning and end fields of the frame are called flags. These fields, made up of a unique series of bits (0111110), can only occur at the beginning and end of the frame—they are never allowed to occur within the bitstream inside the frame through a process that we will describe momentarily. These flags are used to signal to a receiving device that a new frame is arriving or that it has reached the end of the current frame,[3] which is why their unique bit pattern can never be allowed to occur naturally within the data itself—it could indicate to the receiver (falsely) that this is the end of the current frame. If the flag pattern does occur within the bitstream, it is disrupted by the transmitting device through a process called bit stuffing or zero-bit insertion, in which an extra zero is inserted in the middle of the flag pattern, based on the following rule set. When a frame of data is created at an originating device, the very last device to touch the frame—indeed, the device that actually adds the flags—is called a *Universal Synchronous/Asynchronous Receiver-Transmitter* (USART). The USART, sometimes called an *Integrated Data Link Controller* (IDLC), is a chipset that has a degree of embedded intelligence. This intelligence is used to detect (among other things) the presence of a false flag pattern in the bitstream around which it builds the frame. Since a flag is made up of a zero followed by six ones and a final zero, the IDLC knows that it can never allow that particular pattern to exist between any two real flags. So, as it processes the incoming bitstream, it looks for that pattern and makes the following decision: *If I see a zero followed by five ones, I will automatically and without question insert a zero into the bitstream at that point.* This is illustrated in Figure 2-36.

This, of course, destroys the integrity of the message, but it doesn't matter. At the receive device, the IDLC monitors the incoming bits. As the frame arrives it sees a *real* flag at the beginning of the frame, an indication that a frame is beginning. As it monitors the bits flowing by, it *will* find the zero followed by five bits, at which point it knows, beyond a shadow of a doubt, that the very next bit is a zero—which it will promptly remove, thus restoring the integrity of the original message.

The receiving device has the ability to detect the extra zero and remove it before the data moves up the protocol stack for interpretation. This bit stuffing process guarantees that a "false flag" will never be interpreted as a final flag and acted upon in error.

---

[3]Indeed, the final flag of one frame is often the beginning flag of the *next* frame.

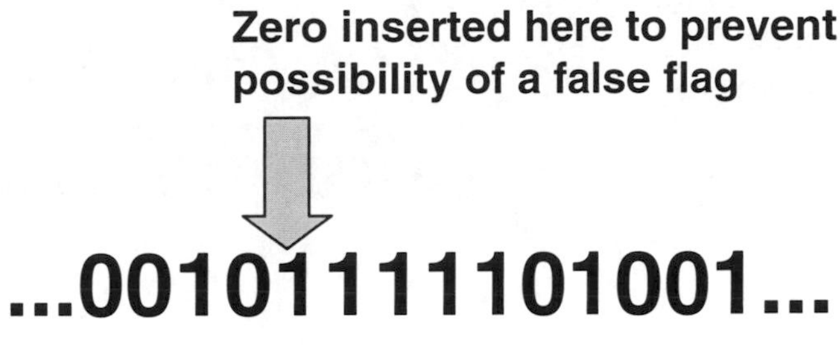

Figure 2-36
Bit stuffing, or
zero-bit
insertion

Zero inserted here to prevent possibility of a false flag

...00101111101001...

The next field in the frame is the *address field*. This field identifies the address of the next switch in the chain to which the frame is directed, and it changes at every node. The only address that remains constant is the destination address, safely embedded in the packet itself.

The third field found in many frames is called the *control field*. It contains supervisory information that the network uses to control the integrity of the data link. For example, if a remote device is not responding to a query from a transmitter, the control field can send a "mandatory response required" message that will allow it to determine the nature of the problem at the far end. It is also used in hierarchical multipoint networks to manage communications functions. For example, a multiplexer may have multiple terminal devices attached to it, all of which routinely transmit and receive data. In some systems, only a single device is allowed to "talk" at a time. The control field can be used to force these devices to take turns. This field is optional; some protocols do not use it.

The final field we will cover is the *Cyclic Redundancy Check*, or CRC, field. The CRC is a mathematical procedure used to test the integrity of the bits within each frame. It does this by treating the zeroes and ones of data as a binary number (which, of course, it is) instead of as a series of characters. It divides the "number," shown in Figure 2-37, by a carefully crafted polynomial value that is designed to *always* yield a remainder following the division process. The value of this remainder is then placed in the CRC field and transmitted as part of the frame to the next switch. The receiving switch performs the same calculation and then compares the two remainders. As long as they are the same, the switch knows that the bits arrived unaltered. If they are different, the received frame is discarded and the transmitting switch is ordered to resend the frame, a process that is repeated until the frame is received correctly. This process can result in transmission delay, because the Data Link Layer will not

Figure 2-37
CRC-based bit-
level error
control

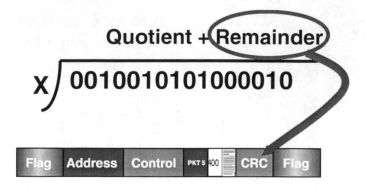

allow a bad frame to be carried through the network. Thus, the Data Link Layer converts errors into delay.

Error Recovery Options

There are a number of techniques in common use that allow receiving devices to recover from bit errors. The simplest of these is *frame discard,* the technique used by frame relay and ATM networks. In frame discard environments, an errored frame is simply discarded—period. No other form of recovery takes place within the network. Instead, the end devices (the originator and receiver) have the end-to-end responsibility to detect that a frame is missing and take whatever steps are necessary to generate a second copy. There are reasons for this strategy that will be discussed in the section on fast-packet services.

A second common technique is called *forward error correction (FEC).* FEC is used when (1) there is no backward channel available over which to request the resend of an errored packet, or (2) the transit delay is so great that a resend would take longer than the application would allow, such as in a satellite hop over which an application is transmitting delay-sensitive traffic. Instead, FEC systems transmit the application data with additional information that allows a receive device to not only determine that an error has occurred, but to fix it. No resend is required.

The third and perhaps most common form of error detection and correction is called *detect and retransmit.* Detect and retransmit systems use the CRC field to detect errors when they occur. The errored frames are then discarded, and the previous switch is ordered to resend the errored frame. This implies a number of things: The frames must be numbered, there must be some form of positive and negative acknowledgment system in place, the transmitter must keep a copy of the frame until its receipt has been acknowledged, and there must be some facility in place to allow the receiver to communicate upstream to the transmitter.

Two recovery techniques are commonly used in synchronous systems. To understand them, we must first introduce a couple of transmission protocols used to meter the transmission of frames between switches.

In early communications systems (1970s), the network was known to be relatively hostile to data transmission. After all, if noise occurred during a voice conversation, no problem—it was a simple matter to ignore it, provided it wasn't too bad. In data, however, a small amount of noise could be catastrophic, easily capable of destroying a long series of frames in a few milliseconds. As a result, early data systems such as IBM's *Binary Synchronous Communications (BISYNC)* used a protocol called *Stop-and-Wait* that would permit only a single frame at a time to be outstanding without acknowledgment from the receiver. This was obviously terribly inefficient, but in those early days was as good as it got. Thus, if a major problem occurred, the maximum number of frames that would ever have to be resent was one.

As time passed and network quality improved, designers got brave and began to allow multiple unacknowledged frames to be outstanding, a process called *pipelining*. In pipelined systems, a maximum *window size* is agreed upon that defines the maximum number of frames that can ever be allowed to be outstanding at any point in time. If the number is reached, the window closes, closing down the pipeline and prohibiting other frames from being transmitted. As soon as one or more frames clear the receiver, that many new frames are allowed into the pipeline by the sliding window. Obviously this protocol is reliant on the ability to number the frames so that the system knows when the maximum window size has been reached.

OK, back to error recovery. We mentioned earlier that there are two common techniques. The first of these is called *selective retransmit*. In selective retransmit environments, if an error occurs, the transmitter is directed to resend *only the errored frame*. This is a complex technique, but is quite efficient.

The second technique is called *Go Back N*. In Go Back N environments, if an error occurs, the transmitter is directed to go back to the errored frame *and retransmit everything from that frame forward*. This technique is less efficient but is far simpler from an implementation point of view.

Let's look at an example. Assume that a transmitter generates five frames, which are numbered one of five, two of five, three of five, etc. Now let's assume that frame three arrives and is found to be errored. In a selective retransmit environment, the transmitter is directed to resend frame three, and *only* frame three. In a Go Back N environment, however, the transmitter will be directed to resend everything from frame three going forward, which means that it will send frames three, four, and five. The receiver will simply discard frames four and five.

So let's review the task of the Data Link Layer. It frames the packet so that it can be checked for bit errors, provides various line control functions so that the network itself can be managed properly, provides addressing information so that the frame can be delivered appropriately; and performs error detection and (sometimes) correction.

Practical Implementations

A number of widely known network technologies are found at the Data Link Layer of the OSI Model. These include the access protocols used in modern local area networks, such Carrier Sense, Multiple Access with Collision Detection (CSMA/CD, used in Ethernet), and Token Ring. CSMA/CD is a protocol that relies on contention for access to the shared backbone over which all stations transmit, while Token Ring is more civilized—stations take turns sharing access to the backbone. This is also the domain of frame relay and ATM technologies, which provide high-speed switching.

Frame relay is a high-speed switching technology that has emerged as a good replacement for private line circuits. It offers a wide range of bandwidth and, while switched, delivers service quality that is equivalent to that provided by a dedicated facility. It is advantageous for the service provider to sell frame relay because it does not require the establishment of a dedicated circuit, thus making more efficient use of network resources. The only downsides of the technology are that it relies on variable-size frames, which can lead to variable delivery delay, and requires careful engineering to ensure that proper quality of service is delivered to the customer. Frame relay offers speeds ranging from 56K to DS3, and is widely deployed internationally.

ATM has become one of the most important technologies in the service provider pantheon today, because it provides the ability to deliver true, dependable, and granular quality of service over a switched-network architecture; thus, it gives service providers the ability to aggregate multiple traffic types on a single network fabric. This means that IP, which is a Network Layer protocol, can be transported across an ATM backbone, allowing for the smooth and service-driven migration to an all-IP network. Eventually, ATM's overhead-heavy quality-of-service capabilities will be replaced by more elegant solutions but, until that time comes, it still plays a central role in the delivery of service.

We will discuss all of these services in later chapters.

The Physical Layer

Once the CRC is calculated and the frame is fully constructed, the Data Link Layer passes the frame down to the Physical Layer, the lowest layer in the networking food chain. This is the layer responsible for the physical transmission of bits, which it accomplishes in a wide variety of ways. The Physical Layer's job is to transmit the bits; this includes the proper representation of zeroes and ones, transmission speeds, and physical-connector rules. For example, if the network is electrical, then what is the proper range of transmitted voltages required to identify whether the received entity is a zero or a one? Is a one in an optical network represented as the presence of light or the absence of light? Is a one represented in a copper-based system as a positive or as a negative voltage, or both? Also, where is information transmitted and received? For example, if pin 2 is identified as the transmit lead in a cable, which lead is data received over? All of these physical parameters are designed to ensure that the individual bits are able to maintain their integrity and be recognized by the receiving equipment.

Many transmission standards are found at the Physical Layer, including T1, E1, SONET, SDH, DWDM, and the many flavors of DSL. *T1* and *E1* are longtime standards that provide 1.544 and 2.048 Mbps of bandwidth respectively; they have been in existence since the early 1960s and occupy a central position in the typical network. *SONET*, the Synchronous Optical Network, and *SDH*, the Synchronous Digital Hierarchy, provide standards-based optical transmission at rates above those provided by the traditional carrier hierarchy. *Dense Wavelength Division Multiplexing*, or DWDM, is a frequency-division multiplexing technique that allows multiple wavelengths of light to be transmitted across a single fiber, providing massive bandwidth multiplication across the strand. It will be discussed in detail later in the chapter on optical networking. And DSL extends the useful life of the standard copper wire pair by expanding the bandwidth it is capable of delivering as well as the distance over which that bandwidth can be delivered.

OSI Summary

We have now discussed the functions carried out at each layer of the OSI Model. Layers six and seven ensure application integrity; layer five ensures security; layer four guarantees the integrity of the transmitted message; layer three ensures network integrity; layer two, data integrity; and layer one, the integrity of the bits themselves. Thus transmission is guaranteed on an end-to-end basis through a series of protocols that are interdependent upon each other and that work closely to ensure integrity at every possible level of the transmission hierarchy.

So let's now go back to our e-mail example and walk through the entire process.

The Eudora e-mail application running on the PC creates a message at the behest of the human user[4] and passes the message to the Application Layer. The Application Layer converts the message into a format that can be universally understood as an e-mail message, in this case X.400. It then adds a header that identifies the X.400 format of the message.

The X.400 message with its new header is then passed down to the Presentation Layer, which encodes it as ASCII, encrypts it using PGP, and compresses it using a British Telecom Lempel-Ziv compression algorithm. After adding a header that details all this, it passes the message to layer five.

The Session Layer assigns a logical session number to the message, glues on a packet header identifying the session ID, and passes the steadily growing message down to the Transport Layer. Based on network limitations and rule sets, the Transport Layer breaks the message into eleven packets and numbers them appropriately. Each packet is given a header with address and quality-of-service information.

The packets now enter the chained layers, where they will first encounter the network. The Network Layer examines each packet in turn, and, based on the nature of the underlying network (connection-

---

[4]I specifically note "human user" here because some protocols do not recognize the existence of the human in the network loop. In IBM SNA environments, for example, "users" are devices or processes that "use" network resources. There are no humans.

oriented? Connectionless?) and congestion status, queues the packets for transmission. After creating the header on each packet, they are handed individually down to the Data Link Layer.

The Data Link Layer proceeds to build a frame around each packet. It calculates a CRC, inserts a Data Link Layer address, inserts appropriate control information, and finally adds flags on each end of the frame. Note that all other layers add a header *only;* the Data Link Layer is the only layer that also adds a trailer.

Once the Data Link frame has been constructed it is passed down to the Physical Layer, which encodes the incoming bitstream according to the transmission requirements of the underlying network. For example, if the data is to be transmitted across a T- or E-Carrier network, the data will be encoded using AMI and will be transmitted across the facility to the next switch at either 1.544 Mbps or 2.048 Mbps, depending on whether the network is T1 or E1.

When the bitstream arrives at the next switch (not the destination), the bits flow into the Physical Layer, which determines that it can read the bits. The Physical Layer hands the bits up to the Data Link Layer, which proceeds to find the flags so that it can frame the incoming stream of data and check it for errors. If we assume that it finds none, it strips off the Data Link frame surrounding the packet and passes the packet up to the Network Layer. The Network Layer examines the destination address in the packet, at which point it realizes that it is not the intended recipient. So, it passes it back to the Data Link Layer, which builds a new frame around it, calculating a new CRC and adding a new Data Link Layer address as it does so. It then passes the frame back to the Physical Layer for transmission. The Physical Layer spits the bits out the facility to the next switch, which, for our purposes we will assume is the intended destination. The Physical Layer receives the bits and passes them to the Data Link Layer, which checks them for errors. If it finds an errored frame it requests a resend, but ultimately receives the frame correctly. It then strips off the header and trailer, leaving the original packet. The packet is then passed up to the Network Layer, which, after examining the packet address, determines that it is in fact the intended recipient of the packet. As a result, it passes the packet up to the Transport Layer, after stripping off the Network Layer header.

The Transport Layer examines the packet and notices that it has received packet three of eleven packets. Because its job is to assemble and pass entire messages up to the Session Layer, the Transport Layer simply places the packet into a buffer while it waits for the other ten packets to arrive. It will wait as long as it has to; it knows that it cannot deliver a partial message because the higher layers are not smart enough to figure out the missing pieces.

Once it has received all eleven packets, the Transport Layer reassembles the original message and passes it up to the Session Layer, which examines the Session header created by the transmitter and notes that this is to be handed to whatever process cares about logical channel number seven. It then strips off the header and passes the message up to the Presentation Layer.

The Presentation Layer reads the Presentation Layer header created at the transmit end of the circuit and notes that this is an ASCII message that has been encrypted using PGP and compressed using BTLZ. It decompresses the message using the same protocol, decrypts the message using the appropriate public key, and, because it is resident in a mainframe, converts the ASCII message to EBCDIC. Stripping off the Presentation Layer header, it hands the message up to the Application Layer. The Application Layer notes that the message is X.400 encoded, and is therefore an e-mail message. As a result, it passes the message to the e-mail application that is resident in the mainframe system.

The process just described happens every time you hit the SEND button. Click.

Other Protocol Stacks

Of course, OSI is not the only protocol model. In fact, for all its detail, intricacy, and definition, it is rarely used in practice. Instead, it serves as a true model for comparing disparate protocol stacks. In that regard, it is unequalled in its value to data communications.

The most commonly deployed protocol stack is that used in the Internet, the so-called *TCP/IP stack*. Instead of seven layers, TCP/IP comprises four. The bottom layer, called the *Network Interface Layer*, includes the functions performed by OSI's Physical and Data Link Layers. It includes a wide variety of protocols as shown in Figure 2-38. We describe them here briefly, because we will describe the functions and relationships of these protocols in a later chapter in great detail.

The *Internet Protocol Layer*, or IP, is roughly functionally equivalent to the OSI Model's Network Layer. It performs routing and congestion control functions and, as the diagram illustrates, includes some of the protocols we mentioned earlier: RIP, OSPF, and a variety of address conversion protocols.

Figure 2-38
The TCP/IP
protocol suite

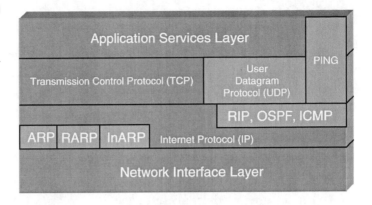

The *Transmission Control Protocol Layer*, or TCP, is responsible for message integrity, similar to the service provided by OSI's Transport Layer. It is extremely capable and has the ability to recover from virtually any network failure imaginable to ensure the integrity of the messages it is designed to protect. For situations in which the high degree of protection provided by TCP (and its attendant overhead) is considered to be overkill, a corollary protocol, called *User Datagram Protocol* (UDP), is also available at this layer. It provides a connectionless network service and is used in situations in which the transported traffic is less critical and in which the overhead inherent in TCP poses a potential problem due to congestion.

The uppermost layer in the TCP/IP stack is called the *Application Services Layer*. This is where the utility of the stack becomes obvious, because this is where the actual applications are found such as HTTP, FTP, Telnet, and the other utilities that make the Internet useful to the user. We will discuss the TCP/IP protocols stack in more detail later in the book, but for now the information provided will suffice.

The point of all these protocols is to give a designer the ability to create a network that will transport the customer's voice, data, video, images, or MP3 files with whatever level of service quality the traffic demands. We now know that data communications protocols make it possible to transport all types of traffic with guaranteed service. Let's turn our attention now to the network itself. In the chapters that follow we will discuss the history of the most remarkable technological achievement on earth—the telephone network. Later we will discuss the anatomy of a typical data network and the technologies found there.

Chapter 2 Questions

1. Why are ASCII and EBCDIC important? Are there other character codes that are still in use? What are they?

2. Convert the number $2{,}387_{10}$ to its base two equivalent.

3. Convert 1101001101010_2 to its base ten equivalent.

4. Give three examples of noise in a telephone network.

5. Explain the difference between noise and distortion.

6. The OSI Model is referred to as a layered protocol model. Why are layered models advantageous?

7. Explain the difference between layers six and seven. Are they both necessary? Why?

8. What is the difference between connectivity and interoperability? Is one more important than the other? Why or why not?

9. What layer of OSI cares most about each of the following: bits, messages, cells, frames, and packets?

10. Explain the difference between a connection-oriented and a connectionless network, including appropriate applications for each.

11. What is the simplest form of error correction?

CHAPTER 3

Telephony

In this chapter we will explore telephony and the telephone network. Described as "the largest machine ever built," the telephone network is an impressive thing. We begin with some history, then explore the network itself, and finally explore telephony service—how it works, how it makes its way across the network, and how it is managed. We will examine the process of voice digitization followed by the multiplexing and high-speed transport of voice through such technologies as PCM, SONET, and SDH.

Some people feel that voice is something of an anachronism and has no place in a book about the new, slick world of telecommunications. I disagree: Without an understanding of how the voice network operates, it is impossible to understand or appreciate how data networks operate. Furthermore, as voice moves inexorably toward an all-IP model, the voice stream becomes nothing more than one more component of the overall corporate data stream. Additionally, since the IT organization manages data in most corporations, voice is rapidly becoming just another corporate data application.

This chapter is designed to bridge the gap between the two, and we begin our tale in New York City.

Miracle on Second Avenue

February 26, 1975 was a business-as-usual day at New York Telephone's lower Manhattan switching center. Located at 2nd Avenue and 13th Street, the massive 11-story building was the telephony nerve center for 12 Manhattan exchanges that provided service to 300 New York City blocks, including 104,000 subscribers and 170,000 telephones. Within the service area were six hospitals, 11 firehouses, three post offices, one police station, nine schools, and three universities. The building was massive, but like most central offices was completely invisible to the public. It was just a big, windowless structure that belonged to the telephone company. No one really knew what went on in there; nobody cared.

When night fell, most of the building's employees went home, leaving a small crew to handle maintenance tasks and minor service problems. The night was quiet; work in the building was carried out routinely. It was going to be a boring evening.

At 12:30, just after midnight, a relatively inconsequential piece of power equipment in the building's subbasement cable vault shorted and spit a few errant sparks into the air around it. It caused no alarms because it didn't actually fail. One of the sparks, however, fell on a piece of insulation and began to smolder. The insulation melted and began to burn, changing from a smoldering spot on the surface of a cable to a full-blown fire. Soon the entire cable vault was aflame.

Cables in a central office leave the cable vault on their way to upper floor switching equipment through 6-inch diameter holes cored in the concrete floors of the office. The fire burned its way up the cables, exited the cored holes, and spread from floor to floor. Soon all of the lower floors of the building were engulfed, and New York Telephone was on its way to hosting the single worst disaster that any American telephone company has ever experienced.

Emergency service vehicles converged on the building. They evacuated everyone inside and began flooding the lower floors with hundreds of thousands of gallons of water. Smoke and steam billowed up and out of the structure (see Figure 3-1), severely damaging equipment on the upper floors that had not, as yet, been affected by the fire.

Figure 3-1
Smoke and steam billow from the 2nd Avenue central office.

Two days later, the fire was finally out and telco engineers were able to reenter the building to assess the damage. On the first floor, the 240-foot main distribution frame was reduced to a melted puddle of iron. Water had ruined the power equipment in the basement. Four panel-switching offices on the lower floors were completely destroyed. Cable distribution ducts between floors were deformed and useless. Carrier equipment on the second floor was destroyed. Switching hardware on the fourth, fifth, and sixth floors were smoke and water damaged and would require massive restoration and cleaning before they could be used. *And 170,000 telephones were out of service.*

Within an hour of discovering the fire, the Bell System mobilized its forces in a massive effort to restore service to the affected area. New York Telephone, AT&T Long Lines, Western Electric (now Lucent), and Bell Laboratories converged on the building and commandeered a vacant storefront across the street to serve as their base of operations. Lee Oberst, New York City Area Vice-President, coordinated the nascent effort that would last just under a month and cost more than $90 million.

The Bell System—and it really was a system—immediately commissioned a host of parallel efforts to restore the central office. Calling upon the organization's widespread resources, Chairman of the Board John deButts put out an urgent demand for personnel, and within hours 4,000 employees from across the globe descended upon New York City and began to work 12-hour shifts, with 2,000 employees in the building on each shift. Meanwhile, other efforts were underway. Central office engineers reviewed the original drawings of the building's complex network to determine what they would need in the way of new equipment to restore service. All other Bell Operating Companies (BOCs) were placed on indefinite equipment holds pending the restoration of the New York office. Nassau Recycle Corporation, a Western Electric subsidiary, moved in to begin the removal and recycling process of the 6,000 tons of debris that came out of the building, much of it toxic. Mobile telephone subsidiaries from across the country sent mobile radio units to the city to provide emergency telephony services. The units were installed all over the affected area and announcements were posted throughout the city to tell residents where they were located.

Simultaneously, Bell Laboratories scientists studied the impact of the fire and crafted the best technique for cleaning the equipment that could be salvaged. 1,350 quarts of specially formulated cleaning fluid were mixed and shipped to the building along with thousands of toothbrushes and hundreds of thousands of Q-Tips. In addition, a high-capacity facility between New York and New Jersey, still in the planning stages, was accelerated, installed, and put into service to carry the traffic that would normally have flowed through the 2nd Avenue office.

Within 24 hours, miracles were underway. Service had been restored to all affected medical, police, and fire facilities. The day after the fire, a new main distribution frame had been located at Western Electric's Hawthorne manufacturing facility and was shipped by cargo plane to New York. Luckily, the third floor of the building had been vacant and was therefore available as a staging area to assemble and install the 240-foot-long iron frame. Under normal circumstances, from the time a frame is ordered, shipped and installed in an office, 6 months elapse. *This frame was ordered, shipped, installed, and wired in 4 days.*

It is almost impossible to understand the magnitude of the restoration effort, but the following numbers may help. Six thousand tons of debris were removed from the building and 3,000 tons of material was installed including 1.2 billion feet of underground wire, 8.6 million feet of frame wire, 525,000 linear feet of exchange cable, and 380 million conductor feet of switchboard cable. Five million underground splices hooked it all together. 30 trucking companies, 11 airlines, and 4,000 people were pressed into service. And just after midnight on March 21, 22 days following the fire, service from the building was restored to 104,000 subscribers. Normally, the job would have taken more than a year. But because of the Bell System's remarkable ability to marshal resources during times of crisis, the building was restored in less than a month. AT&T Chairman John DeButts:

> In the last couple of weeks, I have had the opportunity to observe at first hand the strength of the organization structure that the Justice Department's antitrust suit seeks to destroy. This service restoration has been called a dramatic accomplishment—rightly. But only in its urgency does the teamwork demonstrated on this occasion differ from the teamwork that characterizes the Bell System's everyday job.

Of course, the antitrust suit, now known universally as Divestiture, went forward as planned, beginning in 1984. The Bell System became a memory, with just as many people hailing its death as mourning it. And while there is much to be said for the innovation, competitive behavior and reduced prices for telecommunications services that came from the breakup of the Bell System, it is hard to read an account such as the previous one without a small lump in the throat. In fact, most technology historians conclude that we could not build the Bell System network today.

To understand the creation of a system that could accomplish something as remarkable as the fire recovery described previously, let's take a few pages to describe the history of this marvelous industry. I think you'll find it rather interesting.

The History of Telephony

As early as 1677, inventor Robert Hooke, who formulated the theory of planetary movement, demonstrated that sound vibrations could be transmitted over a piece of drawn iron wire, received at the other end, and interpreted correctly. His work proved to be something of a bellwether, and in 1820, 143 years later, Charles Wheatstone transmitted interpretable sound vibrations through a variety of media including thin glass and metal rods. In 1831, Michael Faraday demonstrated that vibrations such as those made by the human voice could be converted to electrical pulses. Needless to say, this was fundamental to the eventual development of the telephone.

Bernard Finn is the Curator of the Smithsonian's National Museum of American History. His knowledge of the world of science and its vanguard during the heady times of telephony's arrival is unparalleled. He observes that America was a bit of a backwater as far as science was concerned. "The inventive climate of that early part of the century was largely [focused on] mechanical inventions," he notes. "Even the early electrical stuff was basically mechanical and the machine shop was the center of inventive activity."

With the invention of the telegraph, which occurred simultaneously in the United States, England, and Germany in the 1830s, it became possible to communicate immediately across significant distances. By the second half of the nineteenth century, three individuals were working on practical applications of Faraday's observations. Alexander Graham Bell, a speech teacher for the hearing impaired and a part-time inventor; Elijah Gray, the inventor of the telegraph and an employee of Western Electric; and Thomas Edison, an inventor and former employee of Western Union were all working on the same problem: how to send multiple simultaneous messages across a single pair of wires. All three invented devices that they called harmonic telegraphs; they were nothing more than frequency division multiplexers, but for the time were remarkably advanced, conceptually.

The original devices that they created performed the multiplexing task mechanically by vibrating metallic reeds at different frequencies. The frequencies would pass down the line and be received correctly at the other end. The only problem with this technique was channel separation: Because the frequency bands were relatively close to each other, they would eventually interfere with one another, causing problems at the receiving end of the circuit. For the telegraph, therefore, harmonic multiplexing was deemed to not be practical.

All three inventors—Gray, Edison, and Bell—were in a dead heat to create the first device that would transport voice across a long-distance network. None of them had really tested their inventions extensively; they had been used in the laboratory only. Nevertheless, on February 14, 1876, Alexander Graham Bell filed a notice of invention on his own untested device. Simultaneously, Elijah Gray filed a notice of invention on his device. There is some argument as to who filed first; many believe that Gray actually filed first. Whatever the case, three days later Bell became the first person to transmit voice electronically across a network. And in 1877, following numerous experiments and "tweaks" to his original device, Bell founded the Bell Telephone Company.

Meanwhile, Thomas Edison continued to work on his own device, patented in 1875, and in 1876 was hired by Western Union. The arrival of the telephone was seen as a major problem by the company but they had no intention of sitting idly by while Gray and Bell destroyed their business with their new-fangled invention. In 1877 Edison crafted a very good transmitter based on compressed carbon powder (still in widespread use today, by the way; see Figure 3-2) that would eventually replace the far less capable transmitter invented by Bell. For the

Figure 3-2
Bell's carbon
microphone,
still in use
today

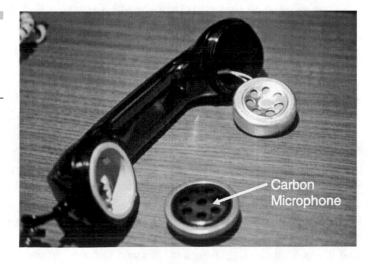

Carbon
Microphone

experimenters among the readers out there, it's a fun exercise to build a compressed carbon microphone. In a mortar and pestle (OK, with a plastic bag and a hammer) grind two charcoal briquettes to a fine powder. Pour the powder into a jar lid until the powder is even with the lip of the lid. Stretch a piece of plastic wrap tightly over the lid and secure it with a rubber band. Next, drill a hole in each side of the lid and insert an insulated wire that has been stripped back about a half inch to expose the copper wire. Next, connect the wires to a speaker and battery. When you speak into the "microphone," the sound waves impinging upon the stretched plastic wrap cause the carbon powder to be compressed and expanded, which in turn changes the resistance between the two conductors inserted into the powder.

Naturally, tensions ran high between the Bell Telephone Company and Western Union during this period. In 1877, Bell offered to sell his patent to Western Union for $100,000. They turned down Bell's offer, deciding instead to buy Gray and Edison's patents and form the American Speaking Telephone Company.

Bell sued Western Union for patent infringement in 1878, and the case was settled in 1879 to everyone's satisfaction. Bell won the rights to Western Union's telephone patents, provided he pay royalties for the duration of the 17-year patent life of each device. Meanwhile, Bell Telephone Company agreed to stay out of the telegraph business and Western Union agreed to stay away from telephony.

Once the legal wrangling had ended, telephone service moved along rather quickly. By the spring of 1880 the United States had 138 exchanges and 30,000 subscribers. By 1887, 146,000 miles of wire had been strung to connect 150,000 subscribers to nearly 750 main offices and 44 branch offices.

As we described in Chapter 2, there were no switches initially, so telephones were sold in pairs, one for each end of the connection. Wires were strung in a haphazard fashion, attached to the outside of buildings, and strung across neighbors' rooftops; they threatened to fill the sky, as shown in Figure 3-3. Obviously, this one-to-one relationship was somewhat limiting. A technique was needed that would allow one phone to connect to many.

The answer came in the form of the central office. Connections were installed from all the phones in a local area to the central exchange, where operators could monitor the lines for service requests. The customer would then tell the operator whom they wanted to speak with, and the operator would set up the call using patch cords. When the parties had finished speaking, the operator would pull down the patch cord, free-

Figure 3-3
Crossbars on pole being prepared to support an awful lot of cable

ing up the equipment for another call. This entire function was the first form of switching. The design soon became problematic, however, because an operator could typically monitor only about 100 positions effectively.

The answer, of course, came with the development of the switch—first with Strowger's invention, and later from a series of switching innovations that led to the technology we have today. The first semiautomatic switch was installed in Newark, New Jersey, in 1914, and the first truly automatic switch went live in Omaha, Nebraska in 1921. For all this innovation, however, direct long-distance dialing did not become a reality until 1951.

Soon the network had evolved to a point where it could handle thousands of simultaneous calls. As the technology evolved, so too did the business. In 1892 Alexander Graham Bell inaugurated a 950-mile circuit between New York and Chicago, ushering in the facile ability to carry out long-distance telephony. Meanwhile the company continued to change; after acquiring Western Electric it changed its name to American Telephone and Telegraph (AT&T) and, in an ongoing attempt to protect its fiefdom, began buying up small competitors as they emerged. In 1909 the company purchased a controlling interest in Western Union, thus garnering the attention of the Interstate Commerce Commission, the federal agency responsible (as of 1909) for the oversight of American wire and radio-based communications. That responsibility would later be passed on to the newly created Federal Communications Commission (FCC).

In 1913, in the first of a series of antitrust decisions, the Kingsbury Commitment forced AT&T to divest its holdings in Western Union, stop buying its competition, and provide free interconnection with other network providers. AT&T did not realize it then, but the Kingsbury Commitment was the harbinger of drastic things to come.

In the years that followed, a number of significant technological events took place in telephony. Load coils were created and installed on telephone lines, extending signal strength dramatically and reducing bandwidth requirements for each customer. In 1915, the first repeaters were installed on long-distance telephone circuits, making it possible to install circuits of essentially unlimited length. In fact, later that same year a circuit was successfully installed between New York and San Francisco.

In 1934, President Roosevelt created the Federal Communications Commission with the signing into law of the Communications Act of 1934. This act moved industry oversight from the Interstate Commerce Commission to the FCC, and recognized the importance of telephone service, mandating that it would be both affordable and universal.

By now, AT&T provided local and long-distance service and offered central office equipment and telephones to more than 90 percent of the market. Not surprisingly, that smacked of monopoly to the federal government, and from that point on they were squarely in the FCC's gun sights.

In 1948, AT&T sued the Hush-a-Phone Corporation for manufacturing a device that physically connected to the telephone network. The device, shown in Figure 3-4, was nothing more than a metal box that fit

Figure 3-4
Hush-a-Phone

over the mouthpiece of the telephone, blocking out extraneous noise and providing a degree of privacy to the person using the phone. It had no electrical component to it; it was merely a box. Hush-a-Phone won the case, and the District of Columbia Court of Appeals ruled that AT&T could not prohibit people from making money from devices that did not electrically connect to AT&T's network.

In 1949, the Justice Department filed an antitrust suit against AT&T and Western Electric. The result was the consent decree of 1956, which allowed AT&T to keep Western Electric, but restricted them to the delivery of common carrier services.

In the early 1960s, inventor Tom Carter created a device called the Carterphone, shown in Figure 3-5. The Carterphone allowed mobile car radios to connect to the telephone network and *did* require electrical connectivity to AT&T. Initially, AT&T prohibited Carter from connecting his

Figure 3-5
Carterphone

device under any circumstances, but when he appealed to the FCC he was given permission, opening the customer-provided equipment (CPE) market for the first time. AT&T's concerns about possible damage to the network were well founded, but by the time *AT&T v. Carter* came to trial the device had been in use for several years and clearly hadn't done any damage whatsoever.

In 1969, the very next year, the FCC gave MCI permission to offer private line services between St. Louis and Chicago over its privately owned microwave system. And in 1971 the court extended its 1956 Consent Decree mandate, ordering AT&T to allow non-Bell companies such as MCI, to connect directly to their network, ending AT&T's stranglehold on the private line market. The decision was based on a far-reaching FCC policy designed to create nationwide, full-blown competition in the private line marketplace.

It didn't end there, however. In 1972 the FCC mandated that satellites could be used to transport voice and compete with AT&T, and that value-added carriers could resell AT&T services. In 1977, in the now famous

Execunet decision, MCI won a legal battle that allowed them to compete directly with AT&T in the switched long-distance services market, AT&T's bastion of revenue. In 1974 they extended their attack, filing an antitrust suit against AT&T and charging them with unfairly restricting competition and dominating the marketplace. That same year, the Justice Department filed an antitrust suit of their own, signaling the beginning of the end for the Bell System.

In 1980, another crack appeared in AT&T's armor when the FCC consciously recognized the difference between basic and enhanced services. *Computer Enquiry II* stipulated that basic services would be regulated and, therefore, tightly controlled. Enhanced services, including CPE, would be deregulated and made completely competitive. AT&T was told that it could sell enhanced services but only through fully separate subsidiaries, to ensure no mixing of revenues between the two sides of the business.

In 1981, the *United States v. AT&T* came to trial in Judge Harold Greene's court. In the months that followed it became painfully clear to AT&T that it would not win this game. On January 8, 1982, the two sides announced that they had reached a mutually acceptable agreement. The agreement, known as the Modified Final Judgment, stipulated that AT&T would divest itself of its 22 local telephone companies, the Bell Operating Companies. The company would retain ownership of Western Electric, Bell Laboratories, and AT&T Long Lines and would be allowed to enter noncarrier lines of business such as computers and the like. Meanwhile, the BOCs were gathered into seven regional BOCs called RBOCs, tasked with providing local service to the regions they served. The RBOCs were further subdivided into Local Access and Transport Areas, or LATAs, which defined the areas within which they were allowed to provide transport. A state like California, for example, is very large; so, even though the traffic between San Francisco and San Diego never leaves Pacific Bell territory, it clearly travels over a long distance and therefore must be handed off to a long-distance carrier for transport between LATAs.

While the best-known impact of divestiture was the breakup of AT&T —one result of which was the liberalization of the telecommunications marketplace in the United States—a second decision that was tightly intertwined with the Bell System's breakup was largely invisible to the public, yet was *at least* as important to AT&T competitors MCI and Sprint as the breakup itself. This decision, known as Equal Access, had one seminal goal: to make it possible for end customers to take advantage of one of the products of divestiture, the ability to select one's long-distance provider from a pool of available service providers—in this case AT&T, MCI, or Sprint. This, of course, was the realization of a truly competitive marketplace in the long-distance market segment.

To understand this evolution, it is helpful to have a high-level understanding of the overall architecture of the network. In the predivestiture world, AT&T was *the* provider for local service, long-distance service and communications equipment. An AT&T central office, therefore, was awash in AT&T hardware—switches, cross-connect devices, multiplexers, amplifiers, repeaters, and myriad other devices.

Figure 3-6 shows a typical network layout in the predivestiture world. A customer's telephone is connected to the service provider's network by a local loop connection (so-called twisted pair wire). The local loop, in turn, connects to the local switch in the central office. This switch is the point at which customers first touch the telephone network, and it has the responsibility to perform the initial call setup, maintain the call while it is in progress, and tear it down when the call is complete. This switch is called a local switch because its primary responsibility is to set up local calls that originate and terminate within the same switch. It has one other responsibility, though, and that is to provide the necessary interface between the local switch and the long-distance switch, so that calls between adjacent local switches (or between far-flung local switches) can be established. The process, then, goes something like this: When a customer lifts the handset and goes off-hook, a switch in the telephone closes, completing a circuit that allows current flow, which, in turn, brings dial tone to the customer's ear. Upon hearing the dial tone, the customer enters the destination address of the call (otherwise known

Figure 3-6
Predivestiture connectivity

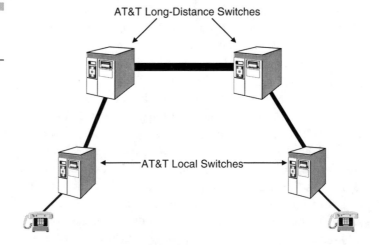

AT&T Long-Distance Switches

←AT&T Local Switches→

as a telephone number). The switch receives the telephone number and analyzes it, determining from the area code and prefix information whether the call can be completed within the local switch or must leave the local switch for another one. If the call is indeed local, it merely burrows through the crust of the switch and then reemerges at the receiving local loop. If the call is a toll or long-distance call, it must burrow through the hard crunchy coating of the switch, pass through the soft chewy center, and emerge again on the other crunchy side on its way to a long-distance switch. Keep in mind that the local switch has no awareness of the existence of customers or telephony capability beyond its own cabinets. Thus, when it receives a telephone number that it is incapable of processing it, hands the number off to a higher-order switch, with the implied message, "Here—I have no idea what to do with this, but I assume that you do."

The long-distance switch receives the number from the local switch, processes the call, establishes the necessary connection and passes the call on to the remote long-distance switch over a long-distance circuit. The remote long-distance switch passes the call to the remote local switch, which rings the destination telephone, and, ultimately, the call is established.

Please note that in this predivestiture example, the originating local loop, local switch, long-distance switch, remote local loop, and all of the interconnect hardware and wiring belong to AT&T. They are mostly manufactured by Western Electric, based on a set of internal manufacturing standards that, were there other manufacturers in the industry, would be considered proprietary. Because AT&T was the only game in town (in the United States at least) prior to divestiture, AT&T created the standard for transmission interfaces.

Fast forward now to January 1, 1984, and put yourself into the mind of MCI's Bill McGowan, whose company's very survival depended upon the successful implementation of Equal Access. Unfortunately, Equal Access had one very serious flaw. Keep in mind that, because the post-1984 network was emerging from the darkness of monopoly control, all of the equipment that made up the network infrastructure was bought at the proverbial company store—and was, by the way, proprietary.

Consider the newly recreated postdivestiture network model shown in Figure 3-7. At the local switch level, precious little has changed—at this point in time there is still only a single local service provider. At the long-distance level, however, there is a significant change. Instead of a single long-distance service provider called AT&T, there are now three—AT&T, MCI, and Sprint. The competitive mandate of Equal Access was designed

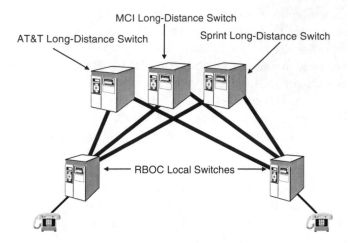

Figure 3-7
Postdivestiture
connectivity

MCI Long-Distance Switch

AT&T Long-Distance Switch   Sprint Long-Distance Switch

RBOC Local Switches

to guarantee that a customer could freely select their long-distance provider of choice. If they wanted to use MCI's service instead of AT&T's, a simple call to the local telephone company's service representative would result in the generation of a service order that would cause the customer's local service to be logically disconnected from AT&T and reconnected to MCI so that long-distance calls placed by the subscriber would automatically be handed off to MCI. The problem of "equal access" to customers for the three long-distance providers was solved.

Since 1984 the telecommunications marketplace has continued to evolve, sometimes in strange and unpredictable ways. Harold Greene oversaw the remarkable transformation of the industry until his death in January of 2000. Over time, the seven RBOCs slowly accreted to a smaller number as SBC, Pacific Bell, and Ameritech joined forces, sucking SNET into their midst in the process; NYNEX and Bell Atlantic danced around each other until they became Verizon, pulling GTE into the fray; and Qwest acquired USWest. Only Bellsouth remains as a standalone suitor from earlier days, and none of them are called RBOCs anymore—they're incumbent local exchange carriers (ILECs). A host of new players emerged from the proverbial woodwork including bypassers, which became competitive access providers (CAPs), which in turn became competitive local exchange carriers (CLECs). Service, now the favored watchword, has given rise to DLECs, BLECs, ISPs, ASPs, and LSPs. Old-timers remember when the world was fine with just BSPs. (Sorry, inside joke.) In 1996 the FCC released the Communications Act of 1996, designed to revamp the Communications Act of 1934 and make it friendlier to the services carried by network providers today. It also was designed to address the requests by ILECs to become long-distance providers through a 14-point checklist that they were required to complete before being considered for entry into the long-distance market. None of them ever complied totally with the demands, and as we will see in the regulatory chapter, significant changes have occurred since 1996.

Overall, the market continues to liberalize. Western Electric has ceased to exist and Lucent has taken its place. The Internet has become the next great technology and service frontier, and both existing and new service providers are leaping at its promises of wealth and riches. Outside the United States, telecom companies once considered primitive rival the level of services offered in the United States, and in spite of the telecom meltdown that plagued the industry for several years, innovation continues and new players spring up like early morning mushrooms.

Let's turn our attention now to the network itself.

The Telephone Network

"Sure, I know how it works. You pick up the phone, dial the numbers, and wait. A little man grabs the words, runs down the line, and delivers them to whomever I'm calling. It seems just about that simple. I mean, come on —how complicated can it be? It's just a telephone call."

Thus was described to me the overall process of placing a telephone call by a fellow on the street, whom I once interviewed for a video I was creating about telephony. His perception of the telephone network, how it works, and what it requires to work is similar to most peoples'. Yet the telephone network is without question the single greatest and most complex agglomeration of technology on the planet. It extends its web seamlessly to every country on earth, providing instantaneous communication for not only voice, but for video, data, television, medical images, sports scores, music, high-quality graphics, secure military intelligence, banking information, and teleconferences. Yet it does so with almost complete transparency and with virtually 100 percent availability. In fact, the only time its presence is noticed is when it isn't there—as happened on 2nd Avenue in New York, in Hinsdale, Illinois following a major central office fire, and in Chicago following a flood that isolated the Mercantile Exchange and placed hundreds of customers out of service.

How the network works is something of a mystery to most people, so we're going to dissect the typical telephone network and examine its anatomy, complete with pictures. This section is not for the squeamish.

The best way to gain an understanding of how the telephone network operates is by studying a modern railroad system. Consider Figure 3-8, which is a route map for the mythical Midland, Amarillo, and Roswell Railroad. Rail yards in each of the three cities are interconnected by high-volume trunk lines. The train yards, also known as switch yards, are used as aggregation, storage, and switching facilities for the cargo that arrives on trains entering each yard. A 90-car train from El Paso, for example, may arrive in Midland as planned. Half the cars are destined for Midland, while the other half are destined for Amarillo. At the Midland yard, the train is stored, disassembled, reassembled for transport, and switched as required to move the Amarillo traffic on to Amarillo. Switches in the yards create temporary paths from one side of the yard to the other. Meanwhile, route bosses in the yard towers analyze traffic patterns and route trains across a variety of alternative paths to ensure the most efficient delivery. For example, traffic from Amarillo to Roswell could be routed through Midland, but it would obviously be more efficient to send it on the direct line that runs between Amarillo and Roswell. Assuming that the direct route is available and is not congested, the route boss might very well choose that alternative.

Notice also that there are short local lines, called feeder lines, which pump local traffic into the switchyards. These lines typically run shorter trains, destined most likely for local delivery. Some of them, however, carry cargo destined for distant cities. In those cases, the cargo would be combined with that of other trains to create a large, efficient train with all cargo bound for the same destination.

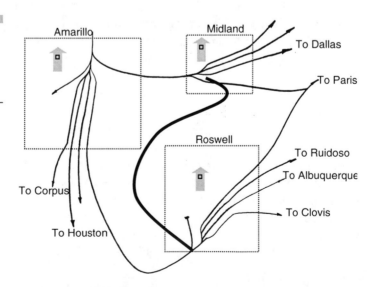

Figure 3-8
Midland,
Amarillo, and
Roswell
Railroad

Trunks, lines, feeders, local access spurs, switches, routers—all terms used commonly in the telecommunications industry. Trunks are high-bandwidth, long-distance transport facilities. Lines, feeders, and local access spurs are local loops. Switches and routers are—well, switches and routers. And the overall function of the two is exactly the same, as is their goal of delivering their transported payload in the most efficient fashion possible.

When Bell's contraption first arrived on the scene it wasn't a lot more complicated than a railroad, frankly. As we noted earlier, there was no initial concept of switching. Instead, the plan was to gradually evolve the network to a full mesh architecture as customers demanded more and more connections. Eventually, operators were added to provide a manual switching function of sorts, and over time they were replaced with mechanical, then electromechanical, then all-electronic switches. Ultimately, intelligence was overlaid in the form of signaling, giving the network the ability to make increasingly informed decisions about traffic handling as well as to provide value-added services.

Now that you understand the overall development strategy that Bell and his cohorts had in mind, as well as the basics of switching, we turn our attention to the inner workings of the typical network, shown in Figure 3-9. I must take a moment here to thank my good friend Dick Pecor, who created this drawing—from his head, I might add. Thanks, Dick.

Scalpel, please.

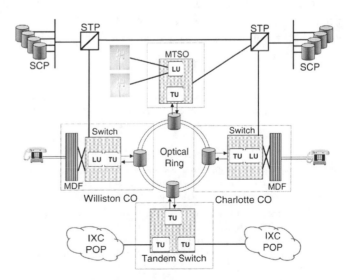

Figure 3-9
Typical network

In the Belly of the Beast

When a customer makes a telephone call, a complex sequence of events takes place that ultimately leads to the creation of a temporary end-to-end, service-rich path. Let's follow a typical phone call through the network. Cristina in South Burlington is going to call Adam in Shelburne. This is a high-level explanation; we'll add more detail later.

When Cristina picks up the telephone in her home, the act of lifting the handset[1] closes a circuit, which allows current to flow from the switch —in the local central office that serves her neighborhood—to her telephone. The switch electronically attaches an oscillator to the circuit called a *dial tone generator*, which creates the customary sound that we all listen for when we wish to place a call.[2] The dial tone serves to notify the caller that the switch is ready to receive the dialed digits.

Cristina now dials Adam's telephone number by pressing the appropriate buttons on the phone. Each button generates a *pair of tones* (listen carefully—you can hear them) that are slightly dissonant. This is done to prevent the possibility of a human voice randomly generating a frequency that could cause a misdial. The tone pairs, shown in Figure 3-10, are carefully selected so that they cannot be naturally generated by the human voice. This technique is called dual tone, multifrequency (DTMF). It is not, however, the only way.

You may have noticed that there is a switch on many phones with two positions labeled "TONE" and "PULSE." When the switch is set to the TONE position, the phone generates DTMF. When it is set on PULSE, it generates the series of clicks that old dial telephones made when they were used to place a call.[3] When the dial was rotated—let's say to the number 3—it caused a switch contact to open and close rapidly three times, sending a series of three electrical pulses (or more correctly, interruptions) to the

---

[1] It doesn't matter whether the phone is corded or cordless. If cordless, pushing the TALK button creates a radio link between the handset and the base station, which in turn closes the circuit.

[2] It's interesting to note that all-digital networks, such as those that use ISDN as the access technology, don't generate dial tone because packets don't make any noise! Instead, the ISDN signaling system (Q.931 for those of you collecting standards) transmits a "ring the phone" packet to the distant telephone. The phone itself generates the dial tone—and the ringing bell!

[3] About 10 years ago I took a trip to Seattle to teach a class, and my family accompanied me. One evening while strolling along Pike Street we came upon a bank of modern pay phones, all of which had dials instead of DTMF buttons. My kids, still little back then, begged me for quarters so that they could call their friends on those cool new phones.

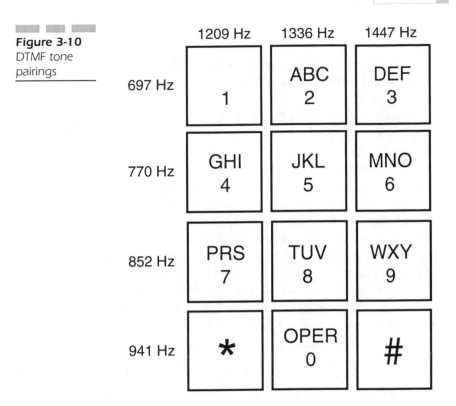

Figure 3-10
DTMF tone
pairings

switch. Want to impress your friends? Try pulse dialing on a DTMF telephone. Simply find a phone that has a real button or buttons that you can push down to hang up the phone. Pick up the handset, then dial the number you wish to call by rapidly pushing and releasing the buttons the appropriate number of times, leaving a second or two of delay between each dialed digit. It takes a little practice but it really does work. Try something simple like Information (411) first.

DTMF has been around since the 1970s, but switches are still capable of being triggered by pulse dialing.

Back to our example. Cristina finishes dialing Adam's number and a digits collector in the switch receives the digits. The switch then performs a rudimentary analysis of the number to determine whether it is served out of the same switch. It does this by looking at the area code (numbering plan area, or NPA) and prefix (NXX) of the dialed number.

A telephone number is made up of three sections, shown here.

(802) 555-7837

The first three digits are the NPA, which identifies the called region. For example, I live in Vermont where the entire state is served by a single NPA—802. Other states, such as California, have a dozen or more NPAs to serve their much denser population bases. Each NPA identifies a calling area.

The second three digits are the NXX, which identify the exchange, or office, that the number is served from, and by extension the switch to which the number is connected. Each NXX can serve 10,000 numbers (555-0000 through 555-9999). Remember the New York fire? Twelve exchanges were lost, under which 104,000 subscribers were operating (out of a possible 120,000). A modern central office switch can typically handle as many as 15 to 20 exchanges of 10,000 lines each.

In our example, Adam is served out of a different switch than Cristina, so the call must be routed to a different central office. Before that routing can take place, however, Cristina's local switch routes a query to the signaling network, known as Signaling System 7 (SS7). SS7 provides the network with intelligence. It is responsible for setting up, maintaining, and tearing down a call, while at the same time providing access to enhanced services such as Custom Local Area Signaling Services (CLASS), 800 Number Portability, Local Number Portability (LNP), Line Information Database (LIDB) lookups for credit card verification, and other enhanced features. In a sense, it makes the local switch's job easier by centralizing many of the functions that formerly had to be performed locally.

The original concept behind SS7 was to separate the actual calls on the public telephone network from the process of setting up and tearing down those calls as a way to make the network more efficient. This had the effect of moving the intelligence out of the PSTN and into a separate network, where it could be somewhat centralized and therefore made available to a much broader population. The SS7 network, shown in Figure 3-11, consists of packet switches (signal transfer points, or STPs) and intelligent database engines (service control points, or SCPs) interconnected to each other and to the actual telephone company switches (service switching points, or SSPs) via digital links, typically operating at 56 to 64 Kbps.

When a customer in an SS7 environment places a call, the following process takes place. The local switching infrastructure issues a software interrupt via the SSP so that the called and calling party information can be handed off to the SS7 network, specifically an STP. The STP, in turn, routes the information to an associated SCP, which performs a database lookup to determine whether any special call handling instruc-

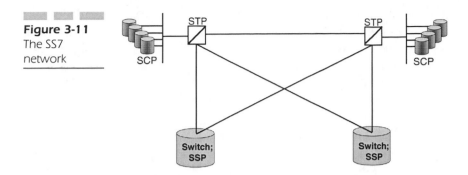

Figure 3-11
The SS7
network

tions apply. For example, if the calling party has chosen to block the delivery of caller ID information, the SCP query will return that fact.

Once the SCP has performed its task, the call information is returned to the STP packet switch, which consults routing tables and then selects a route for the call. Upon receipt of the call, the destination switch will cooperate with SS7 to determine whether the called party's phone is available, and if it is, it will ring the phone. If the customer's number is not available due to a busy condition or some other event, a packet will be returned to the source indicating the fact, and SS7 will instruct the originating SSP to put busy tone or reorder in the caller's ear.

At this point, the calling party has several options, one of which is to invoke one of the many CLASS services such as *automatic ringback*. With automatic ringback, the network will monitor the called number for a period of time, waiting for the line to become available. As soon as it is, the call will be cut through, and the calling party will be notified of the incoming call via some kind of distinctive ringing.

Thus, when a call is placed to a distant switch, the calling information is passed to SS7, which uses the caller's number, the called number, and SCP database information to choose a route for the call. It then determines whether there are any special call handling requirements to be invoked such as CLASS services, and instructs the various switches along the way to process the call as appropriate.

These features are made up of a set of services known as the *Advanced Intelligent Network* (AIN), a term coined by Telcordia (formerly Bellcore, sold to Science Applications International Corporation, recently sold again to Providence Equity and Warburg Pincus). The SSPs (switches) are responsible for basic calling, while the SCPs manage the enhanced services that ride atop the calls. The SS7 network, then, is responsible for the signaling required to establish and tear down calls and to invoke supplementary or enhanced services. It is critically important, even more so today as the network begins the complex process of migrating from a circuit-switched model to a VoIP packet-based model. More on that later in the book.

Network Topology

Before we enter the central office, we should pause to explain the topology of the typical network (please refer to Figure 3-9). The customer's telephone, and most likely, their PC and modem, are connected via *house wiring* (sometimes called inside wire) to a device on the outside of the house or office building called a *protector block* (see Figure 3-12). It is nothing more than a high-voltage protection device (a fuse, if you will) designed to open the circuit in the event of a lightning strike or the presence of some other high-voltage source. It protects both the subscriber and the switching equipment.

The protector connects to the network via *twisted pair wire*. Twisted pair is literally that—a pair of wires that have been twisted around each other to reduce the amount of crosstalk and interference that occur between wire pairs packaged within the same cable. The number of twists per foot is carefully chosen.

The twisted pair(s) that provides service to the customer arrives at the house on what is called the *drop wire*. It is either aerial, as shown in

Figure 3-12
Protector block,
shown at right

Figure 3-13, or buried underground. We note that there may be multiple pairs in each drop wire because today the average household typically orders a second line for a home office, computer, fax machine, or teenager.

Once it reaches the edge of the subscriber's property, the drop wire typically terminates on a terminal box, such as that shown in Figure 3-14. There, all of the pairs from the neighborhood are cross-connected to the main cable that runs up the center of the street. This architecture is used primarily to simplify network management. When a new

Figure 3-13
Aerial drop
wire

Figure 3-14
Terminal box,
sometimes
known as a "B-
Box"

neighborhood is built today, network planning engineers estimate the number of pairs of wire that will be required by the homes in that neighborhood. They then build the houses, install the network and cross-connect boxes along the street, and cross connect each house's drop wire to a single wire pair. Every pair in the cable has an appearance at every terminal box along the street, as shown in Figure 3-15. This allows a cable pair to be reassigned to another house elsewhere on the street, should the customer move. It also allows cable pairs to be easily replaced, should a pair go bad. This design dramatically simplifies the installation and maintenance process, particularly given the high demand for additional cable pairs today. This design also results in a challenge for network designers. These multiple appearances of the same cable pair in each junction box are called *bridged taps*. They create problems for digital services because electrical signal echoes can occur at the point where the wire terminates at a pair of *unterminated* terminal lugs. Suppose, for example, that cable pair number 117 is assigned to the house at #A2 Blanket Flower Circle. It is no longer necessary, therefore, for that cable pair to have an appearance at other terminal boxes because it is now assigned. Once the pair has been cross-connected to the customer's local loop drop wire, the technician *should* remove the appearances at other locations along the street by terminating the open wire appearances at each box. This eliminates the possibility of a signal echo occurring and creating errors on the line, particularly if the line is digital. ISDN is particularly susceptible to this phenomenon.

When outside plant engineers first started designing their networks, they set them up so that each customer was given a cable pair from their house all the way to the central office. The problem with this model was cost. It was very expensive to provision a cable pair for each customer.

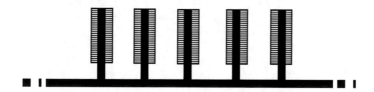

Figure 3-15
Along the street, terminal boxes (sometimes called B-Boxes) are installed. Each B-Box hosts an appearance of every pair of wires in the cable, which may contain as many as 500 pairs. With this design, every pair in the cable is potentially available to every home or business along the street, making installation extremely flexible.

With the arrival of time-division multiplexing, however, the "one dog, one bone" solution was no longer the only option. Instead, engineers were able to design a system under which customers could share access to the network as shown in Figure 3-16. This technique, known as a *subscriber loop carrier* (SLC), uses a collection of T1 carriers to combine the traffic from a cluster of subscribers and, thus, reduce the amount of wire required to interconnect them to the central office. The best known carrier system is called the SLC-96 (pronounced *SLICK*), originally manufactured by Western Electric/Lucent, which transports traffic from 96 subscribers over four, four-wire T-Carriers (plus a spare). A remote SLC terminal is shown in Figure 3-17. Thus 96 subscribers require only 20 wires between their neighborhood and the central office instead of the 192 wires that would otherwise be required. The only problem with this model is that customers are by and large restricted to the 64 Kbps of bandwidth that loop carrier systems assign to each subscriber. That means that subscribers wishing to buy *more* than 64 Kbps—such as those that want DSL—are out of luck. And since it is estimated that as many as 70 percent of all subscribers in the United States are served from loop carriers, this poses a problem that service providers are scrambling to overcome. New versions of loop carrier standards and technologies such as GR-303 and optical remotes that use fiber instead of copper for the trunk line between the remote terminal and the central office terminal go a long way toward solving this problem by making bandwidth allocation far more flexible. There is still quite a ways to go, however.

Typically, as long as a customer is within 12,000 feet from a central office (CO) they will be given a dedicated connection to the network, as shown in Figure 3-18. If they are farther than 12,000 feet from the CO, however, they will normally be connected to a subscriber loop carrier system of one type or another.

Figure 3-16
Subscriber loop carrier system —Customers share access to the network via a collection of multiplexed facilities to reduce outside plant cost.

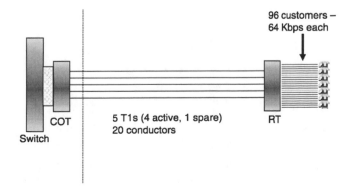

96 customers — 64 Kbps each

COT

Switch

5 T1s (4 active, 1 spare)
20 conductors

RT

Figure 3-17
A remote loop
carrier terminal

Figure 3-18
Carrier Serving
Area (CSA)
architecture

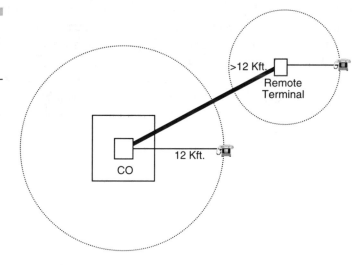

Either way, the customer's local loop, whether as a standalone pair of wires or as a time slot on a carrier system, makes its way over the physical infrastructure on its way to the network. It may be aerial, as shown in Figure 3-19, or underground, as shown in Figure 3-20. If it has to

Figure 3-19
Aerial plant—
Telephony is
the lowest set
of facilities on
the pole.

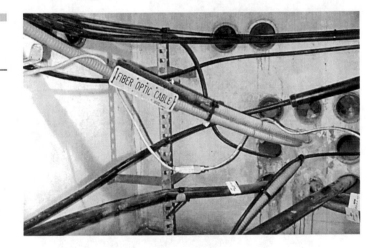

Figure 3-20
Underground
plant

travel a significant distance, it may encounter a load coil along the way, which is a device that "tunes" the local loop to the range of frequencies required for voice transport and extends the distance over which the signals can travel. A *load pot*, shown in Figure 3-21, comprises multiple load coils and performs loading for all the cable pairs in a cable that require it. It may also encounter a repeater if it is a digital loop carrier; the repeater receives the incoming signal, now weakened and noisy from the distance it has traveled, and reconstructs it, amplifying it before sending it on its way.

One last item of interest: Though we call them telephone poles, the wooden poles that support telephone cables are, in fact, shared among telephone, power, and cable providers, as shown in Figure 3-22. Telephony plant is the lowest, followed by cable; power is the highest, high enough that the tallest technician working on a phone or cable problem could not accidentally touch the open power conductors. Normally the telephone company owns the pole and leases space on it to other utilities.

As a cable approaches the central office, its pairs are often combined with those of other approaching cables in a splice case (see Figure 3-23)

Figure 3-21
A load pot containing load coils

Load Pot

Figure 3-22
Shared pole for aerial plant

Power

Cable

Telephone

Figure 3-23
Splice cases
(two on left,
one on right)

Figure 3-24
In the cable
vault, the large
cables shown
on the left are
broken down
into the smaller
cables on the
right for
distribution
throughout the
office.

to create a larger cable. For example, four 250-pair cables may be combined to create a 1,000 pair cable, which in turn may be combined with others to create a 5,000 pair cable that enters the central office. Once inside the office the cables are broken out for distribution. This is done in the *cable vault*; an example is shown in Figure 3-24. The large cable on the left side of the splice case is broken down into the collection of smaller cables, exiting on the right.

Into the Central Office

We could say a great deal more about the access segment of the network, and will later. For now, though, let's begin our tour of the central office.

By now our cable has traveled over aerial and underground traverses and has arrived at the central office. It enters the building via a buried conduit that feeds into the cable vault, the lowest subbasement in the office. (Remember, this is where the New York City fire began.) Because the cable vault is exposed to the underground conduit system, there is a always a danger that noxious and potentially explosive gases (methane, mostly) could flow into the cable vault and be set afire by a piece of sparking electrical equipment. These rooms are, therefore, closely monitored by the electronic equivalent of the canary in a coal mine, and will cause alarms to sound if gas is detected at dangerous levels.

The cables that enter the cable vault are large and encompass hundreds if not thousands of wire pairs, as shown in Figure 3-25. Their security is obviously critical, because the loss of a single cable could mean loss of service for hundreds or thousands of customers. To help maintain the integrity of the outside plant, large cables are pressurized with very dry air, fed from a pressurization pump and manifold system in the cable vault (see Figure 3-26). The pressure serves two purposes: It keeps moisture from leaking into a minor breach in the cable, and serves as an alarm system in the event of a major cable failure. Cable plant technicians can analyze the data being fed to them from pressure transducers along the cable route and very accurately determine the location of the break.

As soon as the large cables have been broken down into smaller pair bundles, they leave the cable vault on their way up to the Main Distribution Frame, or MDF.

Before we leave the basement of the CO, we should discuss power, a major consideration in an office that provides telecommunications services to hundreds of subscribers.

Figure 3-25
Wire pairs

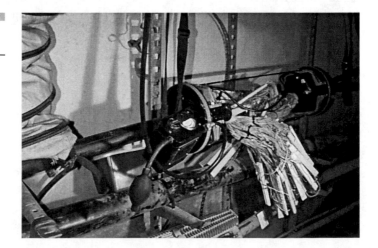

Figures 3-26
Pressure
transducers

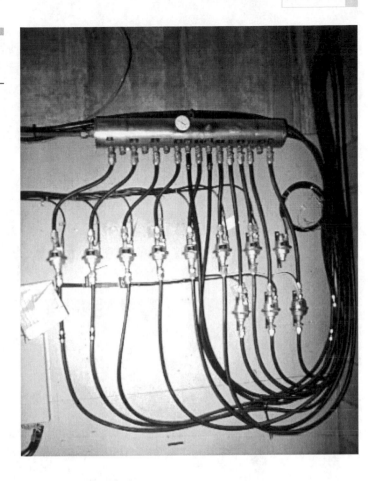

When you plug in a laptop computer, the power supply does not actually power the laptop. The power supply's job is to charge the battery (assuming the battery is installed in the laptop); the battery powers the computer. That way, if commercial power goes away, the computer is unaffected because it was running on the battery to begin with.

Central offices work the same way. They are powered by massive wet cell battery arrays, shown in Figure 3-27, that are stored in the basement. The batteries are quite large—about two feet tall—and are connected together by massive copper buss bars. Meanwhile, commercial power, shown in Figure 3-28 (OK, not really), is fed to the building from several sources on the power grid to avoid dependency on a single source of power. The AC power is fed into an inverter and filter that converts it to 48 volts DC, which is then trickled to the batteries. The voltage

Figure 3-27
Central office
battery array

Figure 3-28
Technicians
working on
central office
commercial
power

maintains a constant charge on them. In the event that a failure of commercial power should occur, the building's *uninterruptible power supply* equipment, or UPS, kicks in and begins a complex series of events designed to protect service in the office. First, it recognizes that the building is now isolated and is on battery power exclusively. The batteries have enough current to power a typical CO for several hours before things start to fail. Second, technicians begin overseeing the process of restoration and of shutting down nonessential systems to conserve battery power. Third, the UPS system initiates startup of the building's turbine, a massive jet engine that can be spun up in about a half hour. The turbine is either in a soundproof room in the basement or in an enclosure on the roof of the building (see Figure 3-29). Once the turbine has spun up to speed, the building's power load can slowly be shed onto it. It will power the building until it runs out of kerosene, but most buildings have several days' supply of fuel in their underground tanks. As Al Bergman, a data center design engineer and friend told me several years ago, "If we're down so long that we run out of fuel, our problems are far worse than an office that's out of service."

The cables leave the cable vault via fireproof foam-insulated ducts and travel upstairs to whatever floor houses the MDF. The MDF, shown in Figure 3-30, is a large iron jungle gym sort of frame that provides physical support and technician access for the thousands of pairs of wire that interconnect customer telephones to the switch. The cables from the cable vault ascend and are hardwired to the *vertical side* of the MDF, shown in Figure 3-31. The ends of the cables are called the "cable heads." The vertical side of the MDF has additional overvoltage protection, shown in the diagram. The arrays of plug-ins are called "carbons;" they are simply carbon fuses that open in the event of an overvoltage

Figure 3-29
Turbine in
rooftop
enclosure

Figure 3-30
Main
Distribution
Frame, or MDF

Figure 3-31
Vertical side of
MDF, showing
protectors

situation and protect the equipment and people in the office. A close-up of carbons is shown in Figure 3-32.

The vertical side of the MDF is wired to the *horizontal side*. The horizontal side, shown in Figure 3-33, is ultimately wired to the central office switch. Notice the mass of wire lying on each shelf of the frame: These are the cable pairs from the customers served by the office. Imagine the complexity of troubleshooting a problem in that hairball. I have spent many hours up to my shoulders in wire, tugging on a wire pair while a technician fifty feet down the frame, also up to his elbows in wire, feels for the wire that is moving so it can be traced. The horizontal side also provides craft/technician access for repair purposes. In Figure 3-34, Dick Pecor has connected a test set to the appearance of a cable pair and is listening for noise.

Figure 3-32
Close-up of carbons

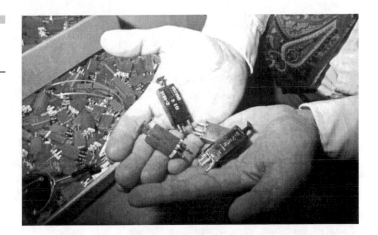

Figure 3-33
Horizontal side of MDF

Figure 3-34
Dick Pecor
testing cable
pair on
horizontal side
of MDF

New technicians are usually given a trial by fire when they arrive for work at a switching office. One of the most common pranks is to ask them to help troubleshoot a cable pair problem. They are asked to reach as far as they possibly can into the mass of wire on the horizontal side of the frame and feel for the wire that is moving, so that they can then wiggle it to help a technician farther down the MDF locate it and trace it to locate a problem. The new tech, now bent over and reaching deep into the frame holding a wire pair, doesn't realize that another technician is now on the vertical side of the frame, reaching into the wire mass with a wire lasso, which is slipped over the new tech's arm and tied tightly to the far side of the frame. The technicians then go to lunch, leaving the new tech wired in place. Naturally, other technicians ignore his pleas for help; they remember their own experiences all too well.

From the MDF, the cable pairs are connected to the local switch in the office. Figure 3-35 shows a Lucent #5ESS local switch. Remember that the job of the local switch is to establish temporary connections between two or more customers who wish to talk, or between two computers that wish to spit bits at each other. The *line units* (LUs) on the switch provide a connection point for every cable pair served by that particular switch in the office. Conceptually, then, it is a relatively simple task for the switch to connect one subscriber in the switch with another subscriber in the same switch. After all, the switch maintains tables—a directory, if you will—so it knows the subscribers to which it is directly connected.

Far more complicated is the task of connecting a subscriber in one switch to a subscriber in another. This is where network intelligence becomes critically important. As we mentioned earlier, when the switch

Figure 3-35
Lucent 5ESS
switch

receives the dialed digits, it performs a rudimentary analysis on them to determine whether the called party is locally hosted. If the number is in the same switch, the call is established. If it resides in another switch, the task is a bit more complex. First, the local switch must pass the call on to the local tandem switch, which provides access to the *points of presence* (POPs) of the various long-distance carriers that serve the area. The tandem switch typically does not connect directly to subscribers; it connects to other switches only. It also performs a database query through SS7 to determine who the subscriber's long-distance carrier of choice is so that it can route the call to the appropriate carrier. The tandem switch then hands the call off to the long-distance carrier, which transports it over the long-distance network to the carrier's switch in the remote (receiving) office. The long-distance switch passes the call through the remote tandem, which in turn hands the call off to the local switch that the called party is attached to.

SS7's influence once again becomes obvious here. One of the problems that occurred in earlier telephone system designs was the following. When a subscriber placed a call, the local switch handed the call off to the tandem, which in turn handed the call off to the long-distance provider. The long-distance provider seized a trunk, over which it transported the dialed digits to the receiving central office. The signaling information, therefore, traveled over the path designed to produce revenue for the telephone company, a process known as *in-band signaling*. As long as the called party was home, and wasn't on the phone, the call would go through as planned and revenue would flow. If they weren't home, however, or if they were on the phone, then no revenue was generated. Furthermore, the network resources that another caller might have used to place a call were not available to them, because they were tied up transporting call setup data—which produces no revenue.

SS7 changes all that. With SS7, the signaling data travels across a dedicated packet network from the calling party to the called party. SS7 verifies the availability of the called party's line, reserves it, and then—*and only then*—seizes a talk path. Once the talk path has been created it rings the called party's phone and places ringing tone in the caller's ear. As soon as the called party answers, SS7 silently bails out until one end or the other hangs up. At that point it returns the path to the pool of available network resources.

Interoffice Trunking

We have not discussed the manner in which offices are connected to one another. As Figure 3-36 illustrates, an optical fiber ring with add-drop multiplexers interconnects the central offices so that interoffice traffic can be safely and efficiently transported. The switches have *trunk units*

Figure 3-36
Typical network showing fiber ring interconnecting offices

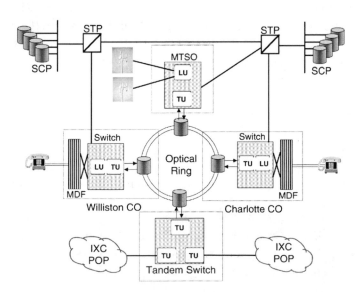

(TUs) that connect the back side, called the trunk side, of the switch to the wide area network; in this case the optical ring.

Trunks that interconnect offices have historically been four-wire copper facilities (a pair in each direction). Today they are largely optical, but are still referred to as "four-wire" because of their two-way nature.

An interesting point about trunks: In the 1960s and early 1970s, most interoffice trunks were analog rather than digital. To signal, they used single frequency tones. Because these trunks did not "talk" directly to customers, there was no reason to worry about a human voice inadvertently emitting a sound that could be misconstrued as a dialing tone. There was, therefore, no reason to use DTMF dialing—instead, trunk signaling was performed using single frequency tones. Specifically, if a switch wished to seize a long-distance trunk, it issued a single-frequency 2,600 Hz tone, which would signal the seizure to take place. Once the trunk seizure had occurred, the dialed digits could be outpulsed and the call would proceed as planned.

In 1972, John Draper, better known by his hacker name "Captain Crunch," determined that the toy plastic bosun's whistle that came packed in boxes of Cap'n Crunch cereal emitted—you guessed it—2600 Hz. Anyone who knew this could steal long-distance service from AT&T by blowing the whistle at the appropriate time during call setup. Before long, word of this capability became common knowledge and Cap'n Crunch cereal became the breakfast food of choice for hackers all over the country.

Soon hackers everywhere were constructing "blue boxes," small oscillators built from cheap parts that would emit 2,600 Hz and make possible the kind of access that Draper and his cohorts were engaged in. In fact, Steve Wozniak and Steve Jobs, both founders of Apple, were early blue box users and, according to legend, used the money they made building blue boxes to fund the company that became Apple.

Draper, a highly skilled hacker, was eventually caught and prosecuted for his activities, but he was given work furlough and was able to continue his software design studies. He is now a successful intrusion detection software designer!

Returning to our example again, then, let's retrace our steps. Cristina dials Adam's number. The call travels over the local loop, across aerial and/or underground facilities, and enters the central office via the cable vault. From the cable vault the call travels up to the main distribution frame and then on to the switch. The number is received by the switch in Cristina's serving central office, which performs an SS7 database lookup to determine the proper disposition of the call and any special service information about Cristina that it should invoke. The local switch routes the call over intraoffice facilities to a tandem switch, which in turn connects the call to the POP of whichever long-distance provider Cristina is subscribed. The long-distance carrier invokes the capabilities of SS7 to determine whether Adam's phone is available, and if it is, hands the call off to the remote local service provider. When Adam answers, the call is cut through and it progresses normally.

There are, of course, other ways that calls can be routed. The local loop could be wireless, and if it is, the call from the cell phone is received by a cell tower (see Figure 3-37) and is transported to a special dedicated cellular switch in a central office called a *Mobile Telephone Switching Office* (MTSO). The MTSO processes the call and hands it off to the wireline network via interoffice facilities. From that point on the call is handled like any other. It is either terminated locally on the local switch or handed to a tandem switch for long-distance processing. The fact of the

Figure 3-37
Cell tower

matter is that the only part of a cellular network that is truly wireless is the local loop—everything else is wired.

Before we wrap this chapter, we should take a few minutes to discuss carrier systems, voice digitization, and multiplexed transport, all of which take place (for the most part) in the central office. In spite of all the hype out there about the Internet and IP magic, plain old telephone service (POTS) remains the cash cow in the industry. In these final sections we'll explain T- and E-Carrier, SONET, SDH, and voice digitization techniques.

Conserving Bandwidth: Voice Transport

The original voice network, including access, transmission facilities, and switching components, was exclusively analog until 1962 when T-carrier emerged as an intra-office trunking scheme. The technology was originally introduced as a short-haul, four-wire facility to serve metropolitan areas and was a technology that customers would *never* have a reason to know about—after all, what customer could ever need a meg-and-a-half of bandwidth? Over the years, however, it evolved to include coaxial cable facilities, digital microwave systems, fiber, and satellite, and of course became a premier access technology that customers knew a great deal about.

Consider the following scenario: A corporation builds a new headquarters facility in a new area just outside the city limits. The building will provide office space for approximately 2,000 employees, a considerable number. Those people will need telephones, computer data facilities, fax lines, videoconferencing circuits, and a variety of other forms of connectivity.

The telephone company has two options that it can exercise to provide access to the building. It can make an assessment of the number of pairs of wire that the building will require and install them, or it can do the same assessment but provision the necessary bandwidth through carrier systems that transport multiple circuits over a shared facility. Obviously, this second option is the most cost-effective and is, in fact, the option that is most commonly used for these kinds of installations. This model should sound familiar: Earlier we discussed loop carrier systems and the fact that they reduce the cost of provisioning network access to far-flung neighborhoods. This is the same concept; instead of a residential neighborhood, we're provisioning a "corporate neighborhood."

The most common form of multiplexed access and transport is T-carrier, or E-carrier outside the United States. Let's take a few minutes to describe them.

Framing and Formatting in T1

The standard T-Carrier multiplexer, shown in Figure 3-38, accepts inputs from 24 sources, converts the inputs to PCM bytes, then time-division multiplexes the samples over a shared four-wire facility, as shown in Figure 3-39. Each of the 24 input channels yields an eight-bit sample, in

Figure 3-38
T-Carrier multiplexer channel banks, showing DS0 cards

Figure 3-39
A time-division multiplexer in action

Regenerators

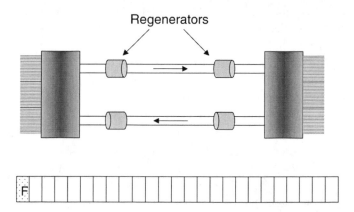

8-bits per sample, 24 samples per frame + frame bit = 193 bits.
8,000 frames are generated per second, yielding 1.544 Mbps.

round-robin fashion, once every 125 microseconds (8,000 times per second). This yields an overall bit rate of 64 Kbps for each channel (eight bits per sample × 8,000 samples per second). The multiplexer gathers one eight-bit sample from each of the 24 channels, and aggregates them into a 192-bit frame. To the frame it adds a frame bit, which expands the frame to a 193-bit entity. The frame bit is used for a variety of purposes that will be discussed in a moment.

The 193-bit frames of data are transmitted across the four-wire facility at the standard rate of 8,000 frames per second, for an overall T1 bit rate of 1.544 Mbps. Keep in mind that 8 Kbps of the bandwidth consists of frame bits (one frame bit per frame, 8,000 frames per second); only 1.536 Mbps belongs to the user.

Beginnings: D1 Framing

The earliest T-Carrier equipment was referred to as D1 and was considerably more rudimentary in function than modern systems (see Figure 3-40). In D1, every eight-bit sample carried seven bits of user information (bits 1 through 7) and one bit for signaling (bit 8). The signaling bits were used for exactly that: indications of the status of the line (on-hook, off-hook, busy, high and dry, etc.), while the seven user bits carried encoded voice information. Since only seven of the eight bits were available to the user, the result was considered to be less than toll quality (128 possible values, rather than 256). The frame bits, which in modern systems indicate the beginning of the next 192-bit frame of data, toggled back and forth between zero and one.

Figure 3-40
D1 framing

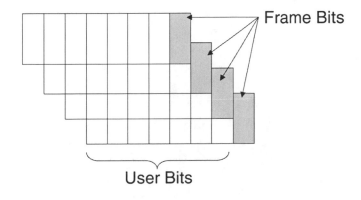

Frame Bits

User Bits

Evolution: D4

As time went on and the stability of network components improved, an improvement on D1 was sought after and found. Several options were developed, but the winner emerged in the form of the D4 or superframe format. Rather than treat a single 193-bit frame as the transmission entity, superframe "gangs together" twelve 193-bit frames into a 2,316-bit entity—shown in Figure 3-41—that obviously includes twelve frame bits. Please note that the bit rate has not changed: We have simply changed our view of what constitutes a frame.

Since we now have a single (albeit large) frame, we clearly don't need twelve frame bits to frame it; consequently, some of them can be redeployed for other functions. In superframe, the six odd-numbered frame bits are referred to as "terminal framing bits," and are used to synchronize the channel bank equipment. The even-numbered framing bits, on the other hand, are called "signal framing bits" and are used to indicate the receiving device *where* robbed-bit signaling occurs.

In D1, the system reserved one bit from every sample for its own signaling purposes, which succeeded in reducing the user's overall throughout. In D4, that is no longer necessary: Instead, we signal less frequently, and only occasionally rob a bit from the user. In fact, because the system operates at a high transmission speed, network designers determined that signaling can occur relatively infrequently and still convey adequate information to the network. Consequently, bits are robbed from the sixth and eighth iteration of each channel's samples, and then only the least significant bit from each sample. The resulting change in voice quality is negligible.

Figure 3-41
Superframe
(D4) framing

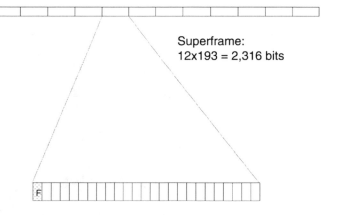

Superframe:
12x193 = 2,316 bits

Back to the signal framing bits: Within a transmitted superframe, the second and fourth signal framing bits would be the same, but the sixth would toggle to the opposite value, indicating to the receiving equipment that the samples in that subframe of the superframe should be checked for signaling state changes. The eighth and tenth signal framing bits would stay the same as the sixth, but would toggle back to the opposite value once again in the twelfth, indicating once again that the samples in that subframe should be checked for signaling state changes.

Today: Extended Superframe

While superframe continues to be widely utilized, an improvement came about in the 1980s in the form of *extended superframe* (ESF), shown in Figure 3-42. ESF groups 24 frames into an entity instead of twelve and, like superframe, reuses some of the frame bits for other purposes. Bits 4, 8, 12, 16, 20, and 24 are used for framing and form a constantly repeating pattern (001011 . . .). Bits 2, 6, 10, 14, 18, and 22 are used as a six-bit cyclic redundancy check (CRC) to check for bit errors on the facility. Finally, the remaining bits—all of the odd-numbered frame bits in the frame—are used as a 4 Kbps facility data link for end-to-end diagnostics and network management tasks.

ESF provides one major benefit over its predecessors: the ability to do unintrusive testing of the facility. In earlier systems, if the user reported trouble on the span, the span would have to be taken out of service for testing. With ESF, that is no longer necessary because of the added functionality provided by the CRC and the facility data link.

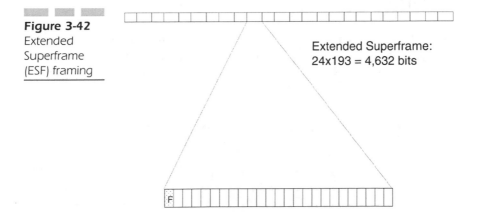

Figure 3-42
Extended
Superframe
(ESF) framing

Extended Superframe:
24x193 = 4,632 bits

The Rest of the World: E1

E1, used for the most part outside of the United States and Canada, differs from T1 on several key points. First, it boasts a 2.048 Mbps facility, rather than the 1.544 Mbps facility found in T1. Second, it utilizes a 32-channel frame rather than 24. Channel 1 contains framing information and a four-bit cyclic redundancy check (CRC-4); Channel 16 contains all signaling information for the frame; and channels 1 through 15, and 17 through 31 transport user traffic. The frame structure is shown in Figure 3-43.

There are a number of similarities between T1 and E1 as well: Channels are all 64 Kbps and frames are transmitted 8,000 times per second. And whereas T1 gangs together 24 frames to create an extended superframe, E1 gangs together 16 frames to create what is known as an ETSI multiframe. The multiframe is subdivided into two sub-multiframes; the CRC-4 in each one is used to check the integrity of the sub-multiframe that preceded it.

A final word about T1 and E1: Because T1 is a departure from the international E1 standard, it is incumbent upon the T1 provider to perform all interconnection conversions between T1 and E1 systems. For example, if a call arrives in the United States from a European country, the receiving American carrier must convert the incoming E1 signal to T1. If a call originates from Canada and is terminated in Australia, the Canadian originating carrier must convert the call to E1 before transmitting it to Australia.

Up the Food Chain: From T1 to DS3 . . . and Beyond

When T1 and E1 first emerged on the telecommunications scene, they represented a dramatic step forward in terms of the bandwidth that service providers now had access to. In fact, they were *so* bandwidth rich

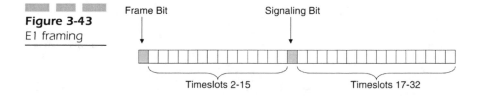

Figure 3-43
E1 framing

that there was no concept that a customer would ever need access to them. What customer, after all, could ever have a use for a million and a half bits per second of bandwidth?

Of course, that question was rendered moot in short order as increasing requirements for bandwidth drove demand that went well beyond the limited capabilities of low-speed transmission systems. As T1 became mainstream, its usage went up, and soon requirements emerged for digital transmission systems with capacity greater than 1.544 Mbps. The result was the creation of what came to be known as the North American Digital Hierarchy, shown in Figure 3-44. The table also shows the European and Japanese hierarchy levels.

From DS1 to DS3

We have already seen the process employed to create the DS1 signal from 24 incoming DS0 channels and an added frame bit. Now we turn

Figure 3-44
North American Digital Hierarchy

Hierarchy Level	Europe	United States	Japan
DS-0	64 Kbps	64 Kbps	64 Kbps
DS-1		1.544 Mbps	1.544 Mbps
E-1	2.048 Mbps		
DS-1c		3.152 Mbps	3.152 Mbps
DS-2		6.312 Mbps	6.312 Mbps
E-2	8.448 Mbps		32.064 Mbps
DS-3	34.368 Mbps	44.736 Mbps	
DS-3c		91.053 Mbps	
E-3	139.264 Mbps		
DS-4		274.176 Mbps	
			397.2 Mbps

our attention to higher bit rate services. As we wander our way through this explanation, pay particular attention to the complexity involved in creating higher rate payloads. This is one of the great advantages of SONET and SDH.

The next level in the North American Digital Hierarchy is called DS2. And while it is rarely seen outside of the safety of the multiplexer in which it resides, it plays an important role in the creation of higher bit rate services. It is created when a multiplexer *bit interleaves* four DS1 signals, inserting as it does so a control bit, known as a C-bit, every 48 bits in the payload stream. Bit interleaving is an important construct here, because it contributes to the complexity of the overall payload. In a bit-interleaved system, multiple bit streams are combined on a bit-by-bit basis as shown in Figure 3-45. When payload components are bit inter-leaved to create a higher-rate multiplexed signal, the system first selects bit 1 from channel 1, bit 1 from channel 2, bit 1 from channel 3, and so on. Once it has selected and transmitted all of the first bits, it goes on to the second bits from each channel, then the third, until it has created the super-rate frame. Along the way it intersperses C-bits, which are used to perform certain control and management functions within the frame.

Once the 6.312 Mbps DS2 signal has been created, the system shifts into high gear to create the next level in the transmission hierarchy. Seven DS2 signals are then bit interleaved along with C-bits after every 84 payload bits to create a composite 44.736 Mbps DS3 signal. The first part of this process, the creation of the DS2 payload, is called *M12 mul-tiplexing;* the second step, which combines DS2s to form a DS3, is called *M23 multiplexing.* The overall process is called *M13,* and is illustrated in Figure 3-46.

The problem with this process is the bit-interleaved nature of the mul-tiplexing scheme. Because the DS1 signal components arrive from dif-ferent sources, they may be (and usually are) slightly off from one another in terms of the overall phase of the signal—in effect, their

Figure 3-45
Bit-interleaved
system

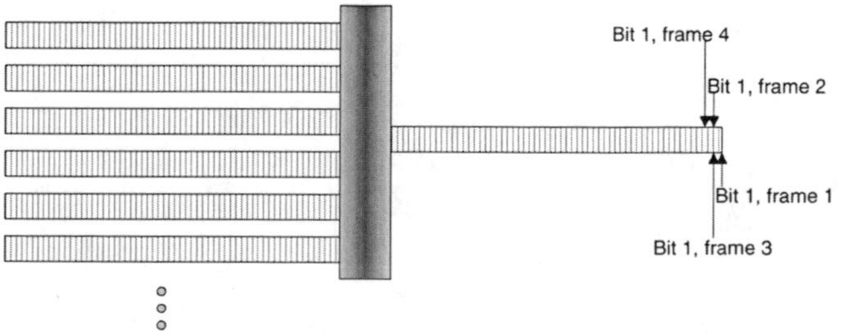

Figure 3-46
M13
multiplexing
process

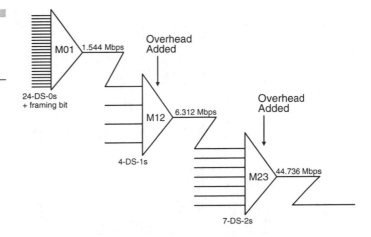

Figure 3-47
M12 frame has
four subframes
and 48-bit
payload
fields—1,176
bits

M0		C1		F0		C1		C1		F1
M1		C2		F0		C2		C2		F1
M1		C3		F0		C3		C3		F1
M1		C4		F0		C4		C4		F1

"speeds" differ slightly. This is unacceptable to a multiplexer which must rate align them if it is to properly multiplex them, beginning with the head of each signal. In order to do this, the multiplexer inserts additional bits, known as "stuff bits," into the signal pattern at strategic places that serve to rate align the components. The structure of a bit-stuffed DS2 frame is shown in Figure 3-47; a DS3 frame is shown in Figure 3-48.

The complexity of this process should now be fairly obvious to the reader. If we follow the left-to-right path shown in Figure 3-49, we see the rich complexity that suffuses the M13 signal-building process. Twenty-four 64 Kbps DS0s are aggregated at the ingress side of the T1[4] multiplexer, grouped into a T1 frame, and combined with a single frame bit to

---

[4]The process is similar for the E1 hierarchy.

Figure 3-48
M13 frame has
seven sub-
frames and 84-
bit payload
fields—4,760
bits

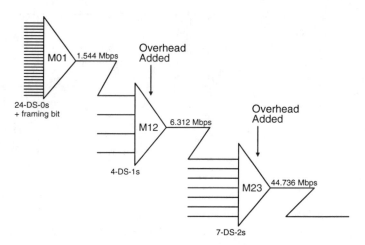

X 1	F 1	C 1	F 0	C 2	F 0	C 3	F 1
X 1	F 1	C 1	F 0	C 2	F 0	C 3	F 1
P 1	F 1	C 1	F 0	C 2	F 0	C 3	F 1
P 2	F 1	C 1	F 0	C 2	F 0	C 3	F 1
M 1	F 1	C 1	F 0	C 2	F 0	C 3	F 1
M 2	F 1	C 1	F 0	C 2	F 0	C 3	F 1
M 3	F 1	C 1	F 0	C 2	F 0	C 3	F 1

Figure 3-49
The M13
multiplexing
process and its
complexity

M01   1.544 Mbps

24-DS-0s
+ framing bit

Overhead
Added

M12   6.312 Mbps

4-DS-1s

Overhead
Added

M23   44.736 Mbps

7-DS-2s

form an outbound 1.544 Mbps signal (I call this the M01 stage; that's my nomenclature, used for the sake of naming continuity). That signal then enters the intermediate M12 stage of the multiplexer, where it is combined (bit interleaved) with three others and a good dollop of alignment overhead to form a 6.312 Mbps DS2 signal. That DS2 then enters the M23 stage of the mux, where it is bit interleaved with six others and another scoop of overhead to create a DS3 signal. At this point, we have a relatively high-bandwidth circuit that is ready to be moved across the wide area network.

Of course, as our friends in the UK are wont to say, there is always the inevitable spanner that gets tossed into the works (those of us on the left side of the Atlantic call it a wrench). Keep in mind that the 28- (do the math) bit-interleaved DS1s may well come from 28 different sources—which means that they may well have 28 different destinations. This translates into the pre-SONET digital hierarchy's greatest weakness, and one of SONET's greatest advantages. In order to drop a DS1 at its intermediate destination, we have to bring the composite DS3 into a set of back-to-back DS3 multiplexers (sometimes called M13 multiplexers). There, the ingress mux removes the second set of overhead, finds the DS2 in which the DS1 we have to drop out is carried, removes its over-head, finds the right DS1, drops it out, and then rebuilds the DS3 frame including reconstruction of the overhead, before transmitting it on to its next destination. This process is complex, time consuming, and expensive. So, what if we could come up with a method for adding and dropping signal components that eliminated the M13 process entirely? What if we could do it as simply as the process shown in Figure 3-50?

We have. It's called SONET in North America, SDH in the rest of the world, and it dramatically simplifies the world of high-speed transport.

SONET brings with it a subset of advantages that makes it stand above competitive technologies. These include mid-span meet, improved OAM&P, support for multipoint circuit configurations, unintrusive facility monitoring, and the ability to deploy a variety of new services. We will examine each of these in turn.

SONET Advantages: Mid-Span Meet

Because of the monopoly nature of early networks, interoperability was a laughable dream. Following the divestiture of AT&T, however, and the realization of Equal Access, the need for interoperability standards became a matter of some priority. Driven largely by MCI, the newly

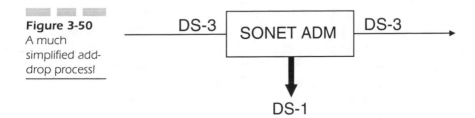

Figure 3-50
A much
simplified add-
drop process!

competitive telecommunications industry fought hard for standards that would allow different vendors' optical multiplexing equipment to inter-operate. This interoperability came to be known as "mid-span meet," SONET's greatest contribution to the evolving industry.

Improved OAM&P

Improved OAM&P is without question one of the most important contri-butions that SONET brings to the networking table. Element and Net-work monitoring, management and maintenance have always been something of a catch-as-catch-can effort because of the complexity and diversity of elements in a typical service provider's network. SONET overhead includes error-checking capability, bytes for network surviv-ability, and a diverse set of clearly defined management messages.

Multipoint Circuit Support

When SONET was first deployed in the network, the bulk of the traffic it carried derived from point-to-point circuits such as T1 and DS3 facili-ties. With SONET came the ability to hub the traffic, a process that com-bines the best of cross-connection and multiplexing to perform a capability known as "groom and fill." This means that aggregated traffic from multiple sources can be transported to a hub, managed as individ-ual components and redirected out any of several outbound paths with-out having to completely disassemble the aggregate payload. Prior to SONET, this process required a pair of back-to-back multiplexers, some-times called an M13 (for "multiplexer that interfaces between DS1 and DS3"). This capability, combined with SONET's discreet and highly capa-ble management features, results in a wonderfully manageable system of network bandwidth control.

Unintrusive Monitoring

SONET overhead bytes are embedded in the frame structure, meaning that they are universally transported alongside the customer's payload. Thus, tight and granular control over the entire network can be realized, leading to more efficient network management and the ability to deploy services on an as-needed basis.

New Services

SONET bandwidth is imminently scalable, meaning that the ability to provision additional bandwidth for customers that require it on an as-needed basis becomes real. As applications evolve to incorporate more and more multimedia content and, therefore, to require greater volumes of bandwidth, SONET offers it by the bucket load. Already, interfaces between SONET and Gigabit Ethernet are being written; interfaces to ATM and other high-speed switching architectures have been in existence for some time.

SONET Evolution

SONET was initially designed to provide multiplexed point-to-point transport. However, as its capabilities became better understood and networks became "mission-critical," its deployment became more innovative, and soon it was deployed in ring architectures as shown in Figure 3-51. These rings represent one of the most commonly deployed network topologies. For the moment, however, let's examine a point-to-point deployment. As it turns out, rings don't differ all that much.

Figure 3-51
Ring architectures used in SONET

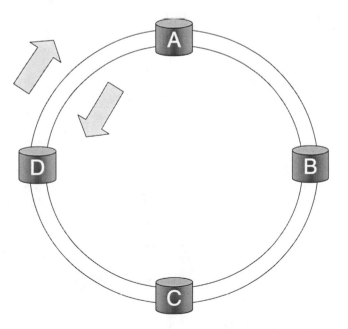

If we consider the structure and function of the typical point-to-point circuit, we find a variety of devices and "functional regions" as shown in Figure 3-52. The components include end devices, multiplexers in this case, which provide the point of entry for traffic originating in the customer's equipment and seeking transport across the network; a full-duplex circuit, which provides simultaneous two-way transmission between the network components; a series of repeaters or regenerators, responsible for periodically reframing and regenerating the digital signal; and one or more intermediate multiplexers, which serve as nothing more than pass-through devices.

When non-SONET traffic is transmitted into a SONET network, it is packaged for transport through a step-by-step, quasi-hierarchical process that attempts to make reasonably good use of the available network bandwidth and ensure that receiving devices can interpret the data when it arrives. The intermediate devices—including multiplexers and repeaters—also play a role in guaranteeing traffic integrity. To that end, the SONET standards divide the network into three "regions": path, line, and section. To understand the differences between the three, let's follow a typical transmission of a DS3, probably carrying 28 T1s, from its origination point to the destination.

When the DS3 first enters the network, the ingress SONET multiplexer packages it by wrapping it in a collection of additional information, called *path overhead*, which is unique to the transported data. For example, it attaches information that identifies the original source of the DS3, so that it can be traced in the event of network transmission problems; a bit-error control byte; information about how the DS3 is actually mapped into the payload transport area (and unique to the payload type); an area for network performance and management information;

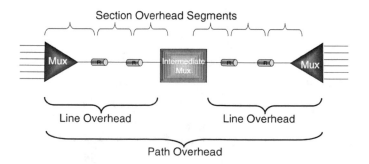

Figure 3-52
SONET network regions and functional devices

and a number of other informational components that have to do with the end-to-end transmission of the unit of data.

The packaged information, now known as a *payload,* is inserted into a SONET frame and, at that point, another layer of control and management information is added, called *Line Overhead.* Line Overhead is responsible for managing the movement of the payload from multiplexer to multiplexer. To do this, it adds a set of bytes that allow receiving devices to find the payload inside the SONET frame. As you will learn a bit later, the payload can occasionally wander around inside the frame, due to the vagaries of the network. These bytes allow the system to track that movement.

In addition to these "tracking bytes," the Line Overhead includes bytes that monitor the integrity of the network and have the ability to effect a switch to a backup transmission span if a failure in the primary span occurs, as well as another bit-error checking byte, a robust channel for transporting network management information, and a voice communications channel that allows technicians at either end of a the line to plug in with a handset (sometimes called a butt-in, or buttinski) and communicate while troubleshooting.

The final step in the process is to add a layer of overhead that allows the intermediate repeaters to find the beginning of, and synchronize, a received frame. This overhead, called the *Section Overhead,* contains a unique initial framing pattern at the beginning of the frame, an identifier for the payload signal being carried, another bit-error check, a voice communications channel, and another dedicated channel for network management information, similar to, but smaller than, the one identified in the Line Overhead.

The result of all this overhead, much of which seems like overkill (and in many peoples' minds *is*), is that the transmission of a SONET frame containing user data can be identified and managed with tremendous granularity from the source all the way to the destination.

So, to summarize, the hard little kernel of DS3 traffic is gradually surrounded by three layers of overhead information, as shown in Figure 3-53, that help it achieve its goal of successfully transiting the network. The Section Overhead is used at every device the signal passes through, including multiplexers and repeaters; the Line Overhead is only used between multiplexers; and the information contained in the Path Overhead is only used by the source and destination multiplexers—the intermediate multiplexers don't care about the specific nature of the payload, because they don't have to terminate or interpret it.

Figure 3-53
Layers of
SONET
overhead
surround the
payload prior
to trans-
mission.

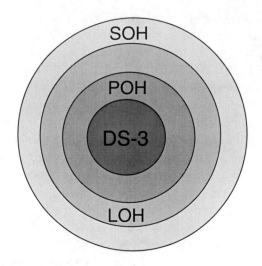

The SONET Frame

Keep in mind once again that we are doing nothing more complicated than building a T1 frame with an attitude. Recall that the T1 frame was made up of 24, eight-bit channels (samples from each of 24 incoming data streams) plus a single bit of overhead. In SONET, we have a similar con-struct—a lot more channel capacity, and a lot more overhead, but the same functional concept.

The fundamental SONET frame is shown in Figure 3-54, and is known as a *Synchronous Transport Signal, Level One,* or STS-1. It is nine bytes tall and 90 bytes wide, for a total of 810 bytes of transported data including both user payload and overhead. The first three columns of the frame are the Section and Line Overhead, known collectively as the *Transport Overhead*. The bulk of the frame itself, to the left, is the *syn-chronous payload envelope* or SPE, which is the "container" area for the user data that is being transported. The data, previously identified as the

Figure 3-54
Fundamental
SONET frame

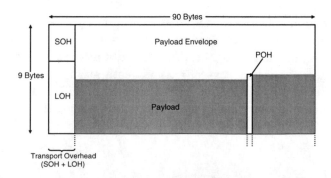

payload, begins somewhere in the payload envelope. The actual starting point will vary, as we will see later. The Path Overhead begins when the payload begins: Because it is unique to the payload itself, it travels closely with the payload. The first byte of the payload is in fact the first byte of the Path Overhead.

A word about nomenclature: Two distinct terms are used, often (incorrectly) interchangeably. The terms are *synchronous transport signal* (STS) and *optical carrier level* (OC). They are used interchangeably because, while an STS-1 and an OC-1 are both 51.84 Mbps signals, one is an electrically framed signal (STS) but the other describes an optical signal (OC). Keep in mind that the signals that SONET transports usually originate at an electrical source such as a T1. These data must be collected and multiplexed at an electrical level before being handed over to the optical transport system. The optical networking part of the SONET system speaks in terms of OC. The SONET frame is transmitted serially on a row-by-row basis. The SONET multiplexer transmits (and therefore receives!) the first byte of row one, all the way to the ninetieth byte of row one, then wraps to transmit the first byte of row two, all the way to the ninetieth byte of row two, and so on, until all 810 bytes have been transmitted.

Because the rows are transmitted serially, the many overhead bytes do not all appear at the beginning of the transmission of the frame—instead, they are peppered along the bitstream like highway markers. For example, the first two bytes of overhead in the Section Overhead are the framing bytes, followed by the single-byte signal identifier. The next 87 bytes are user payload, followed by the next byte of section overhead —in other words, there are 87 bytes of user data between the first three Section Overhead bytes and the next one! The designers of SONET were clearly thinking the day they came up with this, because each byte of data appears just when it is needed. Truly a remarkable thing!

Because of the unique way that the user's data is mapped into the SONET frame, the data can actually start pretty much anywhere in the payload envelope. The payload is always the same number of bytes, which means that if it starts "late" in the payload envelope, it may well run into the payload envelope of the next frame! In fact, this happens more often than not, but it's OK—SONET is equipped to handle this odd behavior. We'll discuss it shortly.

SONET Bandwidth

The SONET frame consists of 810 eight-bit bytes, and like the T1 frame is transmitted once every 125 μsec (8,000 frames per second). Doing the math, this works out to an overall bit rate of

810 bytes/frame $\times$ 8 bits/byte $\times$ 8,000 frames/second = 51.84 Mbps

the fundamental transmission rate of the SONET STS-1 frame.

That's a lot of bandwidth—51.84 Mbps is slightly more than a 44.736 Mbps DS3, a respectable carrier level by anyone's standard. What if more bandwidth is required, however? What if the user wants to transmit multiple DS3s, or perhaps a single signal that requires more than 51.84 Mbps, such as a 100 Mbps Fast Ethernet signal? Or, for that matter, a payload that requires *less* than 51.84 Mbps! In those cases, we have to invoke more of SONET's magic.

The STS-N Frame

In situations in which multiple STS-1s are required to transport multiple payloads, all of which fit in an STS-1's payload capacity, SONET allows for the creation of what are called STS-N frames, where "N" represents the number of STS-1 frames that are multiplexed together to create the frame. If three STS-1s are combined, the result is an STS-3. In this case, the three STS-1s are brought into the multiplexer and *byte interleaved* to create the STS-3, as shown in Figure 3-55. In other words, the multiplexer selects the *first* byte of frame one, followed by the *first* byte of frame two, followed by the *first* byte of frame three. Then it selects the *second* byte of frame one, followed by the *second* byte of frame two, followed by the *second* byte of frame three, and so on, until it has built an interleaved frame that is now three times the size of an STS-1: 9×270

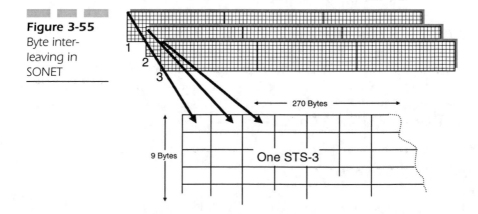

Figure 3-55
Byte interleaving in SONET

bytes instead of 9×90. Interestingly (and impressively!), the STS-3 is still generated 8,000 times per second.

The technique I have just described is called a *single-stage multiplexing process,* because the incoming payload components are combined in a single step. There is also a two-stage technique that is commonly used. For example, an STS-12 can be created in either of two ways. Twelve STS-1s can be combined in a single-stage process to create the byte interleaved STS-12; alternatively, four groups of three STS-1s can be combined to form four STS-3s, which can then be further combined in a second stage to create a single STS-12. Obviously, two-stage multiplexing is more complex than its single-stage cousin, but both are used.

Special note: *The overall bit rate of the STS-N system is N×STS-1. However, the maximum bandwidth that can be transported is STS-1 — but N of them can be transported! This is analogous to a channelized T1.*

The STS-Nc Frame

Let's go back to our Fast Ethernet example mentioned earlier. In this case, 51.84 Mbps is inadequate for our purposes, because we have to transport the 100 Mbps Ethernet signal. For this we need what is known as a *concatenated signal.* One thing you can say about SONET: It doesn't hurt for polysyllabic vocabulary.

On the long, lonesome stretches of outback highway in Australia, unsuspecting car drivers often encounter a devilish vehicle known as a "road train." Imagine an 18-wheel tractor-trailer (see top drawing, Figure 3-56, for a remarkable illustration) barreling down the highway at 80 miles per hour, but now imagine that it has six trailers—in effect, a 98-wheeler. These things give passing a whole new meaning. If a road train is rolling down the highway pulling three 50-foot trailers (middle drawing), then it has the ability to transport 150 feet of cargo—but only if the cargo is segmented into 50-foot chunks.

But what if the trucker wants to transport a 150-foot-long item? In that case, a special trailer must be installed that provides room for the 150-foot payload (bottom drawing).

If you understand the difference between the second and third drawings, then you understand the difference between an STS-N and an

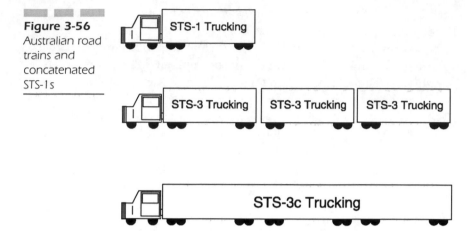

Figure 3-56
Australian road
trains and
concatenated
STS-1s

STS-Nc. The word "concatenate" means "to string together," which is
exactly what we do when we need to create what is known as a *super-rate
frame*—in other words, a frame capable of transporting a payload that
requires more bandwidth than an STS-1 can provide, such as our 100
Mbps Fast Ethernet frame. In the same way that an STS-N is analogous
to a channelized T1, an STS-Nc is analogous to an *unchannelized* T1. In
both cases, the customer is given the full bandwidth that the pipe provides; the difference lies in how the bandwidth is parceled out to the user.

Overhead Modifications in STS-Nc Frames

When we transport multiple STS-1s in an STS-N frame, we assume that
they may arrive from different sources. As a result, each frame is
inserted into the STS-N frame with its own unique set of overhead.
When we create a concatenated frame, though, the data that will occupy
the combined bandwidth of the frame derives from the same source—if
we pack a 100 Mbps Fast Ethernet signal into a 155.53 Mbps STS-3c
frame, there's only one signal to pack. It's pretty obvious, then, that we
don't need three sets of overhead to guide a single frame through the
maze of the network. For example, each frame has a set of bytes that
keep track of the payload within the synchronous payload envelope.
Since we only have one payload, therefore, we can eliminate two of them.
And the Path Overhead that is unique to the payload can similarly be
reduced, since there is a column of it for each of the three formerly individual frames. In the case of the pointer that tracks the floating payload,

the first pointer continues to perform that function; the others are changed to a fixed binary value that is known to receiving devices as a *concatenation indication*. The details of these bytes will be covered later in the overhead section.

Transporting Subrate Payloads: Virtual Tributaries

Let's now go back to our Australian road train example. This time he is carrying individual cans of Fosters Beer. From what I remember about the last time I was dragged into an Aussie pub (and it isn't much), the driver could probably transport about six cans of Fosters per 50-foot trailer. So now we have a technique for carrying payloads smaller than the fundamental 50-foot payload size. This analogy works well for understanding SONET's ability to transport payloads that require less bandwidth than 51.84 Mbps, such as T1 or traditional 10 Mbps Ethernet.

When a SONET frame is modified for the transport of subrate payloads, it is said to carry *virtual tributaries*. Simply put, the payload envelope is chopped into smaller pieces that can then be individually used for the transport of multiple lower-bandwidth signals.

Creating Virtual Tributaries

To create a "virtual tributary-ready" STS, the synchronous payload envelope is subdivided. An STS-1 comprises 90 columns of bytes, four of which are reserved for overhead functions (Section, Line, and Path). That leaves 86 for actual user payload. To create virtual tributaries, the payload capacity of the SPE is divided into seven 12-column pieces, called *virtual tributary (VT) groups*. Math majors will be quick to point out that $7 \times 12 = 84$, leaving two unassigned columns. These columns, shown in Figure 3-57, are indeed unassigned, and are given the rather silly name of *fixed stuff*.

Now comes the fun part. Each of the VT groups can be further subdivided into one of four different VTs to carry a variety of payload types, as shown in Figure 3-58. A VT1.5, for example, can easily transport a 1.544 Mbps signal within its 1.728 Mbps capacity, with a little room left over. A VT2, meanwhile, has enough capacity in its 2.304 Mbps structure to carry a 2.048 Mbps European E1 signal, with a little room left over. A VT3 can transport a DS1C signal, while a VT6 can easily accommodate a DS2, again, each with a little room left over.

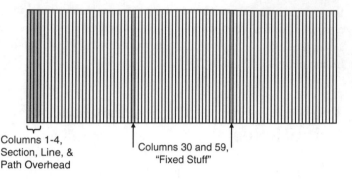

Figure 3-57
SONET fixed stuff

Columns 1-4,
Section, Line, &
Path Overhead

Columns 30 and 59,
"Fixed Stuff"

Figure 3-58
SONET virtual tributary payload transport

VT Type	Columns/VT	Bytes/VT	VTs/Group	VTs/SPE	VT Bandwidth
VT1.5	3	27	4	28	1.728
VT2	4	36	3	21	2.304
VT3	6	54	2	14	3.456
VT6	12	108	1	7	6.912

One aspect of virtual tributaries that must be mentioned is the mix-and-match nature of the payload. Within a single SPE, the seven VT groups can carry a variety of different VTs. However, each VT group can carry only one VT type.

That "little room left over" comment (earlier) is, by the way, one of the key points that SONET and SDH detractors point to when criticizing them as legacy technologies, claiming that in these times of growing competition and the universal drive for efficiency, they are inordinately wasteful of bandwidth, given that they were designed when the companies that delivered them were monopolies and less concerned about such things than they are now. We will discuss this issue in a later section of the book. For now, though, suffice it to say that one of the elegant aspects of SONET is its ability to accept essentially any form of data signal, map it into standardized positions within the SPE frame, and transport it efficiently and at very high speed to a receiving device on the other side of town or the other side of the world.

Creating the Virtual Tributary Superframe

You will recall that when DS1 frames are transmitted through modern networks today, they are typically formatted into extended superframes in order to eke additional capability out of the comparatively large percentage of overhead space that is available. When DS1 or other signals are transported, via an STS-1 formatted into VT groups, four consecutive STS-1s are "ganged together" to create a single VT superframe, as shown in Figure 3-59. To identify the fact that the frames are behaving as a VT superframe, certain of the overhead bytes are modified to indicate the change.

SONET Synchronization

SONET relies on a timing scheme called "plesiochronous timing." As I implied earlier, the word sounds like one of the geological periods that we

Figure 3-59
VT superframe

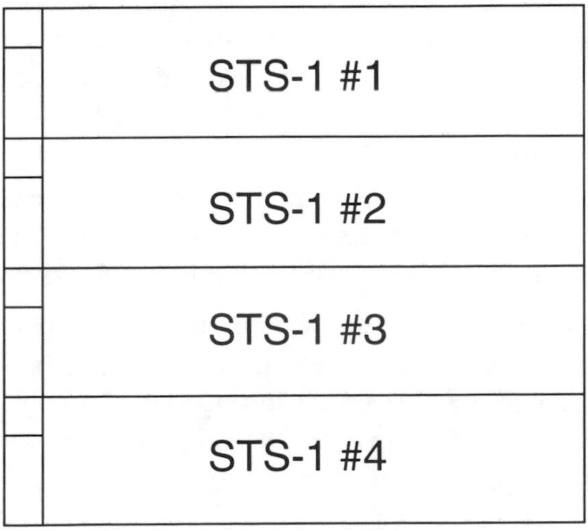

| STS-1 #1 |
| STS-1 #2 |
| STS-1 #3 |
| STS-1 #4 |

all learned in geology classes (Jurassic, Triassic, Plesiochronous, Plasticene). "Plesiochronous" derives from Greek, and means "almost timed." Other words that are commonly tossed about in this industry are *asynchronous* (not timed), *isochronous* (constant delay in the timing), and *synchronous* (timed). SONET is plesiochronous in spite of its name (*Synchronous* Optical Network) because the communicating devices in the network rely on multiple timing sources and are, therefore, allowed to drift slightly relative to each other. This is fine, because SONET has the ability to handle this with its pointer adjustment capabilities.

The devices in a SONET network have the luxury of choosing from any of five timing schemes to ensure accuracy of the network. As long as the schemes have Stratum 4 accuracy or better, they are perfectly acceptable timing sources. The five are discussed later.

■ *Line Timing:* Devices in the network derive their timing signal from the arriving input signal from another SONET device. For example, an add-drop multiplexer that sits out on a customer's premises derives its synchronization pulse from the incoming bit stream, and might provide further timing to a piece of CPE that is out beyond the ADM.

■ *Loop Timing:* Loop timing is somewhat similar to Line Timing; in loop timing, the device at the end of the loop is most likely a terminal multiplexer.

■ *External Timing:* The device has the luxury of deriving its timing signal directly from a Stratum 1 clock source.

■ *Through-Timing:* Similar to line timing, a device that is through-timed receives its synchronization signal from the incoming bit stream, but then forwards that timing signal on to other devices in the network. The timing signal then passes "through" the intermediate device.

■ *Free Running:* In free running timing systems, the SONET equipment in question does not have access to an external timing signal, and must derive its timing from internal sources only.

One final point about SONET should be made. When the standard is deployed over ring topologies, two timing techniques are used. Either external timing sources are depended upon to time network elements, or one device on the ring is internally timed (free running) while all the others are through-timed.

One Final Thought: Next-Generation SONET

As we have discussed, SONET and SDH were introduced in the 1980s as an optical solution for high-bandwidth vendor interoperability in optical networks. SONET transmission rates have steadily increased to OC-768 (40 Gbps), and the protocol was developed primarily to transport voice in fixed timeslots through the network, in retrospect making SONET inefficient for the transport of the increasing volumes of bursty traffic.

Beginning in the 1990s, the Internet arrived, and with it a profusion of data transferred over the existing network infrastructure. STS-1 (51.84 Mbps) is the basic rate at which SONET operates, and higher-bandwidth services are delivered as multiples of STS-1. This technique, however, is less than ideal for the transport of multimedia data payloads, and when it *is* used, a variety of challenges appear. These challenges include: the need to identify techniques to "rescue" the stranded bandwidth that occurs when SONET is used for data transport without having to upgrade existing SONET ADMs and cross-connect boxes; identification of the most effective way to carry data traffic over SONET; the creation and deployment of a reasonable technique for dynamically and efficiently resizing allocated bandwidth without traffic loss; methods and procedures to ensure survivability of link groups in the event of a link loss; and processes to support multiple traffic types across the same physical infrastructure. The solution to this plethora of challenges came in the form of *Next-Generation SONET* (NGS), and with NGS came a new set of terms: *virtual concatenation*, *link capacity adjustment scheme* (LCAS), and *generic framing procedure* (GFP).

Virtual concatenation is based on the concept that a group of smaller containers can be concatenated to form a larger container. The containers, typically STS-1s, are then routed across the network towards the same destination where the individual STS-1s are realigned and sorted to form the original payload. In OSI terms this technique sounds like a layer four process, doesn't it?

Virtual concatenation has been defined for VT1.5, VT2, VT3, VT6, STS-1 SPE and STS-3c SPE containers. A virtually concatenated group is referred to as a *Base Container-nv*, where "n" is a number that indicates the number of containers of the base container type. For example, VT1.5-4v indicates that the group consists of four VT1.5s that are virtually concatenated. This process of "payload channelization" is performed at the end points in the network.

There are two different forms of virtual concatenation. If the channel rate is STS-1 and greater for a channel in a group, then *high-order virtual concatenation* (HOVC) is used. If the channel rate is below STS-1, then *low-order virtual concatenation* (LOVC) is used.

The LCAS is a complementary technology to virtual concatenation. It facilitates the adjustment of the size of a group of channels that have been virtually concatenated. LCAS synchronizes transmission between the sender and receiver so that the size of a virtually concatenated circuit can be increased or decreased without corrupting the data signal.

GFP is an encapsulation technique that allows traffic from a fibre channel-based storage area network (SAN) to be carried directly onto the SONET network efficiently and cost-effectively.

Previously, storage traffic had to be converted to an intermediary protocol such as ATM, frame relay, or IP before being placed on a SONET network. This added overhead and issues related to security, resiliency, latency, and performance. GFP is an optical switch interface that lets companies use their SONET facilities more effectively. GFP can be deployed at the edge of a leased SONET network and can then be used to allocate portions of the circuit using virtual concatenation (VCAT). VCAT extends the utility of SONET transport by letting bandwidth be allocated in multiples of 50 Mbps, thus allocating only the bandwidth required by a particular application.

In combination with VCAT, the cost effectiveness of GFP is significantly improved. Enterprises can transport data using optimal bandwidth allocations, and service providers can maximize the efficiency of their overall network. Furthermore, enterprises can deploy private storage over SONET solutions by purchasing storage-specific equipment for SONET and leasing SONET circuits from a service provider.

SONET Summary

Clearly, SONET is a complex and highly capable standard designed to provide high-bandwidth transport for legacy and new protocol types alike. The overhead that it provisions has the ability to deliver a remarkable collection of network management, monitoring, and transport granularity.

The European Synchronous Digital Hierarchy, or SDH, shares many of SONET's characteristics, as we will now see. SONET, you will recall, is a limited North American standard, for the most part. SDH, on the other hand, provides high-bandwidth transport for the rest of the world.

Most books on SONET and SDH cite a common list of reasons for their proliferation, including a recognition of the importance of the global marketplace and a desire on the parts of manufacturers, therefore, to provide devices that will operate in both SONET and SDH environments; the global expansion of ring architectures; a greater focus on network management and the value that it brings to the table; and massive, unstoppable demand for more bandwidth. To those, add these: increasing demand for high-speed routing capability to work hand-in-glove with transport; deployment of DS1, DS3, and E1 interfaces directly to the enterprise customer as access solutions; growth in demand for broadband access technologies such as cable modems, the many flavors of DSL, and two-way satellite connectivity; the ongoing replacement of traditional circuit-switched network fabrics with packet-based transport and mesh architectures; a renewed focus on the SONET and SDH overhead with an eye toward using it more effectively; and convergence of multiple applications on a single, capable, high-speed network fabric.

SDH Nomenclature

Before launching into a functional description of SDH, it would be good to first cover the differences in naming convention between the two. This will help to dispel confusion (hopefully!).

The fundamental SONET unit of transport uses a nine-row by 90-column frame that comprises three columns of section and line overhead, one column of path overhead, and 87 columns of payload. The payload, which is primarily user data, is carried in a payload envelope that can be formatted in various ways to make it carry a variety of payload types. For example, multiple SONET STS-1 frames can be combined to create higher-rate systems for transporting multiple STS-1 streams, or a single higher-rate stream created from the combined bandwidth of the various multiplexed components. Conversely, SONET can transport subrate payloads, known as virtual tributaries, which operate at rates slower than the fundamental STS-1 SONET rate. When this is done, the payload envelope is divided into virtual tributary groups, which can in turn transport a variety of virtual tributary types.

In the SDH world, similar words apply, but they are different enough that they should be discussed. As you will see, SDH uses a fundamental transport "container" that is three times the size of its SONET counterpart. It is a nine-row by 270-column frame that can be configured into one of five container types, typically written C-n (where "C" means "container"). N can be 11, 12, 2, 3, or 4; they are designed to transport a variety of payload types.

When an STM-1 is formatted for the transport of virtual tributaries (known as "virtual containers" in the SDH world), the payload pointers must be modified. In the case of a payload that is carrying virtual containers, the pointer is known as an Administrative Unit type 3 (AU-3). If the payload is *not* structured to carry virtual containers, but is instead intended for the transport of higher rate payloads, then the pointer is known as an Administrative Unit type 4. Generally speaking, an AU-3 is typically used for the transport of North American Digital Hierarchy payloads; AU-4 is used for European signal types.

The SDH Frame

To understand the SDH frame structure, it is first helpful to understand the relationship between SDH and SONET. Functionally, they are identical: In both cases, the intent of the technology is to provide a globally standardized transmission system for high-speed data. SONET is indeed optimized for the T1-heavy North American market, while SDH is more applicable to Europe; beyond that, however, the overhead and design considerations of the two are virtually identical. There are, however, some key differences.

Perhaps the greatest difference between the two lies in the physical nature of the frame. A SONET STS-1 frame has 810 total bytes, for an overall aggregate bit rate of 51.84 Mbps—perfectly adequate for the North American 44.736 Mbps DS3. An SDH STM-1 frame, however, designed to transport a 139.264 Mbps European E4 or CEPT-4 signal, must be larger if it is to accommodate that much bandwidth—it clearly won't fit in the limited space available in an STS-1. An STM-1, then, operates at a fundamental rate of 155.52 Mbps, enough for the bandwidth requirements of the E4. This should be where the déjà vu starts to kick in: Perceptive readers will remember that 155.52 Mbps number from our discussions of the SONET STS-3, which offers *exactly* the same bandwidth. An STM-1 frame is shown in Figure 3-60. It is a byte-interleaved, 9-row-by-270-column frame, with the first nine columns devoted to overhead and the remaining 261 devoted to payload transport.

A comparison of the bandwidth between SONET and SDH systems is also interesting. The fundamental SDH signal is exactly *three times* the bandwidth of the fundamental SONET signal, and this relationship continues all the way up the hierarchy.

Figure 3-60
An SDH STM-1
frame

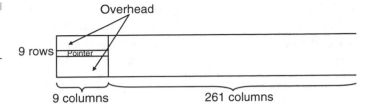

9 rows

Overhead

Pointer

9 columns   261 columns

STM Frame Overhead

The overhead in an STM frame is very similar to that of an STS-1 frame, although the nomenclature varies somewhat. Instead of Section, Line, and Path Overhead to designate the different regions of the network that the overhead components address, SDH uses *Regenerator Section, Multiplex Section,* and Path Overhead, as shown in Figure 3-61. The Regenerator Section (RSOH) occupies the first three rows of nine bytes, the Multiplex Section (MSOH) the final five. Row four is reserved for the pointer. As in SONET, the Path Overhead floats gently on the payload tides, rising and falling in response to phase shifts. Functionally, these overhead components are identical to their SONET counterparts.

Overhead Details

Because an STM-1 is three times as large as an STS-1, it has three times the overhead capacity—nine columns instead of three (plus Path Overhead). The first row of the RSOH is its SONET counterpart, with the exception of the last two bytes, which are labeled as being reserved for national use and are specific to the PTT administration that

Figure 3-61
SDH Overhead

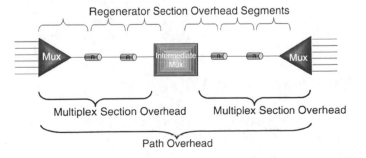

Regenerator Section Overhead Segments

Mux   R   R   Intermediate Mux   R   R   Mux

Multiplex Section Overhead   Multiplex Section Overhead

Path Overhead

implements the network. In SONET they are not yet assigned. The second row is different from SONET in that it has three bytes reserved for media-dependent implementation (differences in the actual transmission medium, whether copper, coaxial, or fiber) and the final two reserved for national use. As before, they are not yet definitively assigned in the SONET realm.

The final row of the RSOH also sports two bytes reserved for media-dependent information, while they are reserved in SONET. All other regenerator section/section overhead functions are identical between the two.

The MSOH overhead in the SDH frame is almost exactly the same as that of the SONET Line Overhead, with one exception: Row nine of the SDH frame has two bytes reserved for national administration use. They are reserved in the SONET world.

The pointer in an SDH frame is conceptually identical to that of a SONET pointer, although it has some minor differences in nomenclature. In SDH the pointer is referred to as an AU pointer, referring to the standard naming convention described earlier.

SONET and SDH were originally rolled out to replace the T1 and E1 hierarchies, which were suffering from demands for bandwidth beyond what they were capable of delivering. Their principal deliverable was voice—and lots of it. Let's take a moment now to describe the process of voice digitization, still a key component of network transport.

Voice Digitization

The goal of digitizing the human voice for transport across an all-digital network grew out of work performed at Bell Laboratories shortly after the turn of the century. That work led to a discrete understanding of not only the biological nature and spectral makeup of the human voice, but also to a better understanding of language, sound patterns, and the sounded emphases that make up spoken language.

The Nature of Voice

A typical voice signal is made up of frequencies that range from approximately 30 Hz to 10 KHz. Most of the speech energy, however, lies between 300 Hz and 3,300 Hz, the so-called voice band. Experiments have shown that the frequencies below 1 KHz provide the bulk of recognizability and intelligibility, while the higher frequencies provide richness, articulation, and natural sound to the transmitted signal.

The human voice has a remarkably rich mix of frequencies, and this richness comes at a considerable price. In order for telephone networks to transmit the voice's entire spectrum of frequencies, significant network bandwidth must be made available to every ongoing conversation. There is a substantial price tag attached to bandwidth; it is a finite commodity within the network, and the more of it that is consumed, the more it costs.

The Network

Thankfully, work performed at Bell Laboratories at the beginning of the twentieth century helped network designers confront this challenge head-on. To understand it, let's take a tour of the telephone network.

The typical network, as shown in Figure 3-62, is divided into several regions: the access plant; the switching, multiplexing, and circuit connectivity equipment (the central office); and the long-distance transport plant. The access and transport domains are often referred to as the *outside plant*; the central office is, conversely, the *inside plant*. The outside plant has the responsibility to aggregate inbound traffic for switching and transport across the long haul, as well as to terminate traffic at a

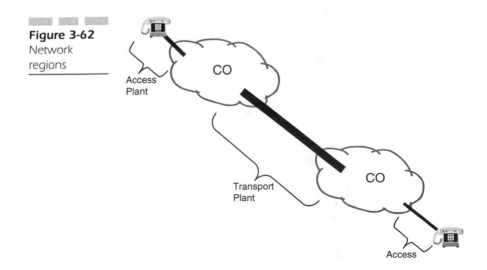

Figure 3-62
Network regions

particular destination. The inside plant, on the other hand, has the responsibility to multiplex incoming traffic streams, switch the streams, and select an outbound path for ultimate delivery to the next central office in the chain or the final destination. Switching, therefore, was centrally important to the development of the modern network.

Multiplexing

Equally important as the development of the central office switch was the concept of multiplexing, which allowed multiple conversations to be carried simultaneously across a single shared physical circuit. The first such systems used frequency-division multiplexing (FDM), a technique made possible by the development of the vacuum tube, in which the range of available frequencies is divided into "chunks" which are then parceled out to subscribers. For example, Figure 3-63 illustrates that subscriber #1 might be assigned the range of frequencies between 0 and 4,000 Hz, while subscriber #2 is assigned 4,000 to 8,000 Hz, #3 8,000 to 12,000 Hz, and so on, up to the maximum range of frequencies available in the channelized system. In FDM, we often observe that users are given "some of the frequency all of the time," meaning that they are free to use their assigned frequency allocation at any time, but may *not* step outside the bounds given to them. Early FDM systems were capable of transporting 24, 4 KHz channels, for an overall system bandwidth of 96 KHz. FDM, while largely replaced today by

Figure 3-63
Frequency-
division
multiplexing
(FDM)

more efficient systems that will be discussed later, is still used in analog cellular telephone and microwave systems, among others.

This model worked well in early telephone systems. Because the lower regions of the 300 to 3,300 Hz voiceband carry the frequency components that allow for recognizability and intelligibility, telephony engineers concluded that while the higher frequencies enrich the transmitted voice, they are not necessary for calling parties to recognize and understand each other. This understanding of the makeup of the human voice helped them create a network that was capable of faithfully reproducing the sounds of a conversation while keeping the cost of consumed bandwidth to a minimum. Instead of assigning the full complement of 10 KHz to each end of a conversation, they employed filters to bandwidth-limit each user to approximately 4,000 Hz, a resource savings of some 60 percent. Within the network, subscribers were frequency-division multiplexed across shared physical facilities, thus allowing the telephone company to efficiently conserve network bandwidth.

Time, of course, changes everything. As with any technology, there were downsides to frequency-division multiplexing. It is an analog technology and, therefore, suffers from the shortcomings that have historically plagued all transmission systems. The wire over which information is transmitted behaves like a longwire antenna, picking up noise along the length of the transmission path and very effectively homogenizing it with the voice signal. Additionally, the power of the transmitted signal diminishes over distance, and if the distance is far enough, the signal will have to be amplified to make it intelligible at the receiving end. Unfortunately, the amplifiers used in the network are not particularly discriminating: They have no way of separating the voice noise; the result is that they convert a weak, noisy signal into a loud, noisy signal—better, but far from ideal. A better solution was needed.

The better solution came about with the development of time-division multiplexing (TDM), which became possible because of the transistor and integrated circuit electronics that arrived in the late 1950s and early 1960s. TDM is a digital transmission scheme, which employs a small number of discrete signal states, rather than the essentially infinite range of values employed in analog systems. While digital systems are as susceptible to noise impairment as their analog counterparts, the discrete nature of their binary signaling makes it relatively easy to separate the noise from the transmitted signal. In digital carrier systems, there are only three valid signal values, one positive, one negative, and zero; anything else is construed to be noise. It is, therefore, a trivial exercise for digital repeaters to discern what is desirable and what is not, thus, eliminating the problem of cumulative noise. The role of the regenerator, as shown is Figure 3-64, is to receive a weak, noisy digital signal; remove the noise; reconstruct the original signal; and amplify it before transmitting the signal on to the next segment of the transmission facility. For this reason, repeaters are also called *regenerators*, because that is precisely the function they perform.

One observation: It is estimated that as much as 60 percent of the cost of building a transmission facility lies in the regenerator sections of the span. For this reason, optical networking, discussed a bit later, has various benefits, not the least of which is the ability to reduce the number of regenerators required on long transmission spans. In a typical network, these regenerators must be placed approximately every 6,000 feet along a span, which means that there is considerable expense involved when providing regeneration along a long-haul network.

Digital signals, often called "square waves," have a very rich mixture of signal frequencies. Not to bring too much physics into the discussion, we must at least mention the Fourier Series, which describes the makeup of a digital signal. The Fourier Series is a mathematical representation of the behavior of waveforms. Among other things, it notes the following fact. If we start with a fundamental signal—such as that shown in Figure 3-65—and mathematically add to it its odd harmonics (a "harmonic" is defined as a wave whose frequency is a whole-number

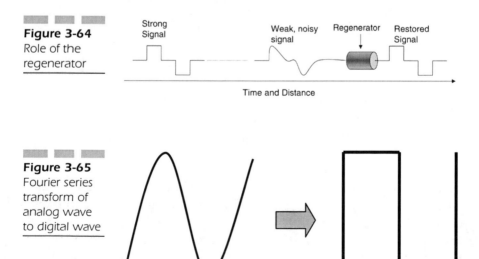

Figure 3-64
Role of the regenerator

Strong Signal   Weak, noisy signal   Regenerator   Restored Signal

Time and Distance

Figure 3-65
Fourier series transform of analog wave to digital wave

multiple of another wave), we see a rather remarkable thing happening: The waveform gets steeper on the sides and flatter on top. As we add more and more of the odd harmonics (there is, after all, an infinite series of them), the wave begins to look like the typical square wave. Now of course, there is no such thing as a true square wave; for our purposes, though, we'll accept the fact.

It should now be intuitive to the reader that digital signals have a mixture of low-, medium-, and high-frequency components, which means that they cannot be transmitted across the bandwidth-limited 4 KHz channels of the traditional telephone network. In digital carrier facilities, the equipment that restricts the individual transmission channels to 4 KHz "chunks" is eliminated, thus giving each user access to the full breadth of available spectrum across the shared physical medium. In frequency-division systems, we observed that we give users "some of the frequency all of the time;" in time-division systems, we turn that around and give users "*all* of the frequency *some* of the time." As a result, high-frequency digital signals can be transmitted without restriction.

Digitization brings with it a cadre of advantages including improved voice and data transmission quality; better maintenance and troubleshooting capability, and therefore reliability; and dramatic improvements in configuration flexibility. In digital carrier systems, the time-division multiplexer is known as a channel bank; under normal circumstances, it allows either 24 or 30 circuits to share a single, 4-wire facility. The 24-channel system is called T-Carrier; the 30-channel system, used in most of the world, is called E-Carrier. Originally designed in 1962 as a way to transport multiple channels of voice over expensive transmission facilities, they soon became useful as data transmission networks as well. That, however, came later. For now, we focus on voice.

Voice Digitization

The process of converting analog voice to a digital representation in the modern network is a logical and straight-forward process. It has four distinct steps: *Pulse Amplitude Modulation* (PAM) sampling, in which the amplitude of the incoming analog wave is sampled every 125 microseconds; *companding*, during which the values are weighted toward those most receptive to the human ear; *quantization*, in which the weighted samples are given values on a nonlinear scale; and finally *encoding*, during which each value is assigned a distinct binary value. Each of these stages of *Pulse Code Modulation* (PCM) will now be discussed in detail.

Pulse Code Modulation (PCM)

Thanks to the work performed by Harry Nyquist at Bell Laboratories in the 1920s, we know that to optimally represent an analog signal as a digitally encoded bitstream, the analog signal must be sampled at a rate that is equal to twice the bandwidth of the channel over which the signal is to be transmitted. Since each analog voice channel is allocated 4 KHz of bandwidth, it follows that each voice signal must be sampled at twice that rate, or 8,000 samples per second. In fact, that is precisely what happens in T-Carrier systems, which we now use to illustrate our example. The standard T-Carrier multiplexer accepts inputs from 24 analog channels, as shown in Figure 3-66. Each channel is sampled, in turn, every one eight-thousandth of a second in round-robin fashion, resulting in the generation of 8,000 pulse amplitude samples from each channel every second. The sampling rate is important. If the sampling rate is too high, too much information is transmitted and bandwidth is wasted; if the sampling rate is too low, then we run the risk of aliasing. "Aliasing" is the interpretation of the sample points as a false waveform, due to the paucity of samples.

This PAM process represents the first stage of PCM, the process by which an analog baseband signal is converted to a digital signal for transmission across the T-Carrier network. This first step is shown in Figure 3-67.

The second stage of PCM, shown in Figure 3-68, is called quantization. In quantization, we assign values to each sample within a constrained range. For illustration purposes, imagine what we now have before us. We have "replaced" the continuous analog waveform of the signal with a series of amplitude samples which are close enough together that we can

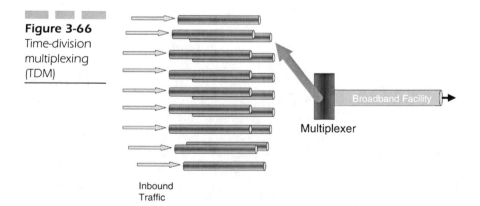

Figure 3-66
Time-division multiplexing (TDM)

Broadband Facility

Multiplexer

Inbound Traffic

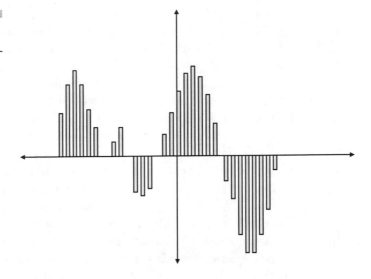

Figure 3-67
PAM samples

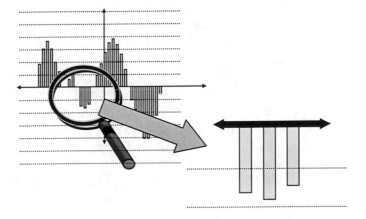

Figure 3-68
Quantizing the
samples

discern the shape of the original wave from their collective amplitudes. Imagine also that we have graphed these samples in such a way that the "wave" of sample points meanders above and below an established zero point on the x-axis, so that some of the samples have positive values and others are negative, as shown.

The amplitude levels allow us to assign values to each of the PAM samples, although a glaring problem with this technique should be obvious to the careful reader. Very few of the samples actually line up *exactly* with the amplitudes delineated by the graphing process. In fact, most of them fall *between* the values, as shown in the illustration. It doesn't take much of an intuitive leap to see that several of the samples will be assigned the same digital value by the coder-decoder that performs this function, yet they are clearly *not* the same amplitude. This inaccuracy in the measurement method results in a problem known as *quantizing noise,* and is inevitable when linear measurement systems, such as the one suggested by the drawing, are employed in coder-decoders (CODECs).

Needless to say, design engineers recognized this problem rather quickly, and equally quickly came up with an adequate solution. It is a fairly well-known fact among psycholinguists and speech therapists that the human ear is far more sensitive to discrete changes in amplitude at low-volume levels than it is at high-volume levels, a fact not missed by the network designers tasked with optimizing the performance of digital carrier systems intended for voice transport. Instead of using a linear scale for digitally encoding the PAM samples, they designed and employed a nonlinear scale that is weighted with much more granularity at low volume levels—that is, close to the zero line—than at the higher amplitude levels. In other words, the values are extremely close together near the x-axis and get farther and farther apart as they travel up and down the y-axis. This nonlinear approach keeps the quantizing noise to a minimum at the low-amplitude levels where hearing sensitivity is the highest, and allows it to creep up at the higher amplitudes, where the human ear is less sensitive to its presence. It turns out that this is not a problem, because the inherent shortcomings of the mechanical equipment (microphones, speakers, the circuit itself) introduce slight distortions at high-amplitude levels that hide the effect of the nonlinear quantizing scale.

This technique of "compressing" the values of the PAM samples to make them fit the nonlinear quantizing scale results in a bandwidth savings of more than 30 percent. In fact, the actual process is called *companding,* because the sample is first compressed for transmission, then expanded for reception at the far end, hence the term.

The actual graph scale is divided into 255 distinct values above and below the zero line. In North America and Japan, the encoding scheme is known as μ-Law (Mu-Law); the rest of the world relies on a slightly different standard known as A-Law.

There are eight segments above the line and eight below (one of which is the shared zero point); each segment, in turn, is subdivided into 16 steps. A bit of binary mathematics now allows us to convert the quantized amplitude samples into an eight-bit value for transmission. For the sake of demonstration, let's consider a negative sample that falls into the thirteenth step in segment five. The conversion would take on the following representation:

1 101 1101

Where the initial "1" indicates a negative sample, "101" indicates the fifth segment, and "1101" indicates the thirteenth step in the segment. We now have an eight-bit representation of an analog amplitude sample that can be transmitted across a digital network, then reconstructed with its many counterparts as an accurate representation of the original analog waveform at the receiving end. This entire process is known as Pulse Code Modulation, or PCM, and the result of its efforts is often referred to as "toll-quality voice."

Alternative Digitization Techniques

While PCM is perhaps the best-known, high-quality voice digitization process, it is by no means the only one. Advances in coding schemes and improvements in the overall quality of the telephone network have made it possible for encoding schemes to be developed that use far less bandwidth than traditional PCM. In this next section, we will consider some of these techniques.

Adaptive Differential Pulse Code Modulation (ADPCM)

Adaptive Differential Pulse Code Modulation (ADPCM) is a technique that allows toll-quality voice signals to be encoded at half-rate (32 Kbps) for transmission. ADPCM relies on the predictability that is inherent in human speech to reduce the amount of information required. The technique still relies on PCM encoding, but adds an additional step to carry out its task. The 64 Kbps PCM-encoded signal is fed into an ADPCM transcoder, which considers the *prior* behavior of the incoming stream to create a prediction of the behavior of the *next* sample. Here's where the magic happens: Instead of transmitting the actual value of the predicted sample, it encodes in four bits and transmits the *difference* between the actual and predicted samples. Since the difference from sample to sample is typically quite small, the results are generally considered to be very close to toll-quality. This four-bit transcoding process, which is based on the known behavior characteristics of human voice, allows the system to transmit 8,000, four-bit samples per second, thus reducing the overall bandwidth requirement from 64 Kbps to 32 Kbps. It should be noted that ADPCM works well for voice, because the encoding and predictive algorithms are based upon its behavior characteristics. It does not, however, work as well for higher bit-rate data (above 4,800 bps), which has an entirely different set of behavior characteristics.

Continuously Variable Slope Delta

Continuously Variable Slope Delta (CVSD) is a unique form of voice encoding that relies on the values of individual bits to predict the behavior of the incoming signal. Instead of transmitting the volume (height or y-value) of PAM samples, CVSD transmits information that measures the changing slope of the waveform. So, instead of transmitting the actual change, it transmits the *rate* of change.

To perform its task, CVSD uses a reference voltage, to which it compares all incoming values. If the incoming signal value is less than the reference voltage, then the CVSD encoder reduces the slope of the curve to make its approximation better mirror the slope of the actual signal. If the incoming value is *more* than the reference value, then the encoder will increase the slope of the output signal, again causing it to approach and, therefore, mirror the slope of the actual signal. With each recurring sample and comparison, the step function can be increased or decreased as required. For example, if the signal is increasing rapidly, then the steps are increased one after the other in a form of step function by the encoding algorithm. Obviously, the reproduced signal is not a particularly exact representation of the input signal: In practice, it is pretty jagged. Filters, therefore, are used to smooth the transitions.

CVSD is typically implemented at 32 Kbps, although it can be implemented at rates as low as 9,600 bps. At 16 to 24 Kbps, recognizability is still possible; down to 9,600, recognizability is seriously affected, although intelligibility is not.

Linear Predictive Coding (LPC)

We mention LPC here only because it has carved out a niche for itself in certain voice-related applications such as voice mail systems, automobiles, aviation, and electronic games that speak to children. LPC is a complex process, implemented completely in silicon, which allows for voice to be encoded at rates as low as 2,400 bps. The resulting quality is far from toll quality, but it is certainly intelligible and its low bit-rate capability gives it a distinct advantage over other systems.

LPC relies on the fact that each sound created by the human voice has unique attributes, such as frequency range, resonance, and loudness, among others. When voice samples are created in LPC, these attributes are used to generate "prediction coefficients." These predictive coefficients represent linear combinations of previous samples, hence the name—*Linear* Predictive Coding.

Prediction coefficients are created by taking advantage of the known *formants* of speech, which are the resonant characteristics of the mouth and throat, which give speech its characteristic timbre and sound. This sound, referred to by speech pathologists as the "buzz," can be described by both its pitch and its intensity. LPC, therefore, models the behavior of the vocal cords and the vocal tract itself.

To create the digitized voice samples, the "buzz" is passed through an inverse filter that is selected based upon the value of the coefficients. The remaining signal, after the buzz has been removed, is called the residue.

In the most common form of LPC, the residue is encoded as either a *voiced* or *unvoiced* sound. Voiced sounds are those that require vocal cord vibration, such as the "g" in "glare," the "b" in "boy," the "d" and "g" in "dog." Unvoiced sounds —such as the "h" in "how," the "sh" in "shoe," the "f" in "frog"—require no vocal cord vibration. The transmitter creates and sends the prediction coefficients, which include measures of pitch, intensity and whatever voiced and unvoiced coefficients are required. The receiver undoes the process: It converts the voice residue, pitch, and intensity coefficients into a representation of the source signal, using a filter similar to the one used by the transmitter to synthesize the original signal.

Digital Speech Interpolation (DSI)

Human speech has many measurable (and therefore predictable) characteristics, one of which is a tendency to have embedded pauses. As a rule, people do not spew out a series of uninterrupted sounds; they tend to pause for emphasis, to collect their thoughts, to reword a phrase while the other person listens quietly on the other end of the line. When speech technicians monitor these pauses, they discover that during considerably more than half of the total connect time, the line is silent.

Digital Speech Interpolation (DSI) takes advantage of this characteristic silence to drastically reduce the bandwidth required for a single channel. Whereas 24 channels can be transported over a typical T1 facility, DSI allows for as many as 120 conversations to be carried over the same circuit. The format is proprietary and requires the setting aside of a certain amount of bandwidth for overhead.

A form of "statistical multiplexing" lies at the heart of DSI's functionality. The standard T-Carrier is a time-division multiplexed scheme in which channel ownership is assured: A user assigned to channel three will *always* own channel three, regardless of whether they are actually using the line. In DSI, channels are not owned. Instead, large numbers of users share a "pool" of available channels. When a user starts to talk, the DSI system assigns an available timeslot to that user and notifies the receiving end of the assignment. This system works well when the number of users is large, because statistical probabilities are more accurate and indicative of behavior in larger populations than in smaller ones.

There is a downside to DSI, of course, and it comes in several forms. *Competitive clipping* occurs when more people start to talk than there are available channels, resulting in someone being unable to talk. *Connection clipping* occurs when the receiving end fails to learn what channel a conversation has been assigned within a reasonable amount of time, resulting in signal loss. Two approaches have been created to address these problems. In the case of competitive clipping, the system intentionally "clips" off the front end of the initial word of the second person who speaks. This technique is not optimal, but does prevent loss of the conversation and also obviates the problem of clipping out the middle of a conversation, which would be more difficult for the speakers to recover from. The loss of an initial syllable or two can be mentally reconstructed far more easily than sounds in the middle of a sentence.

A second technique used to recover from clipping problems is to temporarily reduce the encoding rate. The typical encoding rate for DSI is 32 Kbps; in certain situations, the encoding rate may be reduced to 24 Kbps, thus freeing up significant bandwidth for additional channels.

Both techniques are widely utilized in DSI systems.

Enter the Modern World: Voice-Over IP

It would be irresponsible to update this book without giving at least passing mention to *voice-over IP* (VoIP) as a major telecommunications phenomenon. It is fundamentally changing the way in which people talk on a daily basis. In this section we'll describe what it is, how it works, and how it is being used by a wide array of both residence and enterprise customers.

Let's examine how a telephone call is carried across an IP network. A customer begins the call using a traditional telephone. The call is carried across the PSTN to an IP telephony gateway, which is nothing more than a special-purpose router designed to interface between the PSTN and a packet-based IP network. As soon as the gateway receives the call, it interrogates an associated gatekeeper device, which provides information about billing, authorization, authentication, and call routing. As soon as the gatekeeper has delivered the information to the gateway, the gateway transmits the call across the IP network to another gateway, which in turn hands the call off to the local PSTN at the receiving end, completing the call.

Let's also address one important misconception. At the risk of sounding Aristotelian, "All Internet telephony is VoIP, but all VoIP is not Internet telephony." Got it? Let me explain. It is quite possible today to make telephone calls across the Internet, which by definition are IP-based calls simply because IP is the fundamental operational protocol of the Internet. However, IP-based calls can be made without involving the Internet at all. For example, a corporation may interconnect multiple locations using dedicated private line facilities, and they may transport IP-based phone calls across that dedicated network facility. This is clearly VoIP, but is *not* Internet telephony. There's a big difference.

VoIP Evolution

There was a time when VoIP was only for the brave: those intrepid explorers among us who were willing to try Internet-based telephony just to prove that it could be done, with little regard for QoS, enhanced services, and the like. In many ways it was reminiscent of the 70s, when we all drove around in our cars and talked to each other on CB radios (Remember? Oh, come on . . . "10-4, Rubber Duck. See you on the flip side." Don't deny it . . .). The quality was about that good, and it was experimental, and new, and somewhat exciting. It faded in and out of the public's awareness, but didn't really catch on in a big way until late 2002 or so, when the level of broadband penetration reached a point where adequate access bandwidth was available, and CODEC technology far enough advanced to make VoIP not only possible but actually quite good. Everyone toyed with the technology; Cisco, Avaya, Lucent, and Nortel, not to mention Ericsson and Siemens, all built VoIP products to one degree or another, because they knew that sooner or later the technology would become mainstream. But as always happens, it took *early intro-ducers* (not early adopters) to make the transition from experiment to service. And while companies like Telus in Canada played enormous roles in the ongoing development of VoIP (they were the first company to install a system-wide IP backbone, and one of the first innovators to offer true VoIP services), it took a couple of somewhat off-the-wall companies to really push it to the front of the public consciousness. The first of these was Vonage; the second was Skype.

Vonage Vonage, "The Broadband Phone Company," was founded in January 2001 in Edison, New Jersey. With 600 employees, Vonage completes more than five million calls per week and has more than 350,000 active lines in service. To date they have completed more than 200 million calls over their network, which is Internet-based VoIP.

Signing up for the service could not be easier, because it is all done at the Vonage Web site—there is no need for "a human in the loop." Would-be customers can read all about the service, get answers to their questions, and if they decide they want to sign up they give Vonage a credit card via a secure connection; the card is billed each month for the service. They then enter their area code (NPA), after which Vonage assigns them a phone number. It is important to note that the number is virtual—i.e., while I live in Vermont, I don't have to have a Vermont area code (there's only one for the entire state). For example, a good friend of mine lives in New Jersey but his parents live in southern California. To prevent them having to pay long-distance charges when they call him, he has an LA area code. I have even heard about professional people asking for a 212 area code so that they would appear as if they were in New York City for prestige reasons!

Once the phone number has been assigned, users can then select a wide variety of supplementary services (caller ID, etc.) at no extra charge. One thing they have to do is configure their voice mail service, which is included as part of the low-cost package. There are two options; one is the traditional model, where a subscriber dials an access number, enters a PIN code and then listens to their messages. Alternatively, subscribers can set it up (all done at the Web site) so that received voice mail messages are converted to .wav files and sent as e-mail attachments!

There are two options for connecting to the Vonage network. One is to use a small terminal adapter that Vonage will mail to you; it is connected to the broadband network, and a phone and laptop are in turn plugged into the box. Alternatively, subscribers can download a SoftPhone and make calls from their laptop.

I have been a Vonage user for quite some time and the service could not be easier to use. When I travel I simply plug my laptop into the broadband connection in my hotel room. The Vonage SoftPhone automatically boots up when I start the machine, and as soon as it's online I can start making (and more importantly, receiving) telephone calls. Think about it: *I can be in a hotel room in Mumbai and if someone dials my Vermont-based Vonage telephone number, the phone will ring in India —and the caller has no idea where I am, nor do they need to.* Users beware: Be sure to shut down your laptop when traveling internationally, because the phone will inevitably ring at 3:30 AM because the caller doesn't realize you're traveling abroad!

Vonage offers remarkable service for a remarkably low price, and most of the time it is indistinguishable from traditional circuit-switched voice. Their service is so good that other companies have come into the marketplace, including AT&T (CallVantage), and Verizon (VoiceWing). The number of players grows daily.

To date Vonage has approximately 250,000 customers. Let's now turn our attention to another purveyor of VoIP, which has a few more— *roughly 20 million subscribers.*

Skype

I knew it was over when I downloaded Skype. When the inventors of KaZaA are distributing for free a little program that you can use to talk to anybody else, and the quality is fantastic, and it's free—it's over. The world will change now inevitably.

—Michael Powell, Chairman, FCC

The idea of charging for calls belongs to the last century. Skype software gives people new power to affordably stay in touch with their friends and family by taking advantage of their technology and connectivity investments.

—Niklas Zennström, CEO & cofounder of Skype

Many readers will recall KaZaA, the file sharing program that became well known during the turbulent Napster controversy over the unauthorized sharing of commercial music. The same two programmers who created KaZaA, Niklas Zennström and Janus Friis, put their heads together and concluded that while peer-to-peer music works well, it is fraught with legal ramifications. So what else could be shared in a peer-to-peer fashion? *Voice.*

To date, more than 45 million people have downloaded the free Skype installer, and as I sit here writing this section in mid-December 2004 there are 1.6 million people online. I have used Skype for over a year and the service is excellent. The only downside is that the original, free Skype is a walled-garden technology—users can only talk with other Skype users. But with over a million people online at any time, and with the application's ability to search for other users by city and country, *you'll find someone to talk with*. Skype also supports messenging and video-conferencing.

Of course, the walled-garden concept is limiting, so in early 2004 Skype announced SkypeOut, a paid service that allows Skype users to place calls to regular telephones in about 20 countries (and growing) for about two cents per minute.

Vonage and Skype are but two of the growing herd of public VoIP providers that are redefining the world of voice services. In the enterprise space, VoIP is having an equally dramatic impact. Let's examine the various ways in which enterprise VoIP can be deployed to the benefit of customers.

VoIP Implementation

IP voice has become a reasonable technological alternative to traditional PCM voice. VoIP gateways are in the late stages of development with regard to reliability, features, and manageability. Consequently, service providers wishing to deploy VoIP solutions have a number of options available to them. One is to accept the current state of the technology and deploy it today, while waiting for enhanced versions to arrive, knowing that while the product may not be 100 percent carrier grade, it will certainly be an acceptable solution.

A second option is a variation of the first: Implement the technology that is available today, but make no guarantees of service quality. This approach is currently being used in Western Europe and in some parts of the United States as a way to provide low-cost long-distance service. The service is actually quite good, and while it is not "toll quality," it is inexpensive. And since most telephony users have become inured to the lower than toll quality of cellular service, they are less inclined to be annoyed by VoIP service that is lower than they might have otherwise anticipated. Needless to say this works in the favor of the VoIP service provider. Furthermore, the companies deploying the service often have no complex billing infrastructure to support because they rely on prepaid billing models, thus they can keep their costs low. SkypeOut, for example, allows subscribers to buy credit online using a credit card.

Carrier Class IP Voice

There is no question that VoIP is here to stay and is a serious contender for voice services in both the residence and enterprise markets. The dollars don't lie: Domestic revenues for VoIP will grow exponentially between 2004 and 2009, from $600 million today to $3 billion.

From a customer's perspective, the principal advantages of VoIP include consolidated voice, data, and multimedia transport; elimination of parallel systems; the ability to exercise PSTN fallback in the event that the IP network becomes overly congested; and reduction of long-distance voice and facsimile costs.

For an ISP or CLEC, the advantages are different but no less dramatic, including efficient use of network investments due to traffic consolidation, new revenue sources from existing clients because of demand for service-oriented applications that benefit from being offered over an IP infrastructure, and the option of transaction-based billing. These can collectively be reduced to the general category of customer service, which service providers such as ISPs and CLECs should be focused on. The challenge they face will be to prove that the service quality they provide over their IP networks will be identical to that provided over traditional, circuit-switched technology.

Major carriers are voting on IP with their own wallets, a sure sign of confidence in the technology.

As we observed earlier, the key to IP's success in the voice provisioning arena lies with its "invisibility." If done correctly, service providers can add IP to their networks, maintaining service quality while dramatically improving their overall efficiency. IP voice (not to be confused with Internet voice) will be implemented by carriers and corporations as a way to reduce costs and move to a multiservice network fabric. Virtually every major equipment manufacturer—including such notables as Lucent, Nortel, and Cisco—has added SS7 and IP voice capability to their router products and access devices, recognizing that their primary customers are looking for IP solutions.

IP-Enabled Call Centers

Ultimately, well-run call centers are nothing more than enormous routers. They receive incoming data, delivered using a variety of media (phone calls, e-mail messages, facsimile transmissions, mail order), and make decisions about handling them based on information contained in each message. One challenge that has always faced call center management is the ability to integrate message types and route them to a particular agent based on specified decision-making criteria such as name, address, telephone number, e-mail address, Automatic Number Identification (ANI) triggers, product purchase history, the customer's geographic location, or language preference. This has resulted in the development of a technical philosophy called *unified messaging*. With unified messaging, all incoming messages for a particular agent, regardless of the media over which they are delivered, are housed centrally and clustered under a single account identifier. When the agent logs into the network, their PC lists all of the messages that have been received for them, giving them the ability to much more effectively manage the information contained in those messages.

Today, unified messaging systems also support road warriors. A traveling employee can dial into a message gateway and download all messages—voice, fax, e-mail—from that one central location, thus dramatically simplifying the process of staying connected while away from the office.

Call centers are undergoing tremendous change as the IP juggernaut hits them. The first of these is a redefinition of the market they serve. For the longest time, call centers have primarily served large corporations because they are expensive to deploy. With the arrival of IP, however, the cost is dropping precipitously, all major corporations are now moving toward an IP-enabled call center model because of the ability to create unified application sets and to introduce enormous flexibility into their calling models.

Integrating the PBX

There is an enormous installed base of legacy PBX equipment, and vendors did not enter the IP game enthusiastically. Early entrants arrived with enhancements to existing equipment that were proprietary and expensive and did very little to raise customer awareness or engender trust in the vendor's migration strategy. Over time, however, PBX manufacturers began to embrace the concept of convergence as their customers' demands for IP-based systems grew, and, soon, products began to appear. Most have heard the message delivered by the customer: Preserve the embedded base to the degree possible as a way of saving the existing investment; create a migration strategy that is seamless and as transparent as possible; and preserve the features and functions that customers are already familiar with to minimize churn.

Major vendors like Lucent Technologies and Nortel Networks have responded with products that do exactly that. They allow voice calls and faxes to be carried over LAN, WANs, the Internet, and corporate intranets and function as both a gateway and gatekeeper, providing circuit-to-packet conversion, security, and access to a wide variety of applications; these include enhanced call features such as multiple line appearances, hunt groups, multiparty conferencing, call forward, hold, call transfer, and speed dial. They also provide access to voice mail, CTI applications, wireless interfaces, and call center features.

An Important Aside: Billing as a Critical Service

One area that is often overlooked when companies look to improve the quality of the services they provide to their customers is billing. And while it is not typically viewed as a strategic competitive advantage, studies have shown that customers view it as one of the top considerations when assessing capability in a service provider.

Billing offers the potential to strengthen customer relationships, improve long-term business health, cement customer loyalty, and generally make businesses more competitive. However, for billing to achieve its maximum benefit and strategic value, it must be fully integrated with a company's other operations support systems, including network and service provisioning systems, installation support, repair, network management, and sales and marketing. If done properly, the billing system becomes an integral component of a service suite that allows the service provider to quickly and efficiently introduce new and improved services in logical bundles; improve business indicators such as service timeliness, billing accuracy, and cost; offer custom service programs to individual customers based on individual service profiles; and transparently migrate from legacy service platforms to the so-called next generation network.

In order for billing as a strategic service to work successfully, service providers must build a business plan and migration strategy that takes into account integration with existing operations support systems; business process interaction; the role of IT personnel and processes; and postimplementation testing to ensure compliance with strategic goals stipulated at the beginning of the project.

VoIP Supporting Protocols

In addition to IP, VoIP relies on a superset of additional protocols to guarantee the level of rich functionality that users have come to expect. The relative importance of the protocols discussed in the sections that follow wax and wane like the stages of the moon, but all are important and readers should be familiar with them.

H.323

H.323 started in 1996 as H.320, an ITU-T standard for the transmission of multimedia content over ISDN. Its original goal was to connect LAN-based multimedia systems to network-based multimedia systems. It defined a network architecture that included gatekeepers, which performed zone management and address conversion; endpoints, which were terminals and gateway devices; and multimedia control units, which served as bridges between multimedia types.

H.323 has been rolled out in four phases. Phase one defined a three-stage call setup process: a precall step, which performed user registration, connection admission, and exchange of status messages required for call setup; the actual call setup process, which used messages similar to ISDN's Q.931; and finally, a capability exchange stage, which established a logical communications channel between the communicating devices and identified conference management details.

Phase two allowed for the use of Real-Time Transport Protocol (RTP) over ATM, which eliminated the added redundancy of IP and also provided for privacy and authentication, as well as greatly demanded telephony features such as call transfer and call forwarding. RTP has an added advantage: When errors result in packet loss, RTP does not request resends of those packets, thus providing for real-time processing of application-related content. No delays result from errors.

Phase three added the ability to transmit real-time fax after establishing a voice connection, and phase four, released in May 1999, added call connection over UDP, which significantly reduced call setup time; inter-zone communications; call hold, park, and pickup; and call and message waiting features. This last phase bridged the considerable gap between "IP voice" and "IP telephony."

Several Internet telephony interoperability concerns are addressed by H.323. These include gateway-to-gateway interoperability, which ensures that telephony can be accomplished between different vendors' gateways; gatekeeper-to-gatekeeper interoperability, which does the same thing for different vendors' gatekeeper devices; and finally, gateway-to-gatekeeper interoperability, which completes the interoperability picture.

Session Initialization Protocol (SIP)

While H.323 has its share of supporters, it is slowly being edged out of the limelight by the IETF's Session Initialization Protocol (SIP). SIP supporters claim that H.323 is far too complex and rigid to serve as a standard for basic telephony setup requirements, arguing that SIP, which is architecturally simpler and imminently extensible, is a better choice. In reality H.323 is an "umbrella standard" that includes (among others) H.225 for call handling, H.245 for call control, G.711 and G.721 for CODEC definitions, and T.120 for data conferencing. Originally created as a technique for transporting multimedia traffic over a local area network, gatekeeper functions have been added that allow LAN traffic and LAN capacity to be monitored so that calls are established only if adequate capacity is available on the network. Later, the Gatekeeper Routed Model was added, which allowed the gatekeeper to play an active role in the actual call setup process. This meant that H.323 had migrated from being purely a peer-to-peer protocol to having a more traditional, hierarchical design.

The greatest advantage that H.323 offers is maturity. It has been available for some time now, and while robust and full-featured, was not originally designed to serve as a peer-to-peer protocol. Its maturity, therefore, is not enough to carry it. It currently lacks a network-to-network interface, nor does it adequately support congestion control. This is not generally a problem for private networks, but it does become problematic for service providers who wish to interconnect PSTNs and provide national service among a cluster of providers. As a result of this, many service providers have chosen to deploy SIP instead of H.323 in their national networks.

SIP is designed to establish peer-to-peer sessions between Internet routers. The protocol defines a variety of server types, including feature servers, registration servers, and redirect servers. SIP supports fully distributed services that reside in the actual user devices and, because it is based on existing IETF protocols, provides a seamless integration path for voice/data integration.

Ultimately, telecommunications, like any industry, revolves around profitability. Any protocol that allows new services to be deployed inexpensively and quickly immediately catches the eye of service providers. Like TCP/IP, SIP provides an open architecture that can be used by any vendor to develop products, thus ensuring multivendor interoperability. And because SIP has been adopted by such powerhouses as Lucent, Nortel, Cisco, Ericsson, and 3Com, and is designed for use in large carrier networks with potentially millions of ports, its success is reasonably assured.

Originally, H.323 was to be the protocol of choice to make this possible. And while H.323 is clearly a capable suite of protocols and is indeed quite good for VoIP services that derive from ISDN implementations, it is still incomplete and is quite complex. As a result, it has been relegated to use as a video-control protocol and for some gatekeeper-to-gatekeeper communications functions.

The intense interest in moving voice to an IP infrastructure is driven by simple and understandable factors: cost of service and enhanced flexibility. However, in keeping with the "Jurassic Park Effect" (Just because you *can*, doesn't necessarily mean you *should*), it is critical to understand the differences that exist between simple voice and full-blown telephony with its many enhanced features. It is the feature set that gives voice its range of capability; a typical local switch—such as Lucent Technologies' 5ESS—offers more than 3,000 features, and more will certainly follow. Of course, these features and services are possible because of the protocols that have been developed to provide them across an IP infrastructure.

Media Gateway Control Protocol (MGCP)—and Friends

Many of the protocols that are guiding the successful development of VoIP efforts today stem from work performed early on by Level 3 and Telcordia, which together founded an organization called the International SoftSwitch Consortium. In 1998, Level 3 brought together a collection of vendors that collaboratively developed and released the Internet Protocol Device Control (IPDC). At the same time, Telcordia created and released the Simple Gateway Control Protocol (SGCP). The two were later merged to form the Media Gateway Control Protocol (MGCP), discussed in detail in RFC 2705.

MGCP allows a network device responsible for establishing calls to control the devices that actually perform IP voice streaming. It permits software call agents and media gateway controllers to control streaming media gateways at the edge of the network. These gateways can be cable modems, set top boxes, PBXs, VTOA gateways, and VoIP gateways. Under this design, the gateways manage the circuit-switch-to-IP voice conversion, while the agents manage signaling and call processing.

MGCP makes the assumption that call control in the network is software based, resident in external, intelligent devices that perform all call-control functions. It also makes the assumption that these devices will communicate with one another in a primary-secondary arrangement, under which the call agents send instructions to the gateways for execution. See Table 3-1 for a list of requirements for multimedia applications.

Table 3-1 Multimedia Application Requirements	**Application**	**Required Bandwidth**	**Sensitivity to Delay**
	Voice	Low	High
	Video	High	Medium to high
	Medical Imaging	Medium to high	Low
	Web Surfing	Medium	Medium
	LAN Interconnection	Low to high	Low
	Electronic Mail	Low	Low

Meanwhile, Lucent created a new protocol called the Media Device Control Protocol (MDCP). The best features of the original three were combined to create a full-featured protocol called the MeGaCo, also defined as H.248. In March 1999, the IETF and ITU met collaboratively and created a formal technical agreement between the two organizations, which resulted in a single protocol with two names. The IETF calls it MeGaCo, the ITU calls it H.GCP.

MeGaCo/H.GCP operates under the assumption that network intelligence is housed in the central office, and therefore replaces the gatekeeper concept proposed by H.323. By managing multiple gateways within a single IP-equipped central office, MeGaCo minimizes the complexity of the telephone network. In other words, a corporation might be connected to an IP-capable central office, but because of the IP-capable switches in the CO which have the ability to convert between circuit-switched and packet-switched voice, full telephony features are possible. Thus the next generation switch converts between circuit and packet, while MeGaCo performs the signaling necessary to establish a call across an IP wide area network. It effectively bridges the gap between legacy SS7 signaling and the new requirements of IP, and supports both connection-oriented and connectionless services.

A Final Thought: Network Management for QoS

Because of the diverse audiences that require network performance information and the importance of *service level agreements* (SLAs), the data collected by network management systems must be malleable so that it can be formatted for different sets of corporate eyes. For the purposes of monitoring performance relative to service level agreements, customers require information that details the performance of the network relative to the requirements of their applications. For network operations personnel, reports must be generated that detail network performance to ensure that the network is meeting the requirements of the SLAs that exist between the service provider and the customer. Finally, for the needs of sales and marketing organizations, reports must be available which allow them to properly represent the company's capabilities to customers, and to allow them to anticipate requirements for network augmentation and growth.

For the longest time, the Telecommunications Management Network (TMN) has been considered the ideal model for network management. As the network profile has changed, however, with the steady migration to IP and a renewed focus on service rather than technology, the standard TMN philosophy has begun to appear somewhat tarnished.

Originally designed by the ITU-T, TMN is built around the OSI Model and its attendant standards, which include the Common Management Information Protocol (CMIP) and the Guidelines for the Development of Managed Objects (GDMO).

TMN employs a model, shown in Figure 3-69, made up of a *network element layer,* an *element management layer*, a *network management layer*, a *service management layer*, and a *business management layer*. Each has a specific set of responsibilities closely related to those of the layers that surround it.

The network element layer defines each manageable element in the network on a device-by-device basis. Thus, the manageable characteristics of each device in the network are defined at this functional layer.

The element management layer manages the characteristics of the elements defined by the network element layer. Information found here includes activity log data for each element. This layer houses the actual

Figure 3-69
The TMN
layered model

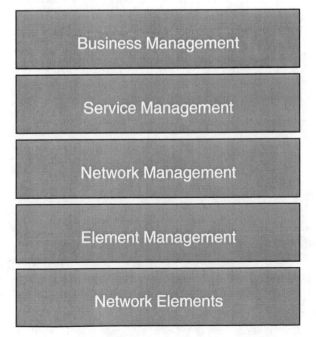

element management systems responsible for the management of each device or set of devices in the network.

The network management layer has the ability to monitor the entire network based upon information provided by the element management layer.

The services management layer responds to information provided by the network management layer to deliver such service functions as accounting, provisioning, fault management, configuration, and security services.

Finally, the business management layer manages the applications and processes that provide strategic business planning and tracking of customer interaction vehicles such as SLAs.

OSI, while highly capable, has long been considered less efficient than IETF management standards, and in 1991, market forces began to affect a shift. That year, the Object Management Group was founded by a number of computer companies—including Sun, Hewlett-Packard, and 3Com —and together they introduced the Common Object Request Broker Architecture (CORBA). CORBA is designed to be vendor independent and built around object-oriented applications. It allows disparate applications to communicate with each other, regardless of physical location or vendor. And while CORBA did not achieve immediate success, it has now been widely accepted, resulting in CORBA-based development efforts among network management system vendors. While this may seem to fly in the face of the OSI-centric TMN architecture, it really doesn't. TMN is more of a philosophical approach to network management, and does not specify technological implementation requirements. Thus, a conversion from CMIP to the Simple Network Management Protocol (SNMP), or the implementation of CORBA, does not affect the overall goal of the TMN.

Summary

IP is here to stay and is profoundly changing the nature of telecommunications at its most fundamental levels. Applications for IP range from carrier-class voice that is indistinguishable from that provided by traditional circuit switching platforms to best-effort services that carry no service quality guarantees. The interesting thing is that all of the various capability levels made possible by the incursion of IP have an application in modern telecommunications, and are being implemented at rapid fire pace. There is still a tremendous amount of hype associated with IP services as they edge their way into the protected fiefdoms of legacy technologies, and implementers and customers alike must be wary of "brochureware" solutions and the downside of the Jurassic Park Effect, which warns that just because you *can* implement an IP telephony solution doesn't necessarily mean you *should*. Hearkening back once again to the old telephone company adage that observes "If it ain't broke, don't fix it," buyers and implementers alike must be cautious as they plan their IP migration strategies. The technology offers tremendous opportunities to offer consolidated services, to make networks and the companies that operate them more efficient, to save cost and pass those savings on to the customer. Ultimately, however, IP's promise lies in its ubiquity, and its ability to tie services together and make them part of a unified delivery strategy. The name of the game is service, and IP provides the bridge that allows service providers to jump from a technology-centric focus to a renewed focus on the services that customers care about.

Chapter Summary

A few years ago, while filming a video about telephony on location in Texas, I interviewed a small town sheriff about his use of telecommunications technology in his job. I wanted to know how it has changed the way he does his job. "Well sir," he began, puffing out his chest and sticking his thumbs into his waistband, "We use telecommunications all the time. We use our radios and cell phones in the cars, and our telephones there to talk here in town and for long distance. And now," he said, patting a fax machine on his desk, "we've got backup." I wasn't sure what he meant by that, so I asked. "We've always had trouble with the phones goin' out around here, and that was a bad thing. Can't very well do our job if we can't talk. But now we've got this here fax machine." I was puzzled—I was missing his point, so I probed a little deeper. "Don't you get it, son? Before, if we lost the phones, we were like fish outta water. Now, I know that if the phones go down, I can always send a fax."

Unfortunately, many peoples' understanding of the inner workings of the telephone network is at about the same level as the understanding of the sheriff (who was a wonderful guy, by the way. I set him straight on his understanding of the telephone network and he treated me to the best chicken fried steak I have ever eaten). Hopefully this chapter, and those that follow, are helping to lift the haze a bit.

We will return to the central office several times as we examine the data transport side of telecommunications, but for now, let's introduce a necessarily painful component of telephony. Next, we delve into the Byzantine world of telecom regulation.

Chapter 3 Questions

1.   Why was it called "The Bell System?"

2.   What is the difference between a BOC, an RBOC, and an ILEC?

3.   Who were the three principal inventors working on the earliest telephones?

4.   What was the Kingsbury Commitment?

5.   What was the Communications Act of 1934? Why was it important?

6.   What was Computer Enquiry II?

7.   The Divestiture of AT&T in 1984 was far reaching. Was it a good idea? Have the results of it been positive?

8.   What is a central office?

9.   What is the difference between an NPA and an NXX?

10.   How many telephone numbers are there in a single NXX? How many in an NPA? Why?

11.   Explain the difference between a local switch and a tandem switch.

12.   "SS7 is perhaps the single most important component of the modern telephone network." Explain.

13.   Is twisted pair the only option for a local loop?

14.   Explain the differences between D1, D4, and ESF.

15.   What is SONET? How important is it in the modern network?

16.   Explain the basic differences between a traditional, circuit-switched network and a VoIP network.

The Byzantine World of Telecom Regulation

It had been a long and painful process, but starting today they would enjoy the fruits of their collective labors. Today, telecomm regulation had a whole new face, one that would redefine the way incumbent service providers, their competitors, and the customers they serve interact. So radically different was it that it had taken 3 years to force the changes through Congress and another two to get the FCC to rally around the changes.

In his newly created cabinet-level role as Secretary of Telecommunications and Information Technology, Roy Marcus had worked hard to bring about a sweeping rewrite of telecom law and a major philosophical shift on the part of the industry at large. Yes, today would be different. Instead of the contrived, artificially competitive model that drove the industry for so long, all incumbent local exchange carriers had agreed to subject themselves to structural separation, resulting in the creation of a wholesale arm and a retail arm for each company. The wholesale arm sells infrastructure; the retail arm sells services. All competitive carriers are free to buy from the wholesale division of any of the carriers at a cost that is the same for all. And to ensure that the services delivered by the infrastructure company are as good as they can be, a well-informed oversight committee ensures that the technology infrastructure is *also* as good as it can be. Of course, such an infrastructure is expensive; after all, the reason that the FCC and Congress were even willing to consider the concept was that the cost of maintaining and upgrading the infrastructure to meet customer demands had become so onerously high that the service providers, faced with declining revenues on all service fronts, could not afford to make the necessary investment. So to offset the cost —this was a stroke of genius, if he did say so himself—Congress had passed a "chip and device tax," which placed a surcharge on every manufactured semiconductor and optoelectronic device as well as on the finished product—a tax similar to the VAT used in Europe. The monies collected will pay for the Universal Service Fund and for guaranteed, scheduled network upgrades. The money also pays to guarantee Quality of Service, technological currency, and what his committee had come to call "Quality of Experience." But now he had to go—he was late for his meeting with the Global League of Service Providers.

"Wake up, Steve—it's time for your pill." As my friend Paul Whalen likes to say, "My mind just had an out-of-body experience, but it's back now." You may laugh at my characterization of the regulatory environment of the future, but it isn't any more far-fetched than some of the other ideas that have come along in the last few years.

Today, writing a definitive section on telecom regulation is like playing that wonderful arcade game from the 1970s, Whack-a-Mole: Every time things seem like they're under control, another one pops up in an entirely new and unexpected place. In this section, we will attempt to create a snapshot of telecom regulation with enough detail to make it useful, but with the understanding that as I sit here typing the environment is shape-shifting around me. I can't emphasize enough how critical it is for decision makers in the tech sector to stay aware of changes in regulatory law and attuned to the impacts that regulation has on all facets of the technology industry.

The "reworking" of the telecommunications industry that began with the divestiture of AT&T in 1984 reaches far beyond the borders of the United States. It has become a global phenomenon as companies move to compete in what has become an enormously lucrative business. Customers have been quick to point out that technology is a wonderful thing, but unless the provider positions it in a way that makes it useful to the would-be customer, it has very little value. Customers expect service-provider representatives to have relevant technical knowledge as well as knowledge about them as customers: business issues, competition, major segment concerns, etc. Furthermore, the level of customer satisfaction or dissatisfaction varies from industry segment to industry segment. ILECs are too rigid and expensive. CLECs must compete on price but must also show a demonstrated ability to respond quickly and offer OSS/NM capabilities—and, in today's marketplace, must show staying power. Ninety-two percent of all business customers want to work with an account team when purchasing services; yet, according to both formal and informal surveys that have been conducted, account teams are not creating long-lasting relationships or loyalty in their day-to-day dealings with customers. Eighty-eight percent of those customers buy their services from the local provider, but only 53 percent say that they would continue to do so if presented with a capable alternative. According to numerous studies, the single most important factor that customers take into consideration when selecting a service provider is *cost*. Yet, the most important factor identified for improvement is *customer service*.

The most daunting challenge that has historically faced service providers is the requirement to provide true universal service, in rural areas as well as in metropolitan and suburban areas. The American Communications Act of 1934, signed into law by President Roosevelt, mandated that telephony service be universal and affordable. To make this happen, AT&T agreed in the 1940s to offer low-cost service with the

tacit approval of the Justice Department to charge a slightly higher price for long distance, some business line services, and value-added services, as a way to offset the cost of deploying high-cost rural service and to support low-income families. These subsidies made it possible for AT&T to offer true universal, low-cost service.

Today, 60 to 70 percent of the local loops in the United States are still subsidized to the tune of anywhere from $3 to $15 per month. As a result, the incumbent service providers often charge significantly less for the service they sell than it actually costs them to provide it.

The problem with this model is that only the ILECs enjoy these subsidies, which means that CLECs believe themselves to be penalized right out of the starting gate. Couple that with the fact that 40 percent of all CLEC revenues go to the ILECs for access charges and we begin to see

the basis for much of the acrimony that exists between the incumbents and the new entrants. As a result of these and other factors, CLECs typically ignore the residence market in favor of the far more lucrative business markets. Business markets are significantly easier and less costly to provision than residence installations because of the dominance of *multitenant buildings* (MTUs). Clearly the metro market is a desirable place to be. Of course, there is an added issue: Because of their overall market dominance, ILECs control more than 60 percent of all metro spending.

The regulatory environment in many countries fails to address the disparity that exists between incumbent providers and would-be competitors. Recent rulings, such as the United States Communications Act of 1996, consisted of a series of decisions designed to foster competition at the local loop level. Unfortunately, by most players' estimations, they have had the opposite effect.

The introduction of new technologies always conveys a temporary advantage to the first mover. Innovation involves significant risk; the temporary first mover advantage allows first movers to extract profits as a reward for taking the initial risk, thus allowing them to recover the cost of innovation. As technology advances, the first mover position usually erodes, so the advantage is fleeting. The relationship, however, fosters a zeal for ongoing entrepreneurial development.

Under regulatory rules in many countries today, incumbent providers are required to open their networks through element unbundling, and to lease their network resources—including new technologies—at a wholesale price to their competitors. In the minds of regulators this creates a

competitive marketplace, but it is artificial. In fact, the opposite often happens: The incumbents lose their incentive to invest in new technology and innovation progresses at "telco time." Under wholesale unbundling requirements, the rewards for innovative behavior are socialized, while the risks undertaken by the incumbents are privatized. Why should they invest and take substantial economic risk, they argue, when they are required by law to immediately share the rewards with their competitors?

Ultimately, the truth comes down to this: Success in the access marketplace, translated as sustainable profits, relies on network ownership —period. During the heady bubble years, many would-be competitors such as E.spire and ICG bought switches and other network infrastructure components, built partial networks, and created business plans that consciously relied on the ILECs for the remainder of their network infrastructure. Due to the behavior of the marketplace and fickle consumers, most CLECs failed to attract a viable customer base and quickly learned that this model could not succeed. Many of them have now disappeared, and among those that remain many have filed for bankruptcy court protection, leaving some manufacturers with unpaid bills totaling in the billions of dollars.

Consider this: When WorldCom bought Metropolitan Fiber Systems (MFS), they weren't after the firm's business; they were after the in-place local network that MFS had built, which would allow WorldCom to satisfy customer demands for end-to-end service without going through the cost of a network build-out. They paid more than six times the value of MFS' in-place assets for the company. The message? Network ownership is key.

Possible Regulatory Solutions

So what are the possible solutions? Two in particular have been discussed for a very long time. The first is to eliminate the local service subsidies and allow the ILECs to raise rates so that they are slightly above the actual cost of provisioning service. Subsidies could continue for high-cost rural areas and low-income families (roughly 18 percent of the total), but would be eliminated elsewhere. New entrants would then have a greater chance of playing successfully in the local access business. The subsidy dollars, estimated to be in the neighborhood of $15 billion, could then be redeployed to finance universal broadband access deployment. The monies could be distributed among ILECs, CLECs, cable companies, ISPs, wireless companies, and DSL providers, to facilitate broadband.

A second solution is to call for the structural separation of the ILECs, which would result in the establishment of retail and wholesale arms. The retail arm would continue to sell to traditional customers, while the wholesale arm would sell unbundled network resources to "all comers," including the retail arm of the ILEC. The result would be enhanced innovation; the downside would undoubtedly be strong resistance from the ILECs because of the cost and complexity required to carry out structural separation.

The FCC, led by Kevin Martin, has indicated that it wants less government intervention in the telecommunications marketplace rather

than more, a good thing in light of the industry's current troubles. One could argue that well-intended regulatory strictures have in fact done damage. Consider the case of WorldCom once again. The company's original plan in the late 1990s was to challenge local service providers all over the world by creating a broadband voice and data IP network through acquisitions and mergers. The regulators, concerned by World-Com's aggressive plans, felt that the intended company looked too much like a monopoly. They forced the divestiture of MCI's internet company to Cable & Wireless and rejected the proposed merger with Sprint because of fears that they would control 80 percent of the long-distance market. This decision was made while long-distance revenues were plummeting due to the influence of such disruptive technologies as multichannel optical transport and IP. The result is two badly weakened companies that have not yet recovered—and may not. Ironically, they could now be prime acquisition targets for the ILECs. (Although, at the time of this writing, competitors are calling for the dismemberment of MCI over arguments that their reentry into the market would give them an unfair competitive advantage, particularly in light of their unethical behavior).

Another example is AT&T itself. There was a huge expectation that AT&T would be a big winner in the local broadband access game following its acquisitions of cable properties for its plans to deliver high-speed Internet and interactive services. Many analysts expected a market cross-invasion between the ILECs and cable providers, but it has only *just* begun to happen. Cable providers concentrated on adding Internet service and additional channels; telephone companies concentrated on penetrating the long-distance market. Furthermore, when talk of open access and loop unbundling began to be targeted at the cable industry in 2000, AT&T's hopes of a competitive advantage through cable ownership were dashed; they sold their cable properties to Comcast.

There were also expectations that ILECs would work hard to penetrate each other's markets, but this never happened, either. Who better than the ILECs knows that network ownership is the most critical factor for success in the local access game? If you control the network, you control the customers. More importantly, if you *don't* have that control, don't get into the game.

In the metro market, which is highly dependent on high-speed access technologies such as DSL (and, to a lesser degree, cable and wireless), these regulatory decisions represent pivot points in the evolution of the metro domain. If the FCC chooses to reduce the regulatory pressure on the incumbents, customers will invest in bandwidth at the edge, driving traffic into the metro—which will, in turn, drive traffic into the network core.

Terminology

Before we discuss the status of current regulatory decisions, let's take a moment to clarify some important terminology that relates to ongoing decisions. As part of the Telecommunications Act of 1996, incumbent local service providers were required to "unbundle" their local loops—in other words, to create a list of all the billable elements that make up a local loop and determine wholesale prices for each element. This exercise would allow ILECs to set reasonable wholesale prices for their local loops to facilitate the creation of competition. Since CLECs were not inclined initially to build their own networks, they instead wanted to simply buy the loop elements for a wholesale price, mark the resulting loops up appropriately and resell them in competition with their ILEC competitors. These were known as *unbundled network elements* (UNEs).

Ultimately, service providers took a different tack by creating what was known as the *unbundled network element-platform* (UNE-P), which was a technique under which CLECs could lease "packages" of elements including switching elements—in effect, an entire service platform.

The unbundled elements included everything from the interface between the customer's inside wire and the outside plant to the switching and multiplexing elements located within the CO and embedded throughout the outside plant facilities. The elements are as follows:

- *Network Interface Device* (NID): Physical interface between the local service provider and the customer's inside wiring
- *Loop Distribution:* Facilities that connect NID to feeder distribution interface
- *Loop Feeder:* Facilities that connect loop distribution to the local switch
- *Local Switch:* Switch that initiates and terminates local calls
- *Tandem Switch:* Switch that interfaces between local, long-distance networks
- *Common Transport*: Transmission facilities used by multiple carriers
- *Dedicated Transport:* Facilities used exclusively by a single carrier
- *Service Control Point*: Database used to provision Caller ID-based enhanced and supplementary services
- *Signaling Transfer Point:* Packet switch for signaling communication
- *Signaling Link Transport:* Facilities between signaling points
- *Concentrator/Multiplexer:* All forms of carrier systems
- *Operator Systems and Operations Support Systems:* Operator services, directory assistance, and element and network management systems for maintaining the health and welfare of the network

Recent Regulatory Decisions

In the early months of 2003, the FCC dramatically rewrote the basics of telecom regulation. Following the 2003 Triennial Review that concluded in February of that year, the FCC released a series of high-impact decisions that pleased everyone—and no one.

In an effort to resolve the question of which UNEs the ILECs had to make available to their competitors, the FCC essentially turned the issue over to the states for local resolution. They also ruled that ILECs no

longer had to provide switching as an unbundled element to CLECs for business customers unless state regulatory agencies could prove within 90 days of the order that this resulted in an overly onerous impairment to their ability to do business.

In the residence market, the FCC outlined specific criteria that each state had to use to determine whether CLECs were impaired without unbundled switching. They also concluded that ILECs did not have to provide competitors with SONET-level transport services (OC-N), but shifted responsibility to the states to determine whether ILECs should be required to unbundle dark fiber and DS3 transport on a route-by-route basis.

For broadband, the FCC's decision was clearer. They ruled that all new broadband buildouts, including both fiber builds and hybrid loops, were exempt from unbundling requirements as well. They also ruled that line sharing would no longer be classified as a UNE, a decision that put considerable pressure on competitive carriers, such as Covad Communications, that rely on line sharing.

Apparently, the decisions reached by the Commission were far from universal. Commissioner Kevin Martin was the only member who agreed with the entire order. Chairman Powell and Commissioner Kathleen Abernathy disagreed with the decisions about line sharing and unbundled switching, while Jonathan Adelstein and Michael Copps disagreed with the broadband decision.

The feedback from various industry segments about the decisions was, as would be expected, varied. CLECs grudgingly agreed for the most part that it was about as good as it could have been, and could have been far worse. ILECs, on the other hand, saw the decision to turn much of the decision making over to the states as a step in the wrong direction, concluding that it would extend the period of time over which definitive decisions would be made.

What we do know is this: While the decisions made by the fragmented FCC were not ideal for all players (how could they possibly be?), they did move the industry forward and shook up the players in a positive way. ILECs now had a renewed incentive to invest in broadband infrastructure, while competitive providers had incentives to invest in alternative technologies such as cable and wireless (both fixed and mobile) in their competitive efforts. DSL rollouts were accelerated, and as penetration climbed, alternative solutions were invoked. So, while the results were not as comfortable as they could be for the industry players, they led to the appropriate marketplace behavior—and that's a good thing.

Consider the following simple scenario. Regulators, after a great deal of wrangling, remove unbundling requirements, providing the appropriate degree of incentive to incumbent providers to accelerate broadband deployment. ILECs publicly commit to universal broadband (DSL) deployment throughout their operating areas. In response, cable and wireless players accelerate their own rollouts, preparing for the price wars that will inevitably come.

The broadband deployment effort involves infrastructure; shortly after the service announcement, ILECs issue RFPs and RFQs for DSLAMs, optical extension hardware, DSL modems, and begin to jockey for content alliances since DSL provides the necessary bandwidth for television signal delivery in addition to voice and high-speed data. Customers, meanwhile, excited by the prospect of higher-speed access and all that it will bring them in the way of enhanced capability, buy upgraded PCs and broadband service packages. Hardware manufacturers and service providers applaud the evolution because they know that the typical broadband user generates 13 times the traffic volume that a dial-up user generates, and while service providers rarely charge by the transported megabyte, the increased traffic volume requires capability upgrades to the network—CAPEX, in other words.

Meanwhile, software manufacturers, mobile appliance makers, and content owners scramble to develop products for wireline and wireless broadband delivery. Upgrades occur; innovation happens; prices come down. Cable and wireless players march along with their own parallel efforts and soon this great, dynamic money engine known as the telecom industry starts to turn again, slowly at first but building rapidly as it feeds on its own self-generated fuel. And that is the ultimate end state— a self-perpetuating industry that evolves and changes in concert with market demand, sustained by a forward-thinking, reasonable regulatory environment.

What this boils down to is Michael Powell's ultimate goal: to force the industry at large to become facilities-based. The fact is that the local loop is a natural monopoly—there's no getting around that. The UNE rules are basically unnatural and simply don't work as well as they were envisioned. Want to compete with the incumbent service providers? Use a different technology, like cable, or broadband wireless. Clearly this strategy is working in the cable industry's favor.

Other Regulatory Activity

On May 12, 2003, the FCC commissioners were presented with a proposal to modify the existing rules that govern media ownership. On the surface, this does not appear to affect the technology sector per se, but in fact it potentially has a significant impact because of a blurring of the lines between the sectors. Under the terms of the proposal, ownership of broadcast rights could change dramatically. Among the changes are the following.

The proposal plans to allow a single company to own television stations that reach as many as 45 percent of U.S. households, today capped at 35 percent. Needless to say, the major networks are in favor of eliminating the existing 35 percent cap. This decision would favor companies like News Corporation (the owner of Fox) and Viacom (the owner of CBS and UPN), which are already in violation of the 35 percent limit because of merger and acquisition activity that put them (slightly) over the top. There are some inviolable restrictions to the decision: The four major networks (CBS, NBC, ABC, and Fox) are prohibited from merging, and ownership of more than eight broadcasting stations in a single market remains prohibited.

Two cross-ownership rules are under scrutiny and proposals to modify them are included in the proposal. One prevents a company from owning a newspaper *and* a radio or television station in the same city; the other limits ownership of both radio and television stations in the same market. Under the terms of the proposal, these two rules would be combined into a single rule that eliminates most of the existing restrictions. Cross-ownership would be allowed in large and medium-size markets, but would be restricted or perhaps even banned in smaller markets.

Under the terms of federal communications law, the FCC is required to consider reasonable changes to the rules it oversees that affect the communications marketplace and the public it serves. Since some of these laws were first written into law over 50 years ago, they are in dire need of revision. In many cases the changes are driven by the growing influence of the cable and Internet-dominated sectors.

Recent Events

In March 2004, an appeals court struck down FCC rules for how regional telephone companies must open their networks to competitors. Federal law originally required regional phone companies to lease parts of their networks (the UNE-P mandates) to competitors at reasonable rates, set by the states. The ILECs have long contended that they have been forced to give competitors rates that are below their actual cost.

In its decision, the appeals court found that the FCC wrongly gave power to state regulators to decide which parts of the telephone network had to be unbundled. The court also upheld an earlier FCC decision that the ILECs are not required to lease their high-speed facilities to competitors at discount rates the way they do their standard phone lines.

In April 2004, regulators rejected AT&T's petition to eliminate the requirement that they pay long-distance fees on calls transported partially over the Internet.

The decision means that AT&T may be required to pay hundreds of millions of dollars in unpaid retroactive fees to the ILECs.

Earlier in 2004 the FCC ruled that calls that originated and terminated on the Internet, such as those made using Skype, Vonage, and other voice-over-Internet service providers, are free from the fees and taxes that traditional phone companies are required to pay, such as support for E911, Universal Service Fund, etc. The FCC said that because calls that travel over the Internet don't provide anything in the way of enhanced features, standard rules apply. Furthermore, in a recent decision, the FCC consolidated all regulatory power over VoIP service at the federal level, wresting it from the states.

The key here is that regulatory agencies are trying to balance the need for a regulated telecommunications marketplace with the need for an unfettered development environment that can technologically innovate without fear of onerous fees and taxes. The FCC wants to engender a spirit of facilities-based competition, rather than the UNE-based environment that has not worked as well as hoped.

Perhaps the reason for this is a realization on the part of the regulators of exactly what their role should be. For the longest time it appeared that regulators had come to believe that their primary responsibility was to create competition. In fact, the primary goals of regulation are two-fold: first and foremost, to ensure that users of the regulated service (whether it be telecommunications, power, or airlines) have the best possible experience while using the service; and second, to "fill in the gaps" where the market fails to create the appropriate environment. Now, regulators can use competition as a way to achieve these goals, but competition is a means to an end, not the end itself.

Furthermore, regulators should not be in the business of regulating technology per se, but rather the application of technologies according to the mandates listed previously. Today, there is significant regulatory attention being paid to VoIP. But consider this: VoIP is a technology, not a service. What comes out of the phone is still the same voice that has always come out of the phone. And while VoIP today enjoys a certain amount of immunity from regulatory forbearance, that is a temporary situation. The time will come, and not all that long from now, when VoIP providers will reach "critical mass" and will find themselves under the same degree of regulatory scrutiny as other, more traditional, providers.

I have now finished writing this chapter; it is 11:15 PM eastern time on December 30, 2004, and I am sure that it is already woefully out of date. Readers, let me leave you with this: Regulation is a complex, often boring, difficult subject to become engaged with. It is, however, one of the most critical subjects going in telecom today, so if you have skillfully avoided it, now is the time that changes. If you are to be informed about the current goings-on in telecom and the industries it serves, you *must* be able to discuss the status of current regulation.

Chapter 4 Questions

1.   What are the primary goals of telecom regulation?

2.   How would you compare the Communications Act of 1934 with the Telecommunications Act of 1996 (TA96)?

3.   How did the ILECs basically control the regulatory process when it came to the introduction of new services and technologies?

4.   What are the 14 requirements under TA96 that the ILECs were required to meet as a prerequisite for entry into long distance?

5. Why did the long-distance carriers not have a similar list of requirements?

6. What is UNE-P?

7. Explain the concept of "Structural Separation." Is it a good idea?

8. Should VoIP be regulated the same way as traditional telephony? Why or why not?

9. Should telecom be regulated exclusively at the federal level, or should there be state-level regulatory oversight as well?

10. "The local loop is a natural monopoly." Explain.

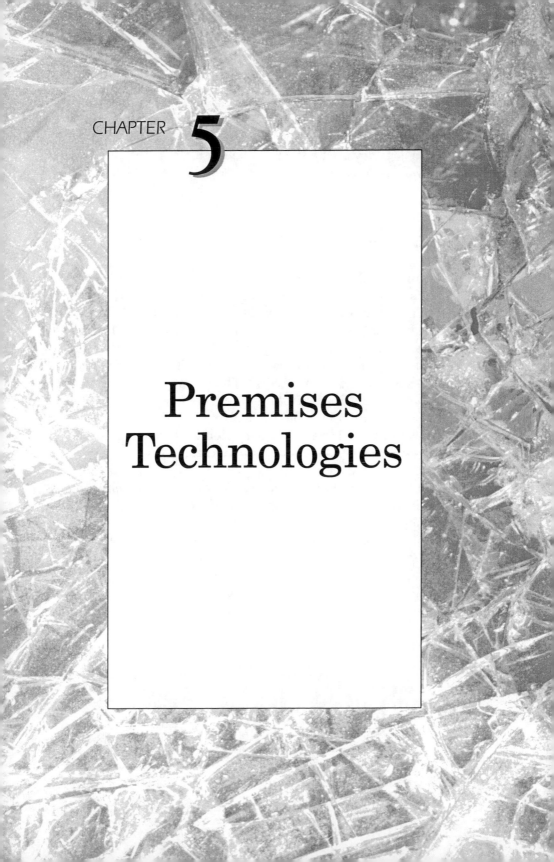

CHAPTER **5**

Premises Technologies

This chapter addresses the network devices found in a typical premises environment including computers, wired and wireless local area networks, and a number of other options. We begin with an examination of a typical computer. In one way or another, the computer is the ultimate premises technology device; it appears, in one form or another, in every device used by a customer to access the network.

The Computer

For all its complexity, the typical computer has only a small number of components, as shown in Figure 5-1. These are the central processing unit (CPU), main memory, secondary memory, input/output (I/O) devices, and a parallel bus that ties all the components together. It also has two types of software that make the computer useful to a human. The first is application software such as word processors, spreadsheet applications, presentation software, and MP3 encoders. The second is the operating system that manages the goings on within the computer including hardware component inventory and file location. In a sense, the operating system is the executive assistant to the computer itself; some mainframe manufacturers refer to their operating system as the EXEC.

The concept of building modular computers came about in the 1940s when Hungarian-born mathematician John Von Neumann applied the work he had done in logic and game theory to the challenge of building large electronic computers (Figure 5-2). As one of the primary contributors to the design of the Electronic Numerical Integrator and Computer (ENIAC), Von Neumann introduced the concept of stored program control and modular computing—the design under which all modern computers are built today. The internals of a typical modern computer are shown in Figure 5-3.

Figure 5-1
Computer
components

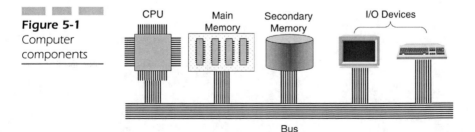

Figure 5-2
A section of the original ENIAC

The CPU

The CPU is the brain of the computer. Its job is to receive data input from the I/O devices (keyboard, mouse, modem, etc.), manipulate the data in some way—based on a set of commands from a resident application—and package the newly gerrymandered data for presentation to a human user at another I/O device (monitor). The CPU has its own set of sub-components. These include a clock, an arithmetic-logic unit (ALU), and

Figure 5-3
PC internals
showing major
components

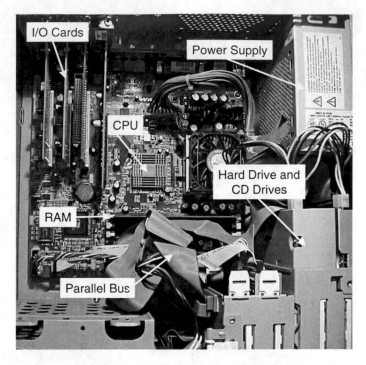

registers. The clock is the device that provides synchronization and timing to all devices in the computer, and is characterized by the number of clock cycles per second it is capable of generating. These cycles are called Hertz; modern systems today operate at a range of speeds as high as 4 Megahertz, or MHz. The faster the clock, the faster the machine can perform computing operations.

The ALU is the specialized silicon intelligence in the CPU that performs the mathematical permutations that make the CPU useful. All functions performed by a computer—word processing, spreadsheets, multimedia, videoconferencing . . . *all functions*—are viewed by the computer as mathematical functions and, therefore, are executed as such. It is the job of the ALU to carry out these mathematical permutations.

Registers are nothing more than very fast memory located close to the ALU for rapid input/output functions during execution cycles.

Main Memory

Main memory, sometimes called *random access memory* (RAM), is another measure of the "goodness" of a computer today. RAM is the segment of memory in a computer used for execution space and as a place to store operating system, data, and application files that are in current use. RAM is silicon-based, solid-state memory and is extremely fast in terms of access speed. It is, however, volatile: When the PC is turned off, or power is lost, whatever information is stored in RAM disappears. When basic computer skills courses tell students to "save often," this is the reason. When a computer user is writing a document in a word processor, or populating a spreadsheet, or manipulating a digital photograph, the file lives in RAM until the person saves, at which time it is written to main memory, which is nonvolatile as we'll see in a moment. Modern systems typically have a minimum of 512 Megabytes (MB) of RAM.

It amazes me to look across my office at my first Mac, which I purchased in 1986. It has a 40 MB hard drive, 640 KB of RAM, and a processor that doesn't run—it dribbles. Yet, it runs the same office applications I use today (although much reduced in terms of bells and whistles) and boots in seconds, compared to my current, very powerful XP machine.

Secondary Memory

Secondary memory has become a very popular line of business in the evolving PC market. It provides a mechanism for long-term storage of data files and is nonvolatile—when the power goes away, the information it stores does not. Secondary memory tends to be a much slower medium in terms of access time than main memory because it is *usually* mechanical in nature, whereas main memory is solid-state. Consider, for example, the hard drive shown in Figure 5-4, which chose a hot July day to fail, right in the middle of writing this book (and yes, I had backed up often). That's why the disk platters are exposed. Notice that the drive comprises three platters and an armature upon which are mounted read/write heads similar to those used in cassette decks of yore. Figure 5-5 is a schematic diagram that more clearly illustrates a bit more how hard drives actually work. The platters, typically made of aluminum and cast to extremely exacting standards, are coated with iron oxide identical to what is found on recording tape. Adjacent to the platters, which are mounted on a spindle attached to a high-speed motor, is an armature upon which is mounted a stack of read-write heads that write information to the disk surfaces and read information *from* the disk surfaces. Remember when we were discussing the importance of saving often when working on a computer? The save process reads information stored temporarily in RAM and transports the information to the disk controller, which in turn, under the guidance of the operating system, transmits the information to the write heads; they, in turn, write the bits to the disk surface by magnetizing the iron oxide surface. Of course, once the bits have been written to the disk, it's important to keep track of where the information is stored on the disk. This is a function of the operating system. Have you ever inserted a floppy into a drive or (far less fun) a new hard drive into a computer as I recently had to do, and had the computer ask you if you're sure you want it to format the drive? In order to keep track of where files are stored on a disk, whether a hard drive, a CD, a zip drive, or a floppy, the operating system marks the disk with a collection of road markers that help it find the "beginning" of the disk so that it can use that as a reference. Obviously there is no "beginning" on a circle, by design; by creating an arbitrary start point, however, the operating system can then find the files it stores on the drive. This start point is called the *file allocation table*, or FAT; if the FAT goes away for some reason the operating system will not be able to find files stored on the drive. When my hard drive failed, my initial indication that something bad was about to happen was an ominous message that appeared during the boot cycle that said "UNABLE TO FIND FAT TABLE."

Figure 5-4
Close-up of a hard drive, showing platters and read-write head armature

Read-Write Heads

Platters

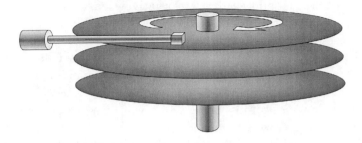

Figure 5-5
Schematic diagram of a hard drive—As the platters spin, the read-write heads access the magnetized surfaces of each platter, writing information to and reading information from them.

When the operating system formats a disk, it logically divides the disk into what are called cylinders, tracks, and sectors, as shown in Figure 5-6. It then uses those delimiters as a way to store and recall files on a drive, using three dimensions. When the *operating system* (OS) writes the file to the disk surface, there is every possibility (in fact it is likely) that the file will not be written to the surface in a contiguous stream of bits. Instead, the file may be broken into pieces, each of which may be written on a different part of the disk array. The pieces are linked together by the operating system using a series of pointers, which tell the OS where to go to get the next piece of the file when the time comes to retrieve it. The pointer, of course, is a cylinder:track:sector indication.

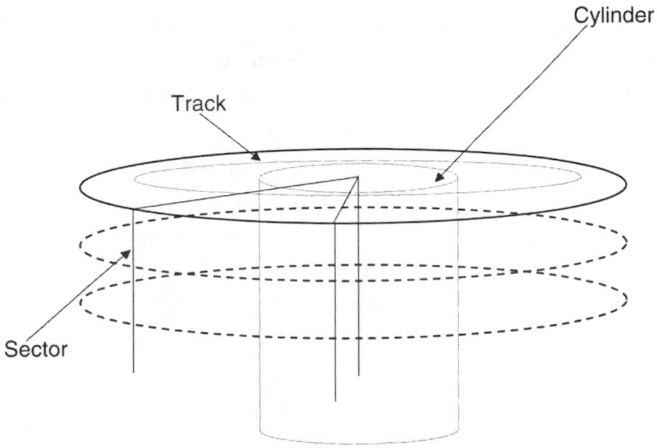

Figure 5-6
Cylinders, tracks, and sectors on a typical disk

Cylinder

Track

Sector

Cylinders, tracks, and sectors are relatively easy to understand. A *track* is a single writeable path on one platter of the disk array. A *cylinder* is a "stack of tracks" on multiple platters. And a sector is a "pie slice" of the disk array. With a little imagination it is easy to see how a file can be stored or located on a disk by tracking those three indicators.

It should also be easy to see now how some of those *wonderful* applications like Norton Utilities work. When you direct your computer to erase a file, it doesn't really erase it; it just removes the pointers so that it is no longer a "registered" file and can no longer be found on the hard drive. What file restore utilities do is remember where the pointers were when you told the computer to erase the file. That way, when you beg the computer to find the file after you've done something stupid (this is always the trick of killing the chicken and lighting candles), the utility has a trivial task: Simply restore the pointers. As long as there hasn't been too much disk activity since the deletion and the file hasn't been overwritten, it can be recovered.

Another useful tool is the disk optimization utility. It keeps track of the files and applications that you, the user, accesses most often while at the same time keeping track of the degree of fragmentation that the files stored on the disk are experiencing. Think about it: When disks start to get full, there is less of a chance that the OS will find a single contiguous piece of writeable disk space and will therefore have to fragment the file before writing it to the disk. Disk optimization utilities perform two tasks. First, they rearrange files on the disk surface so that those accessed most frequently are written to new locations close to the spindle, which spins faster than the outer edge of the disk and, therefore, is accessible more quickly. Second, they rearrange the file segments and attempt to reassemble files or at least move them closer to each other so that they can be more efficiently managed.

OK, enough about hard drive anatomy. Other forms of secondary memory include those mentioned earlier—writeable CDs, DVDs, zip disks, floppies, and nonvolatile memory arrays. As I mentioned, this has become a big business because these products make it relatively easy for users to back up files and keep their systems running efficiently.

Input/Output (I/O) Devices

I/O devices are those that provide an interface between the user and the computer and include mice, keyboards, monitors, printers, scanners, modems, speakers, and any other devices that take data in or spit data out of the computer.

The Bus

Interconnecting the components of the computer is a cable known as a *parallel bus*. It is called "parallel" because the bits that make up one or more eight-bit bytes travel down the bus beside each other on individual conductors, rather than one after the next as occurs in a serial cable on a single conductor. Both are shown schematically in Figure 5-7. The advantage of a parallel bus is speed: By pumping multiple bits into a device in the computer simultaneously, the device—such as a CPU—can process them faster. Obviously, the more leads there are in the bus, the more bits can be transported. It should come as no surprise, then, that another differentiator of computers today is the width of the bus. A 32-bit bus is four times faster than an eight-bit bus; as long as the internal device to which the bus is transporting data has as many input/output leads as the bus, it can handle the higher volume of traffic. The parallel

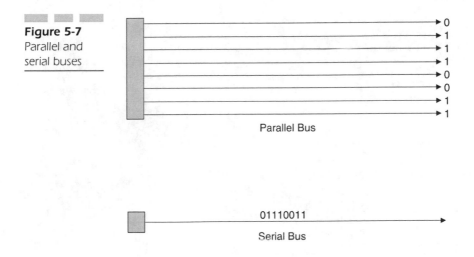

Figure 5-7
Parallel and
serial buses

bus does not have to be a flexible gray cable; for those devices that are physically attached to the main printed circuit board (often called the "motherboard"), the parallel bus is extended as wire leads that are etched into the surface of the motherboard. Devices, such as I/O cards, attach to the board via edge connectors.

Enough about computer internals. A brief word about computer history and evolution is now in order. The work performed by Von Neumann and his contemporaries led to the development of the modern mainframe computer (Figure 5-8) in the 1960s and its many succeeding machine design generations. Targeted primarily at corporate applications, the mainframe continues to provide computing services required by most corporations: access to enormous databases, security, and support for hundreds of simultaneous users. These systems host enormous disk pools (Figure 5-9) and tape libraries (Figure 5-10), and are housed in self-contained windowless computer centers. Their components are interconnected by large cables that require they be installed on raised floor (Figure 5-11). They produce so much heat that they are fed enormous amounts of chilled air and water and require constant attention and monitoring. It makes sense: These are very powerful creatures.

Over time, computer technology advanced, and soon a new need arose. Mainframes were fine for the computing requirements of large, homogeneous user communities, but as the technology became cheaper and more ubiquitous, the applications for which computers could be used became more diverse. Soon a need arose for smaller machines that could be used in more specialized departmental applications, and in the 1970s, thanks to companies like Xerox, Digital Equipment Corporation (DEC), and

Figure 5-8
Mainframe computers in a "clean room" environment— No people work on this floor; the machines are controlled from another area.

Figure 5-9
Mainframe disk pool, sometimes called a "Direct Access Storage Device (DASD) Farm"

Figure 5-10
Tape library— Many data centers now use cartridges that look like the old eight-track tapes and hold significantly more data than the reels shown here.

Data General, the minicomputer was born (Figure 5-12). The minicomputer made it possible for individual departments in a corporation, or even small corporations, to take charge of their own computer destinies and not be shackled to the centralized data centers of yore. It carried with it a price, of course: Not only did these companies or organizations lose their dependency on the data center, they also lost the centralized support that came with it. So there was a downside to this evolution.

The real evolution, of course, came with the birth of the personal computer. Thanks to Bill Gates and his concept of a simple operating system (DOS) and to Steve Jobs with his vision of "computing for the masses," truly ubiquitous computing became a reality. From the perspective of the individual user, this evolution was unparalleled. The revolution began in

Figure 5-11
Raised floor,
showing cables
below

Figure 5-12
Minicomputers

January 1975 with the announcement of the MITS ALTAIR (Figure 5-13). Built by Micro Instrumentation and Telemetry Systems (MITS) in Albuquerque, New Mexico, the ALTAIR was designed around the Intel 8080 microprocessor and a specially designed 100-pin connector. The machine ran a BASIC operating system developed by Bill Gates and Paul Allen; in effect, MITS was Microsoft's first customer.

The ALTAIR was really a hobbyist's machine, but soon the market shifted to a small business focus. Machines like the Apple, Osborne, and Kaypro were smaller and offered integrated keyboards and video displays. In 1981, of course, IBM entered the market with the DOS-based personal PC, and soon the world was a very different place. Apple followed with the Macintosh and the first commercially available *graphical user interface* (GUI), and the rest, as they say, is history. Soon corporations embraced the PC; "a chicken in every pot, a computer on every desk" seemed to be the rallying cry for IT departments everywhere.

There was a downside to this evolution, of course. The arrival of the PC heralded the arrival of a new era in computing that allowed every individual to have his own applications, her own file structures, and his own data. The good news was that each person now controlled his or her own individual computer resources; the bad news was that each person now controlled his or her own individual computer resources. Suddenly the control was gone: Instead of having *a copy* of the database, there were now as many copies as there were users. This led to huge problems. Furthermore, PC proliferation led to another challenge: connectivity, or lack of it. Whereas before every user had electronic access to every other user

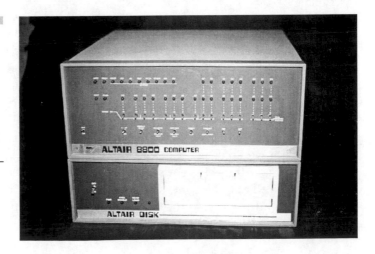

Figure 5-13
The ALTAIR 8800 computer (Photo courtesy Jim Willing, The Computer Garage)

via the mainframe or minicomputer-based network that hooked everyone together, the PC did not have that advantage. Furthermore, PCs eliminated the efficiency with which expensive network resources such as printers could be shared. Some of you may remember the days when being the person in the office with *The Laser Printer* attached to your machine was akin to approaching a state of Nirvana. You were able to do your work and print anytime you wanted—except, of course, for the disruption caused by all of those people standing behind you with floppy disks in their hands promising you that "It's only one page—it'll just take a second." Thus was born the term "Sneakernet;" in order to print something, you had to put the document on a diskette (probably a 5¼-inch floppy, back when floppies really were floppy), walk over to the machine with the directly attached printer, and beg and wheedle for permission to print the file—not the most efficient technique for sharing a printer. Something else was needed. That *something* was the *local area network* (LAN).

LAN Basics

A LAN is exactly what the name implies: A physically small network, typically characterized by high-speed transport, low error rate, and private ownership, which serves the data transport needs of a geographically small community of users. Most LANs provide connectivity, resource sharing, and transport services within a single building, although they can operate within the confines of multiple buildings on a campus.

When LANs were first created, the idea was to design a network option that would provide a low-cost solution for transport. Up until their arrival, the only transport option available was dedicated private line from the telephone company or X.25 packet switching. Both were expensive, and X.25 was less reliable than was desired. Furthermore, the devices being connected together were relatively low-cost devices; it simply didn't make sense to interconnect them with expensive network resources. That would defeat the purpose of the LAN concept.

All LANs, regardless of the access mechanism, share certain characteristics. They all rely on some form of transmission medium that is shared among all the users on the LAN; all use some kind of interrupt and contention protocol to ensure that all devices get an equal opportunity to use the shared medium; and all have some kind of software called a *network operating system* (NOS) that controls the environment.

All LANs have the same basic components as shown in Figure 5-14. A collection of devices such as PCs, servers, and printers serve as the interface between the human user and the shared medium. Each of these machines hosts a device called a *network interface card* (NIC), which provides the physical connectivity between the user device and the shared medium. The NIC is either installed inside the system or, less commonly, as an external device. In laptop machines, the NIC is either a PC card that plugs into a slot in the machine, shown in Figure 5-15, or, more commonly today, built into the system. In many machines today Wi-Fi is also built in; it's basically a wireless version of Ethernet.

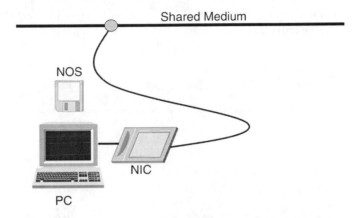

Figure 5-14
Typical LAN components

Figure 5-15
A PC card network interface card (NIC)

The NIC device implements the access protocol that devices wishing to access the shared medium use on their particular LAN. These access schemes will be discussed shortly. The NIC also provides the connectivity required to attach a user device to the shared network.

Topologically, LANs differ greatly. The earliest LANs used a bus architecture—shown in Figure 5-16—so-called because they were literally a long run of twisted-pair wire or coaxial cable to which stations were periodically attached. Attachment was easy; in fact, early coax systems relied on a device called a "vampire tap," which poked a hole in the insulation surrounding the center conductor in order to suck the digital blood from the shared medium. Later designs, such as IBM's Token Ring, used a contiguous ring architecture like that shown in Figure 5-17. Both have their advantages and will be discussed later in this chapter.

Later designs combined the best of both topologies to create star-wired LANs (Figure 5-18), also discussed later.

Figure 5-16
A bus-based
LAN

Figure 5-17
A ring-based
LAN

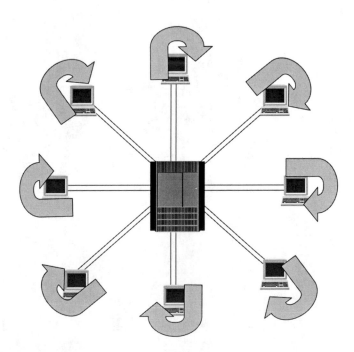

Figure 5-18
A star-wired
LAN

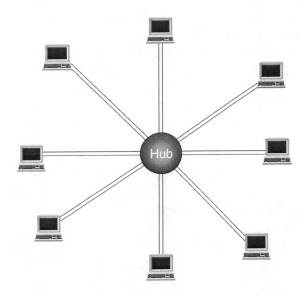

LAN Access Schemes

LANs have traditionally fallen into two primary categories characterized by the manner in which they access the shared transmission medium (shared among all the devices on the LAN). The first, and most common, is called *contention,* and the second group is called *distributed polling.* I tend to refer to contention-based LANs as the Berkeley Method, whereas I view distributed-polling LANs as users of the Harvard Method. I'll explain in a moment.

Contention-Based LANs

Dear Mr. Metcalf:

We're not sure how to break this to you, but we have discovered that your claim of a patent for the invention of Ethernet must be denied after the fact due to its existence prior to the date of your claim of invention. There is a small freshwater fish, Gymnarchus Niloticus, that uses an interesting technique for locating mates, food, and simply communicating with peers. The fish's body is polarized, with a cathode on its head and an anode on its tail. Using special electric cells in its body similar to those employed by the electric eel or the California electric ray, Gymnarchus emits nominal 300 Hz, 10 volt pulses, which reflect back and inform the fish about its immediate environment.

In the event that two Gymnarchus are in the same area, their emissions interfere with one another (intersymbol interference?), rendering their detection mechanisms ineffective. But, being the clever creatures that they are, Gymnarchus has designed a technique to deal with this problem. When the two fish "hear" each other's transmissions, they both immediately stop pulsing. Each fish waits a measurably random period of time while continuing to swim, after which they begin to transmit again, but this time at slightly different frequencies to avoid interference.

We hope that you understand that under the circumstances we cannot in good conscience grant this patent.

Sincerely yours,

U.S. Patent Office

I don't think a fish can hold a patent, but if it could, *Niloticus* would hold the patent for a widely used technique called "Carrier Sense, Multiple Access with Collision Detection." Please read on.

Perhaps the best-known, contention-based medium access scheme is Ethernet, a product developed by 3Com founder and Xerox PARC veteran Bob Metcalfe. In contention-based LANs, devices attached to the network vie for access using the technological equivalent of gladiatorial combat. "If it feels good, do it" is a good way to describe the manner in which they share access (hence the Berkeley Method). If a station wants to transmit, it simply does so, knowing that the possibility exists that the transmitted signal may collide with the signal generated by another station that transmits at the same time. Even though the transmissions are electrical and are occurring on a LAN, there is still some delay between the time that both stations transmit and the time that they both realize that someone else has transmitted. This realization is called a collision, and it results in the destruction of both transmitted messages as shown in Figure 5-19. In the event that a collision occurs as the result of simultaneous transmission, both stations back off by immediately stopping their transmissions, wait a random amount of time, and try again. This technique has the *wonderful* name of *truncated binary exponential back-off*. It's one of those phrases you just *have* to commit to memory because it sounds *so good* when you casually let it roll off the tongue in conversation.

Ultimately, each station *will* get a turn to transmit, although how long they may have to wait is based on how busy the LAN is. Contention-based systems are characterized by what is known as *unbounded delay,*

Figure 5-19
A collision on a
contention-
based LAN

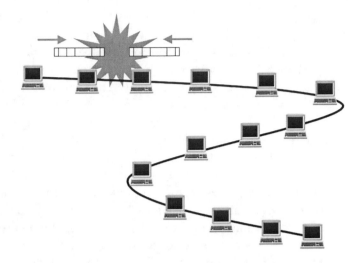

because there is no upward limit on how much delay a station can incur as it waits to use the shared medium. As the LAN gets busier and traffic increases, the number of stations vying for access to the shared medium —which only allows a single station at a time to use it, by the way—also goes up, which naturally results in more collisions. Collisions translate into wasted bandwidth, so LANs do everything they can to avoid them. We will discuss techniques for this in the contention world a bit later in this chapter.

The protocol that contention-based LANs employ is called *carrier sense, multiple access with collision detection* (CSMA/CD). In CSMA/CD, a station observes the following guidelines when attempting to use the shared network. First, it listens to the shared medium to determine whether it is in use or not—that's the "carrier sense" part of the name. If the LAN is available (not in use), it begins to transmit, but continues to listen while it is transmitting, knowing that another station could also choose to transmit at the same time—that's the "multiple access part." In the event that a collision is detected, usually indicated by a dramatic increase in the signal power measured on the shared LAN, both stations back off and try again. That's the "collision detection" part.

Ethernet is the most common example of a CSMA/CD LAN. Originally released as a 10 Mbps product based on IEEE standard 802.3, Ethernet rapidly became the most widely deployed LAN technology in the world.

As bandwidth-hungry applications such as E-Commerce, Enterprise Resource Planning (ERP), and Web access evolved, transport technologies advanced, and bandwidth availability (and capability) grew, 10 Mbps Ethernet began to show its age. Today, new versions of Ethernet have emerged that offer 100 Mbps (Fast Ethernet) and 1,000 Mbps (Gigabit Ethernet) transport, with plans afoot for even faster versions. Furthermore, in keeping with the demands being placed on LANs by convergence, standards are evolving for LAN-based voice transport mechanisms that guarantee quality of service for mixed traffic types.

Gigabit Ethernet has become a fundamentally important technology as its popularity and level of deployment have climbed. Emerging applications certainly make the case for Gigabit Ethernet's bandwidth capability: LAN telephony, server interconnection, and video to the desktop all demand low-latency solutions, and Gigabit Ethernet may be positioned to provide it. Many vendors have entered the marketplace including Alcatel, Lucent Technologies, Nortel Networks, and Cisco Systems. The other aspect of the LAN environment that began to show weaknesses was the overall topology of the network itself. LANs are broadcast environments, which means that when a station transmits, every station on the LAN segment hears the message (Figure 5-20). While this is a simple implementation scheme, it is also wasteful of bandwidth, since stations hear broadcasts that they have no reason to hear. In response to this, a technological evolution occurred. It was obvious to LAN implementers that the traffic on most LANs was somewhat domain-oriented; that is, it tended to cluster into communities of interest based on the work groups using the LAN. For example, if employees in sales shared a LAN with shipping and order processing, three discernible traffic groupings emerged, according to what network architects call "The 80:20 Rule." The 80:20 Rule simply states that 80 percent of the traffic that originates in a particular work group tends to stay in that work group, an observation that makes network design distinctly simpler. If the traffic naturally tends to segregate itself into groupings, then the topology of the network could change to reflect those groupings. Thus was born the *bridge*.

Bridges are devices with two responsibilities: They filter traffic that does not have to propagate in the forward direction, and they forward traffic that does. For example, if the network described previously were to have a bridge inserted in it (Figure 5-21), all of the employees in each of the three work groups would share a LAN segment, and each segment would be attached to a port on the bridge. When an employee in sales transmits a message to another employee in sales, the bridge is

When one station transmits, all stations hear the message. This can result in significant waste of bandwidth.

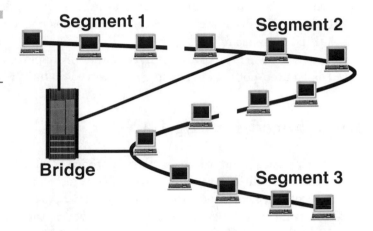

Using a bridge to segment a LAN

intelligent enough to know that the traffic does not have to be forwarded to the other ports. Similarly, if the sales employee now sends a message to someone in shipping, the bridge recognizes that the sender and receiver are on different segments and, thus, forwards the message to the appropriate port, using address information in a table that it maintains (the filter/forward database). Bridges operate at Layer Two of the OSI Model and, as such, are frame switches.

Following close on the heels of bridging is *LAN switching*. LAN switching qualifies as "bridging on steroids." In LAN switching, the filter/forward database is distributed—that is, a copy of it exists at each port, which implies that different ports can make simultaneous traffic handling decisions. This allows the LAN switch to implement full-duplex transmission, reduce overall throughput delay, and, in some cases, implement per-port rate adjustment. The first 10 Mbps Ethernet LAN switches emerged in 1993, followed closely by Fast Ethernet (100 Mbps) versions in 1995, and Gigabit Ethernet (1,000 Mbps) switches in 1997.

Fast Ethernet immediately stepped up to the marketplace bandwidth challenge and was quickly accepted as the next generation of Ethernet. LAN switching also helped to propagate the topology called "star wiring." In a star-wired LAN, all stations are connected by wire runs back to the LAN switch or a hub that sits in the geographical center of the network, as shown in Figure 5-22. Any access scheme (contention-based or distributed polling) can be implemented over this topology because it defines a wiring plan, not a functional design. Because all stations in the LAN are connected back to a center point, management, troubleshooting and administration of the network is simplified. Today, high-speed LAN switches have become the architectural norm for network deployment in multitenant office buildings and corporate facilities because of low cost, high utility, and ease of management.

Contention-based LANs are the most commonly deployed LAN topologies. Distributed-polling environments, however, do have their place.

Distributed-Polling LANs

In addition to the gladiatorial combat approach to sharing access to a transmission facility, there is a more civilized technique known as distributed polling, or as it is more commonly known, *token passing*. IBM's Token-Passing Ring is perhaps the best known of these products, followed closely by FDDI, a 100 Mbps version occasionally seen in campus and metropolitan area networks (although the sun seems to be setting on FDDI).

In token-passing LANs, stations take turns with the shared medium, passing the right to use it from station-to-station by handing off a "token" that gives the bearer the one-time right to transmit while all other stations remain quiescent. Thus, the Harvard Approach. This is a much

Figure 5-22
LAN switching

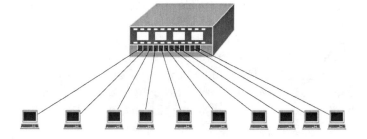

fairer way to share access to the transmission medium than CSMA/CD, because while every station has to wait for its turn, it is absolutely guaranteed that it will get that turn. These systems are therefore characterized by bounded delay, because there is a maximum amount of time that any station will ever have to wait for the token.

Token-passing rings work as shown in Figure 5-23. When a station wishes to transmit a file to another station on the LAN, it must first wait for the "token"—a small and unique piece of code that must be "held" by a station to validate the frame of data that is created and transmitted. Let's assume for a moment that a station has secured the token because a prior station has released it. The station places the token in the appropriate field of the frame it builds (actually, the *medium access control scheme*, called a MAC scheme and implemented on the NIC card that builds the frame), adds the data and address, and transmits the frame to the next station on the ring. The next station—which also has a frame it wishes to send—receives the frame, notes that it is not the intended recipient, and also notes that the token is busy. It does not transmit, but instead passes the frame of data from the first station on to the next station. This process continues, station by station, until the frame arrives at the intended recipient on the ring. The recipient validates that it is the intended recipient, at which time it makes a copy of the received frame,

Figure 5-23
A token-passing, distributed-polling LAN

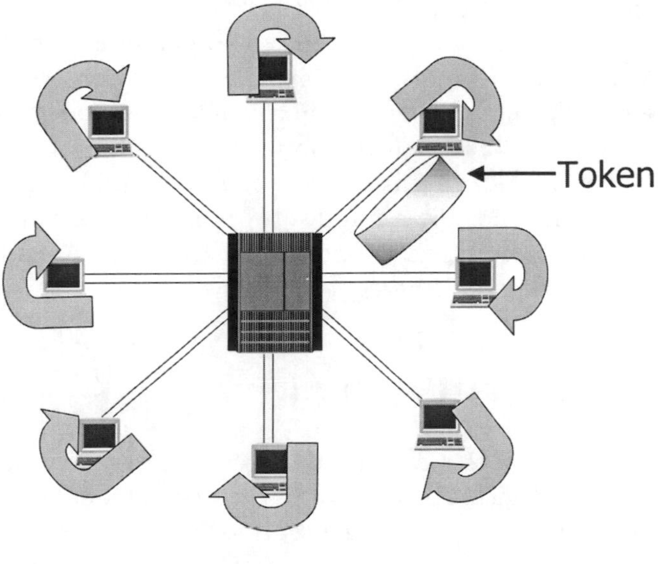

Token

sets a bit in the frame to indicate that it has been successfully received, *leaves the token set as busy,* and transmits the frame on to the next station on the ring. Because the token is still shown as busy, no other station can transmit. Ultimately, the frame returns to the originator, at which time it is recognized as having been received correctly. The station, therefore, removes the frame from the ring, frees the token, and passes it on to the next station (it is not allowed to send again just because it is in possession of a free token).

This is where the overall fairness scheme of this technique shines through. *The very next station to receive a free token* is the station that first indicated a need for it. It will transmit its traffic, after which it will pass the token on to the next station on the ring, followed by the next station, and so on.

This technique works very well for situations in which high traffic congestion on the LAN is the norm. Stations will always have to wait for what is called *maximum token rotation time,* that is, the amount of time it takes for the token to be passed completely around the ring, but they will *always* get a turn. Thus, for high-congestion situations, a token-passing environment may be better.

Traditional token ring LANs operate at two speeds—4 and 16 Mbps. Like Ethernet, these speeds were fine for the limited requirements of text-based LAN traffic that was characteristic of early LAN deployments. However, as demand for bandwidth climbed, the need to eliminate the bottleneck in the token ring domain emerged and fast token ring was born. In 1998, the IEEE 802.5 committee (the oversight committee for token ring technology) announced draft standards for 100 Mbps high-speed token ring (HSTR 802.5t). A number of vendors stepped up to the challenge and began to produce high-speed token ring equipment including Madge Networks and IBM.

Gigabit token ring is on the horizon as draft standard 802.5v, and while it may become a full-fledged product, many believe that it may never reach commercial status because of competition from the far less expensive Gigabit Ethernet.

Logical LAN Design

One other topic that should be covered before we conclude our discussion of LANs is logical design. Two designs for LAN data management have emerged over the years. The first is called peer to peer. In a peer-to-peer LAN, shown in Figure 5-24, all stations on the network operate at the

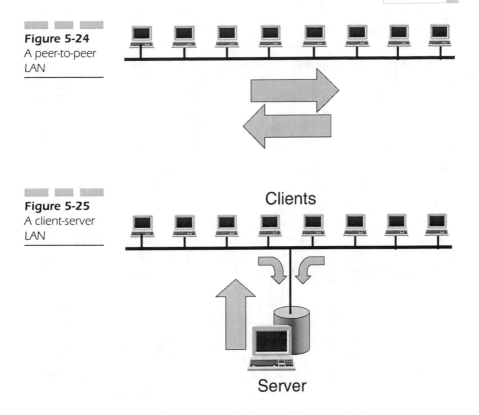

Figure 5-24
A peer-to-peer LAN

Figure 5-25
A client-server LAN

Clients

Server

same protocol layer, and all have equal access *at any time* to the shared medium and other resources on the network. They do not have to wait for any kind of permission to transmit; they simply do so. Traditional CSMA/CD is an example of this model. It is simple, easy to implement, and does not require a complex operating system to operate. It does, however, result in a free-for-all approach to networking, and in large networks can result in security and performance problems.

The alternative and far more commonly seen technique is called *client server*. In a client-server LAN, all data and application resources are archived on a designated server that is attached to the LAN and accessible by all stations (user PCs) with appropriate permissions, as illustrated in Figure 5-25. Because the server houses all of the data and application resources, client PCs do not have to be particularly robust. When a user wishes to execute a program, such as a word processor, he or she goes through the same keystrokes he or she would on a standalone PC. In a client-server environment, however, the application actually executes on the server, giving the user the appearance of local execution. Data files modified by the user are also stored on the server, resulting in a significant improvement in data management, cost control, security, and "software harmonization" compared to the peer-to-peer design. This also means that client devices can be relatively inexpensive, because they need very little in the way of onboard computing resources. The server, on the other hand, is really a PC with additional disk, memory, and processor capacity so that it can handle the requests it receives from all the users that depend on it. Needless to say, client-server architectures are more common than peer-to-peer architectures in corporate environments today.

Deployment

So when should Ethernet be used as opposed to token ring? Both have their advantages and disadvantages, both have solid industry support, and both are manufactured by any number of respectable, well-known players. CSMA/CD (for all intents and purposes, Ethernet) is far and away the most widely deployed LAN technology because it is simple, inexpensive, and capable of offering very high bandwidth to any marketplace, including residential. I am writing this in my home office on a PC that is connected to a 100 Mbps Ethernet LAN that ties together three PCs and a couple of printers, and the total cost of the network—including the router and firewall that protects the machines from intrusion due to the always-on connection through the cable modem—was less than $200.

Most businesses use Ethernet today because most businesses have normal traffic flows—office automation traffic and the like. For businesses that experience constant, bandwidth-intensive traffic such as that found in engineering firms, architectural enterprises or businesses with other graphics-heavy traffic, token ring may be a better choice, although there are those who will argue. Businesses that already have a large installed base of IBM hardware may also be good candidates for token ring, since it integrates well (for obvious reasons) into IBM environments. Even still, Ethernet is ruling the roost.

Content

It would be irresponsible of me to write a book like this without at least mentioning the content that rides on today's networks. The section that follows explores multimedia and all its many flavors. Multimedia is the primary driving force for bandwidth today; an understanding of what it is and where it is going is important. Besides, it's fascinating stuff.

The World of Multimedia

In the last decade a revolution has taken place in visual applications. Starting with simple, still image-based applications such as gray-scale facsimile, the technology has diverged into a collection of visually oriented applications that include video and virtual reality. Driven by aggressive demands from sophisticated, applications-hungry users and fueled by network and computer technologies capable of delivering such bandwidth-and-processor-intensive services, the telecommunications industry has undergone a remarkable metamorphosis as industry players battle for the pole position.

Why this rapid growth? Curt Carlson, vice president of information systems at the David Sarnoff Research Institute in Princeton, New Jersey, observes that more than half of the human brain is devoted to vision-related functions—an indication that vision is our single most important sense. He believes that this rapid evolution in image-based systems is occurring because these are the systems that people actually *need*.

"First we invented radio," he observes, "then we invented television. Now we are entering what we call the age of interactivity, in which we will . . . merge all of those technologies and add the element of user interaction. Vision is one of the key elements that allow us to create these exciting new applications." Indeed, many new applications depend on the interactive component of image-based technologies. Medical imaging, interactive customer service applications, and multimedia education are but a few.

Still Images

Still-image applications have been in widespread use for quite some time. Initially, there were photocopy machines. They did not provide storage of documents nor did they offer the ability to electronically transport them from one place to another. That capability arrived on a limited basis with the fax machine.

While a fax transmission allows a document to be moved from one location to another, what actually moves is not document content but rather an *image* of the document's content. This is important, because there is no element of flexibility inherent in this system that allows the receiver to make immediate and easy changes to the document. The image must be converted into machine-readable data, a capability that is just now becoming possible with *optical character recognition* (OCR) software.

As imaging technology advanced and networks grew more capable, other technological variations emerged. The marriage of the copy machine and the modem yielded the scanner, which allows high-quality images to be incorporated into documents or stored on a machine-accessible medium such as a hard drive.

Other advances followed. The emergence of *high-quality television* (HQTV) coupled with high-bandwidth, high-quality networks led to the development and professional acceptance of medical imaging applications, with which diagnosis-quality X-ray images can be used for remote teleradiology applications. This made possible the delivery of highly specialized diagnostic capabilities to rural areas, a significant advancement and extension of medicine.

Equally important are imaging applications that have emerged for the banking, insurance, design, and publishing industries. Images convey enormous amounts of information. By digitizing them, storing them online, and making them available simultaneously to large numbers of users, the applications for which the original image was intended are enhanced. Distance ceases to be an issue, transcription errors are eliminated, and the availability of expertise becomes a nonproblem.

There are downsides, of course. Image-based applications require expensive end-user equipment. Image files tend to be large, so storage requirements are significant. Furthermore, because of the bandwidth-intensive nature of transmitted image files, network infrastructures must be reexamined.

The Arrival of Compression

To deal with the storage and transmission issues associated with image-based applications, corollary technologies such as digital compression have emerged. The main technique used today for still image compression is JPEG, developed by the Joint Photographic Experts Group, a cooperative effort between the International Organization for Standardization (ISO), ITU-T, and the International Electrotechnical Commission (IEC).

JPEG, discussed briefly in an earlier chapter, works as follows. Digital images, composed of thousands of picture elements (pixels), are "dissolved" into a mosaic of 16-pixel-by-16-pixel blocks. These blocks are then reduced to eight-by-eight blocks by removing every other pixel value. JPEG software then calculates an average brightness and color value for each block, which is stored and used to reconstruct the original image during decompression.

Today, still-image technologies remain in widespread use, and will continue to play a key role in the application of visual technologies. But others have emerged as well. One of the most promising is video.

Video

The video story begins in 1951 at RCA's David Sarnoff Research Institute. During a celebration dinner, Brigadier General David Sarnoff, chairman of RCA and the founder of NBC, requested that the institute work on three new inventions, one of them what he called a "videograph." In his mind, a videograph was a device capable of capturing television signals on some form of inexpensive medium, such as tape.

Remember the time frame that we're talking about. Thanks to Philo T. Farnsworth, who invented the predecessor of today's *cathode ray tube* (CRT—and yes, that's his real name), electronic television became a reality in the 1920s. By the early 1950s, black and white television was widespread in America. The gap between the arrival of television and the demand for video, therefore, was fairly narrow.

Work on the videograph began almost immediately and a powerful cast of characters was assembled. There was one unlikely member of this cast who served as a catalyst: Bing Crosby. Keenly interested in broadcast technologies, Crosby wanted to be able to record his weekly shows for later transmission. The Bing Crosby Laboratories played a key role in the development and testing of video technology.

The Sarnoff Institute called upon the capabilities of several companies to reach its goal of creating what Sarnoff dubbed the "Hear-See machine." One of them was Ampex, developer of the first commercial audiotape recorder. The Sarnoff team believed that audiotape technology could be applied to video recording.

To a certain extent, they were correct. Marvin Camras, a Sarnoff team member and scientist who developed the ability to record audio signals on steel wire used during WWII, soon discovered that the video signal was dramatically broader than the relatively narrow spectrum of the audio signal. Early audiotape machines typically moved the tape along at a stately 15 inches per second (ips). To meet the bandwidth requirements of the video signal, the tape had to be accelerated to somewhere between 300 and 400 ips—roughly 25 miles per hour.

To put this into perspective, a tape that would accommodate an hour's worth of audio in those days would hold *one minute* of video—which did not take into account the length of the leader that had to be in place to allow the recorder to reach its ridiculously high tape transport speed. To hold 15 minutes of video, a reel of quarter-inch tape would have had to be three feet in diameter—not exactly portable. Put another way, a one-hour show would require 25 miles of tape!

To get around this problem, Camras invented the spinning record head. Instead of moving the tape rapidly past the recording head, he moved the tape slowly, and rapidly spun the head. By attaching the head to a 20,000 rpm Hoover vacuum cleaner motor (stolen, by the way, from his wife's vacuum cleaner), he was able to use two-inch tape and reduce the tape transport speed to 30 to 40 inches per second—a dramatic improvement.

The first video demonstrations were admirable, but rather funny. First of all, the resolution of the television was only 40 lines per inch, compared to more than 250 on modern systems. The images were so poor that audiences required a narrator to tell them what they were seeing on the screen.

Luckily, other advances followed. The original video systems rendered black and white images, but soon a color system was developed. It recorded five tracks (red, blue, green, synch, and audio) on half-inch tape and ran at the original speed of 360 ips. The system was a bit unwieldy: It required a *mile* of color tape to capture a 4-minute recording.

Obviously, mile-long tapes were unacceptable, especially if they would only yield 4-minute programs. As a result, the Sarnoff/Ampex team reexamined the design of the recording mechanism. Three scientists—Charles Ginsburg, Alex Maxey, and Ray Dolby (later to be known for his work in audio)—redesigned the rotating record head, rotating it about 90 degrees so that the video signal was now written on the tape in a zigzag design. This redesign, combined with FM instead of AM signal modulation, allowed the team to reduce the tape speed to a remarkable 17.5 inches per second. For comparison's sake, modern machines consume tape at about 2.5 inches per second. This, by the way, is why the record head in your home VCR sits at a funny angle. (It's that big silver cylinder you see when you open the slot where the tape is inserted.)

By 1956, Sarnoff and Ampex had created a commercially viable product. They demonstrated the Mark IV to 200 CBS affiliate managers in April of that year. When David Sarnoff walked out on the stage and stood next to his own prerecorded image playing on the television next to him, the room went berserk. In 4 days, Ampex took $5 million in video machine orders.

Modern Video Technology

Today's palm-size videotape recorders are a far cry from the washer-dryer-size Mark IV of 1956. But physical dimensions are only a piece of the video story.

The first video systems were analog and relied on a technique called *composite video*. In composite video, all of the signal components—such as color, brightness, audio, and so on—are combined into a single multiplexed signal. Because of the interleaved nature of this technique, a composite signal is not particularly good and suffers from such impairments as clarity loss between generations (in much the same way an analog audio signal suffers over distance) and color bleeding. Unfortunately, bandwidth at the time was extremely expensive and the cost to transport five distinct high-bandwidth channels was inordinately high. Composite video, therefore, was a reasonable alternative.

As the cost of bandwidth dropped in concert with advances in transmission technology, *component video* emerged, in which the signal components are transported separately, each in its own channel. This eliminated many of the impairments that plagued composite systems.

Several distinct component formats have emerged including RGB, YUV (for luminance [Y], hue [U], and saturation [V]), YIQ, and a number of others. An interesting aside: Luminance is analogous to the brightness of the video signal, whereas hue and saturation make up the chrominance (color) component.

All of these techniques accomplish the task of representing the red, green, and blue components needed to create a color video signal. In fact, they are mathematical permutations of one another.

One final observation is that RGB, YUV, and the others previously mentioned are *signal formats*—a very different beast from *tape formats* such as D1, D2, Betacam, VHS, and S-VHS. Signal formats describe the manner in which the information that represents the captured image is created and transported; tape formats define how the information is encoded on the storage medium.

Of course, these signal formats are analog and, therefore, still suffer from analog impairments such as quality loss, as downstream generations are created. Something better is needed.

Digital Video

In the same way that digital data transmission was viewed as a way to eliminate analog signal impairments, digital video formats were created to do the same for video. Professional formats such as D1, D2, and Digital Betacam (the latter often discounted because it incorporates a form of compression), and even High-8 and MiniDV, virtually eliminate the problem of generational loss.

One downside is that even though these formats are digital, they still record their images sequentially on videotape. Today, video and computer technologies are being married as nonlinear video systems for editing and management ease. Because the video signal can be digitized ("bits is bits"), it can be stored on a large hard drive just as easily as a text or image file can be stored. Of course, these files tend to be large, and therefore require significant amounts of hard drive space. For example, a full-motion, full-color, TV-screen-size, two-hour movie requires 26 MB per second of storage capacity—a total of nearly 24 Gigabytes of storage.

Today, the market demands an inexpensive, high-quality solution to the storage of video—CD-ROM is a popular "target" for such files. While this is an appealing concept, let's explore what it takes to put video on a CD-ROM.

The Video Process

The signal that emerges from a video camera, while it can be either analog or digital, is typically analog. The signal is laid down on either analog (Betacam, VHS, S-VHS) or digital tape (D1, D2, Digital Betacam). Digital tape is a quite expensive medium, but it eliminates generational loss and, therefore, is popular in commercial video.

To create a digital representation of analog video, the analog signal is typically sampled at three to four times the bandwidth of the video channel (unlike audio, which typically relies on a two times sample). As a result, the output bandwidth for digital-tape machines is quite high— 411 Mbps for D1, and 74 Mbps for D2.

A single CD-ROM drive will hold approximately 650 MB of data. The best numbers available today, including optimal sampling and compression rates (discussed a bit later) indicate that VHS-quality recording requires roughly 5 MB of storage per second of recorded movie. That's 7,200 seconds for a two-hour movie, or roughly 30 CD-ROM disks. Finally, the maximum transfer rate across a typical bus in a PC is 420 kilobytes per second, somewhat less than that required for a full-motion movie.

What the Market Wants, the Market Gets: Compression

Growth in desktop video applications for videoconferencing, on-demand training, and gaming is fueling the growth in digital video technology, but the problems mentioned earlier still loom large. Recent advances have had an impact; for example, storage and transport limitations can often be overcome with compression.

The most widely used compression standard for video is MPEG, created by the Moving Pictures Expert Group, the joint ISO/IEC/ITU-T organization that oversees standards development for video. MPEG is relatively straightforward. In its initial compression step, MPEG creates what is called a "reference frame," an actual copy of the original frame of video. It intersperses these so-called I-Frames every 15 frames in the transmission. Because they are used as a reference point they are only minimally compressed.

MPEG assumes (correctly) that a relatively small amount of the information in a series of video frames changes from frame to frame. The background, for example, stays relatively constant in most movies, for long periods of time. Therefore, only a small amount of the information in each frame needs to be recaptured, recompressed, and re-stored.

Two additional frame types are created in the MPEG process. Predicted frames use information contained in past frames to "predict" the content of the next frame in the series, thus reducing the amount of information required. These predicted frames experience medium to significant compression.

Finally, MPEG may create bidirectional interpolated frames that are interspersed between the original I-Frames and the predicted frames. They require input from both, and because they constitute the bulk of the transmitted frame stream, are subject to the most compression. In fact, compression ratios of 200:1 are not uncommon in MPEG. An 18 Gigabyte movie can be reduced to a somewhat more manageable 90 MB.

There are other compression schemes for video. Motion JPEG, for example, is an intraframe compression technique (unlike MPEG, which is interframe) that compresses and stores each and every image as a separate entity. This is different from MPEG. Compression ratios with motion JPEG are considerably lower than MPEG's 200:1 numbers; ratios of 20:1 are about the limit, assuming equivalent image quality.

Other techniques exist, but they are largely proprietary. These include PLV, Indeo, RTV, Compact Video, AVC, DV, DVCAM, and a few others.

Television Standards

It's interesting to note that in the drive toward digitization, the ultimate goal is to create a video storage and transport technology that will yield as good a representation of the original image as analog transmission does.

Two primary governing organizations dictate television standards: One is the National Television System Committee, or NTSC (sometimes said to stand for "Never Twice the Same Color" based on the sloppy color management that characterizes the standard); the other, used primarily in Europe, is the Phased Alternate Line (PAL) system. NTSC is built around a 525-line-per-frame, 30-frame-per-second standard, while PAL uses 625 lines per frame and 25 frames per second. While technically different, both address the same concerns and rely on the same characteristics to guarantee image quality.

Video Quality Factors

Four factors influence the richness of the video signal. They are frame rate, color resolution, image quality, and spatial resolution.

Frame rate is a measure of the refresh rate of the actual image painted on the screen. The NTSC video standard is 30 frames per second, meaning that the image is updated 30 times every second. Each frame consists of odd and even fields. The odd field contains the odd-numbered screen lines, while the even field contains the even-numbered screen lines that make up the picture.

Television sets paint the screen by first painting the odd field, then the even. They repeat this process at the rate of 60 fields—or 30 frames—per second. The number 60 is chosen to coincide with the frequency of electricity in the United States. By the same token, PAL relies on a scan rate that is very close to 50 Hz, the standard in Europe.

This odd–even alternation of fields is called *interlaced video.*

Many monitors, on the other hand, use a technique called *progressive scan,* in which the entire screen is painted 30 times per second, from top to bottom. This is referred to as *noninterlaced video.* Noninterlaced systems tend to demonstrate less flicker than their interlaced counterparts.

Computers often rely on variable graphics array (VGA) monitors, which are much sharper and clearer than television screens. This is due to the density of the phosphor dots on the inside of the screen face that yield color when struck by the deflected electron beams, as well as a number of other factors. The scan rate of VGA is much higher than that of traditional television and, therefore, can be noninterlaced to reduce screen flicker.

Another quality factor is *color resolution.* Most systems resolve color images using a technique called RGB—for the red, green, and blue primary colors. While video does rely on RGB, it also uses a variety of other resolution techniques, including YUV and YIQ. YUV is a color scheme used in both PAL and NTSC. As noted previously, Y represents the luminance component; U and V—hue and saturation, respectively—make up the color component. Varying the hue and saturation components changes the color.

Image quality plays a critical role in the final outcome, and the actual resolution varies by application. For example, the user of a slow scan, desktop, videoconferencing application might be perfectly happy with a half-screen, 15-frame-per-second, eight-bit image, whereas a physician using a medical application might require a full frame, 768-by-484 (the NTSC standard screen size) pixel image, with 24-bit color for perfect accuracy. Both frame rate (frames per second) and color density (bits per pixel) play a key role.

Finally, *spatial resolution* comes into the equation. Many PCs have displays that measure 640-by-480 pixels. This is considerably smaller than the NTSC standard of 768-by-484, or even the slightly different European PAL system. In modern systems, the user has great control over the resolution of the image, because he or she can vary the number of pixels on the screen. The pixels seen on the screen are simply memory representations of the information displayed. By selecting more pixels, and therefore better resolution, the graininess of the screen image is reduced. Some VGA adapters, for example, have resolutions as dense as 1024-by-768 pixels.

The converse, of course, is also true. By selecting fewer pixels, and therefore increasing the graininess of the image, special effects can be created, such as "pixelization" or "tiling."

Video Summary

Let's review the factors and choices involved in creating video.

The actual image is captured by a camera that is either analog or digital. The resulting signal is then encoded on either analog or digital tape. If a VHS tape is to be the end result, it will be created directly from the original tape.

If desktop video or CD-ROM video is the end result, then additional factors come into play. Because of the massive size of the original "file," it must be modified in some way to make it transportable to the new medium. It might be compressed, in which case a 200:1 reduction in size (or greater) could occur, with a resulting loss of quality (rain, for example, because it is a random event, tends to disappear in MPEG-compressed movies); it might be sampled less frequently, thus lowering the total number of frames, but causing the animation-like jerkiness that results from lower sampling rates; or, the screen size could be reduced, thus reducing the total number of pixels that need to be encoded.

During the last decade, video has achieved a role of some significance in a wide variety of industries. It has found a home in medicine, travel, engineering, and education, and provides not only a medium for the presentation of information, but combined with telecommunications technology it also makes possible such applications as distance learning, video teleconferencing, and desktop video. The ability to digitize video signals has brought about a fundamental change in the way video is created, edited, transported, and stored.

Its emergence has also changed the players in the game. Once the exclusive domain of filmmakers and television studios, video is now fought over by creative companies and individuals who want to control its content, cable and telephone companies who want to control its delivery, and a powerful market that wants it to be ubiquitous, richly featured, and cheap. Regulators are in the mix as well, trying to make sense of a telecommunications industry that, once designed to transport voice, now carries a broad mix of fundamentally indistinguishable data types.

Summary

Premises technologies—including LANs, wireless solutions, and the content that rides on both of them—are the most visible components of the overall network to a customer. In the next chapter we dive into the network itself and explore the access technologies that provide the bridge between premises technologies and the transport network.

Chapter 5 Questions

1. What are the five principal components of a computer? Describe the functions of each.

2. What factors affect a CPU's processing speed?

3. What does "4.0 MHz" actually mean?

4. What is a bit? A byte? A megabyte?

5. What are the functions of the ALU?

6. Explain the differences between main and secondary memory. Give examples of each.

7. Explain the differences between a token-passing and a contention-based LAN.

8. Why is Ethernet the preferred LAN scheme?

9. What is the difference between 802.3 and Ethernet?

10. What is the difference between a peer-to-peer network and a client-server network?

Access Technologies

For the longest time, "access" described the manner in which customers reached the network for the transport of voice services. In the last 20 years, however, that definition has changed dramatically. In 1981—*20 years ago*—IBM changed the world when it introduced the PC, and in 1984 the Macintosh arrived, bringing well-designed and organized computing power to the proverbial masses. Shortly thereafter, hobbyists began to take advantage of emergent modem technology and created online databases—the first bulletin board systems that allowed people to send simple text messages to each other. This accelerated the modem market dramatically, and before long data became a common component of local loop traffic. At that time there was no concept of instant messaging or of the degree to which e-mail would fundamentally change the way people communicate and do business. At the same time, the business world found more and more applications for data, and the need to move that data from place to place became a major contributor to the growth in data traffic on the world's telephone networks.

In those heady, early days, data did not represent a problem for the bandwidth-limited local loop. The digital information created by a computer and intended for transmission through the telephone network was received by a modem, converted into a modulated analog waveform that fell within the 4 KHz voice band, and fed to the network without incident. As we mentioned in an earlier chapter, the modem's job was (and is) quite simple: Invoke the Wizard of Oz Protocol—when a computer is doing the talking, the modem must make the network think it is talking to a telephone. "Pay no attention to that man behind the curtain!"

Over time, modem technology advanced, allowing the local loop to provide higher and higher bandwidth. This increasing bandwidth was made possible through clever signaling schemes that allowed a single signaling event to transport more than a single bit. These modern modems, often called "Shannon-Busting modems" because they defy the limits of signaling defined by Shannon, are commonplace today. They allow baud levels to reach unheard-of extremes and permit the creation of very high bit-per-signal rates.

The analog local loop is used today for various voice and data applications in both business and residence markets. The new lease on life it enjoys thanks to advanced modem technology as well as a focus by installation personnel on the need to build clean, reliable, outside plants has resulted in the development of faster access technologies designed to operate across the analog local loop, including traditional high-speed modem access and such options as DSL.

Marketplace Realities

According to a number of demographic studies, there are approximately 100 million households today that host home-office workers, and the number is growing rapidly. These numbers include both telecommuters and those who are self-employed and work out of their homes. They require the ability to connect to remote LANs and corporate databases, retrieve e-mail, access the Web, and, in some cases, conduct videoconferences with colleagues and customers. The traditional bandwidth-limited local loop is not capable of satisfying these requirements with traditional modem technology. Dedicated private-line service, which would solve the problem, is far too expensive as an option, and because it is dedicated is not particularly efficient. Other solutions are required, and these have emerged in the form of access technologies that take advantage of either a conversion to end-to-end digital connectivity (ISDN) or expanded capabilities of the traditional analog local loop (DSL, 56K modems). In some cases, a whole new architectural approach is causing excitement in the industry (wireless local loop, or WLL). Finally, cable access has become a popular option as the cable infrastructure has evolved to a largely optical, all-digital system with high-bandwidth, two-way capability. We will discuss each of these options in the pages that follow.

56 Kbps Modems

With the accelerating proliferation of broadband access underway, it almost feels medieval to write about 56 Kbps modems. They are still out there, however, and a remarkably large number of them are still in use. In fact, I rely on the one in my laptop far more often than I care to, particularly when I travel internationally. All too often I check into a hotel that doesn't have broadband of any flavor—wired or wireless. So, while this seems like a technology that is on its way out, it isn't—and I feel compelled to include it in the second edition of the book.

One of the most important words in telecommunications is "virtual." It is used in a variety of ways, but in reality only has one meaning. If you see the word "virtual" associated with a technology or product, you should immediately say to yourself, "It's a lie."

A 56 Kbps modem is a good example of a virtual technology. These devices have attracted a great deal of interest since they were introduced a few years ago. Under certain circumstances they do offer higher access speeds designed to satisfy the increasing demands of bandwidth-hungry applications and increasingly graphics-oriented Web pages. The problem they present is that they do not really provide true 56K access, even under the best of circumstances.

56K modems provide asymmetric bandwidth, with 56 Kbps delivered downstream toward the customer (sometimes), and significantly less bandwidth (33.6 Kbps) in the upstream direction. While this may seem odd, it makes sense given the requirements of most applications today that require modem access. A Web session, for example, requires very little bandwidth in the upstream direction to request that a page be downloaded; the page itself, however, may require significantly more, since it may be replete with text, graphics, Java applets, and even small video clips. Since the majority of modem access today is for Internet surfing, asymmetric access is adequate for most users.

The limitations of 56K modems stem from a number of factors. One of them is that under current FCC regulations (specifically Part 68), line voltage supplied to a communications facility is limited such that the maximum achievable bandwidth is 53 Kbps in the downstream direction. Another limitation is that these devices require that only a single analog-to-digital conversion occur between the two end points of the circuit. This typically occurs on the downstream side of the circuit, and usually at the interface point where the local loop leaves the central office. Consequently, downstream traffic is less susceptible to the noise created during the analog-to-digital conversion process, while the upstream channel is affected by it and is therefore limited in terms of the maximum bandwidth it can provide. In effect, 56K modems, in order to achieve their maximum bandwidth, require that one end of the circuit, typically the central office end, be digital.

The good news with regard to 56K modems is that even in situations in which the 56 Kbps speed is not achievable, the modem will fall back to whatever maximum speed it can fulfill. Furthermore, no premises wiring changes are required, and since this device is really nothing more than a faster modem, the average customer is comfortable with migration to the new technology. This is certainly demonstrated by sales volume: As was mentioned before, most PCs today are automatically shipped with a 56K modem.

The Integrated Services Digital Network (ISDN) has been the proverbial technological roller coaster since its arrival as a concept in the late 1960s. Often described as "the technology that took 15 years to become an overnight success," ISDN's level of success has been all over the map.

Internationally, it has enjoyed significant uptake as a true, digital local loop technology. In the United States, however, because of competing and often incompatible hardware implementations, high cost, and spotty availability, its deployment has been erratic at best. In market areas where providers have made it available at reasonable prices, it has been quite successful. In fact, one application that has made its mark with ISDN is videoconferencing, which we'll discuss shortly.

ISDN Technology

The typical non-ISDN local loop is analog. Voice traffic is carried from an analog telephone to the CO using a frequency-modulated carrier; once at the CO the signal is typically digitized for transport within the digital network cloud. On the one hand, this is good because it means that there is a digital component to the overall transmission path. On the other hand, the loop is still analog and, as a result, the true promise of an end-to-end digital circuit cannot be realized. The circuit is only as good as the weakest link in the chain, and the weakest link is clearly the analog local loop.

In ISDN implementations, local-switch interfaces must be modified to support a digital local loop. Instead of using analog frequency modulation to represent voice or data traffic carried over the local loop, ISDN digitizes the traffic at the origination point, either in the voice set itself or in an adjunct device known as a *terminal adapter* (TA). The digital local loop then uses time-division multiplexing to create multiple channels over which the digital information is transported, which provide for a wide variety of truly integrated services.

The Basic Rate Interface (BRI)

There are two well-known implementations of ISDN. The most common (and the one intended primarily for residence and small-business applications) is called the Basic Rate Interface, or BRI. In BRI, the two-wire local loop supports a pair of 64 Kbps digital channels—known as B-Channels—as well as a 16 Kbps D-Channel, which is primarily used for signaling but can also be used by the customer for low-speed (up to 9.6 Kbps) packet data. The B-Channels can be used for voice and data, and in some implementations can be bonded together to create a single 128 Kbps channel for videoconferencing or other higher bandwidth applications.

Figure 6-1 shows the layout of a typical ISDN BRI implementation. In this diagram, the LE is the local exchange, or switch. The NT1 is the network termination device that serves as the demarcation point between the customer and the service provider; among other things, it converts the two-wire local loop to a four-wire interface on the customer's premises. The TE1 (terminal equipment, type 1) is an ISDN-capable device such as an ISDN telephone. This simply means that the phone is a digital device and, therefore, is capable of performing the voice digitization itself. A TE2 (terminal equipment, type 2) is a non-ISDN-capable device, such as a POTS telephone. In the event that a TE2 is used, a TA must be inserted between the TE2 and the NT1 to perform analog-to-digital conversion and rate adaptation.

The reference points mentioned earlier identify circuit components between the functional devices just described. The U reference point is the local loop; the S/T reference point sits between the NT1 and the TEs; the R reference point is found between the TA and the TE2.

BRI Applications

While BRI does not offer the stunning bandwidth that other more recent technologies (such as DSL) do, its bondable 64 Kbps channels provide reasonable capacity for many applications. The two most common today are remote LAN and Internet access. For the typical remote worker the bandwidth available through BRI is more than adequate, and new video compression technology even puts reasonable quality videocon-

Figure 6-1
The ISDN Basic
Rate Interface
(BRI)

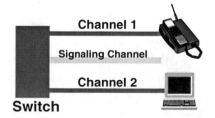

Channel 1

Signaling Channel

Channel 2

Switch

ferencing within the grasp of the end user at an affordable price. Sixty-four Kbps makes short shrift of LAN-based text file downloads and reduces the time required for graphics-intensive Web page downloads to reasonable levels.

The Primary Rate Interface (PRI)

The other major implementation of ISDN is called the *Primary Rate Interface*, or PRI. The PRI is really nothing more than a T-Carrier in that it is a four-wire local loop, uses AMI and B8ZS for ones-density control and signaling, and provides twenty-four 64 Kbps channels that can be distributed among a collection of users as the customer sees fit (see Figure 6-2). In PRI, the signaling channel operates at 64 Kbps (unlike the 16 Kbps D-Channel in the BRI) and is not accessible by the user. It is used solely for signaling purposes—that is, it cannot carry user data. The primary reason for this is service protection: In the PRI, the D-Channel is used to control the 23 B-Channels and, therefore, requires significantly more bandwidth than the BRI D-Channel. Furthermore, the PRI standards allow multiple PRIs to share a single D-Channel, which makes the D-Channel's operational consistency all the more critical.

The functional devices and reference points are not appreciably different from those of the BRI. The local loop is still identified as the U reference point. In addition to an NT1, we now add an NT2, which is a service distribution device, usually a PBX, which allocates the PRI's 24 channels to customers. This makes sense since PRIs are typically installed at businesses that employ PBXs for voice distribution. The S/T reference point is now divided; the S reference point sits between the

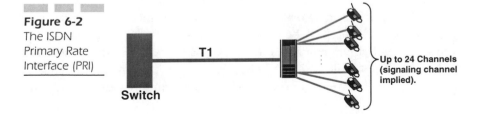

Figure 6-2
The ISDN
Primary Rate
Interface (PRI)

T1

Switch

Up to 24 Channels
(signaling channel
implied).

NT2 and TEs, while the T reference point is found between the NT1 and the NT2.

PRI service also has the ability to provision B-Channels as super-rate channels to satisfy the bandwidth requirements of higher bit rate services. These are called "H-Channels," and are provisioned as shown in Table 6-1.

Table 6-1

H-Channels

Channel	Bandwidth
H0	384 Kbps (6B)
H10	1.472 Mbps (23B)
H11	1.536 Mbps (24B)
H12	1.920 Mbps (30B)

PBX Applications

The PRI's marketplace is the business community, and its primary advantage is pair gain—that is, to conserve copper pairs by multiplexing the traffic from multiple user channels onto a shared, four-wire circuit. Inasmuch as a PRI can deliver the equivalent of 23 voice channels to a location over a single circuit, it is an ideal technology for a number of applications including interconnection of a PBX to a local switch, dynamic bandwidth allocation for higher-end videoconferencing applications, and interconnection between an ISP's network and that of the local telephone company.

Some PBXs are ISDN-capable on the line (customer) side, meaning that they have the ability to deliver ISDN services to users that emulate the services that would be provided over a direct connection to an ISDN-provisioned local loop. On the trunk (switch) side, the PBX is connected to the local switch via one or more T1s, which in turn provide access to the telephone network. This arrangement results in significant savings, faster call setup, more flexible administration of trunk resources, and the ability to offer a diversity of services through the granular allocation of bandwidth as required.

▬▬ Videoconferencing

Videoconferencing—an application that now enjoys widespread acceptance in the market thanks to service providers like Proximity Systems and affordable, effective video CODECs from companies like PolyCom and Tandberg—has moved from the boardroom to the home office and everywhere in between. While BRI provides adequate bandwidth for casual conferencing on a reduced-size PC screen, PRI is preferable for high-quality, television-scale sessions. Its ability to provision bandwidth-on-demand makes it an ideal solution. All major videoconferencing equipment manufacturers have embraced PRI as an effective connectivity solution for their products and, consequently, have designed their products in accordance with accepted international standards—specifically H.261 and H.320.

A typical corporate-level videoconferencing unit incorporates an inverse multiplexer (IMUX) that combines the bandwidth available in three or more ISDN BRI lines to create a super-rate, two-way channel for transporting the video and audio signals, as illustrated in Figure 6-3. Three ISDN BRI lines provide six 64 Kbps channels for an aggregate channel of 384 Kbps, perfectly adequate for high-quality videoconferencing. These devices range in price from as low as $6,000 to as high as

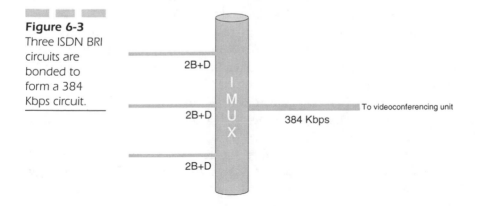

Figure 6-3
Three ISDN BRI circuits are bonded to form a 384 Kbps circuit.

$80,000. And while many people would turn up their noses at the thought of using ISDN, let it be known that far and away the vast majority of all videoconferencing rooms use ISDN for network connectivity, as opposed to some other protocol.

Automatic Call Distribution (ACD)

Another significant application for PRI is call routing in the call center environment. Ultimately, call centers represent critical decision points within a corporate environment, and the degree to which they are successful at what they do translates directly into visible manifestations of customer service. If calls are routed quickly and accurately based on some well-designed internal decision-making process, customers are happy, and the call center provides an effective image of the corporation and its services before the customer.

Automatic call distribution, or ACD, is a popular switch feature that allows a customer to create custom routing tables in the switch so that incoming calls can be handled most effectively on a call-by-call basis. The ACD feature often relies on information culled from both corporate sources and the SS7 network's SCP databases, but the ultimate goal is to provide what appears to each caller to be customized answering. For example, some large call centers handle incoming calls from a variety of countries and, therefore, potentially different language groups. Using caller ID information, an ACD can filter calls as they arrive and route them to language-appropriate operators. Similarly, the ACD can change the assignment of voice channels on a demand basis. For example, if the call center is large and receives calls from multiple time zones simultaneously, there may be a need to add incoming trunks to one region of the call center based on time-of-day call volumes, while reducing the total coverage in another. By reallocating the 64K channels of the PRI(s), bandwidth can be made available as required. Furthermore, it can be done automatically, triggered by time of day or some other preidentified event.

ISDN is only one of the so-called twisted-pair solutions that extend the capacity and lifetime of the local loop. Another is DSL, a technology family that has achieved significant attention in the last couple of years. The attention is due to both successes and failings on the part of the technology. When it works, it works extremely well. When it doesn't, it tends to fail loudly and publicly.

DSL

The access technology that has enjoyed the greatest amount of attention in recent times is the digital subscriber line, or DSL. It provides a good solution for remote LAN access, Internet surfing, and access for telecommuters to corporate databases.

DSL came about largely as a direct result of the Internet's success. Prior to its arrival, the average telephone call lasted approximately four minutes, a number that central office personnel used while engineering switching systems to handle expected call volumes. This number was arrived at after nearly 125 years of experience designing networks for voice customers. They knew about Erlang theory, loading objectives, peak-calling days/weeks/seasons, and had decades of trended data to help them anticipate load problems. So in 1993, when the Internet—and with it, the World Wide Web—arrived, network performance became unpredictable as callers began to surf, often for hours on end. The average four-minute call became a thing of the past as online service providers such as AOL began to offer flat rate plans that did not penalize customers for long connect times. Then, the unthinkable happened: Switches in major metropolitan areas, faced with unpredictably high call volumes and hold times, began to block during normal business hours, a phenomenon that had only occurred in the past during disasters or on Mothers Day.

A number of solutions were considered including charging different rates for data calls than for voice calls, but none of these proved feasible until a technological solution was proposed. That solution was DSL.

It is commonly believed that the local loop is incapable of carrying more than the frequencies required to support the voice band. This is a misconception. ISDN, for example, requires significantly high bandwidth to support its digital traffic. When ISDN is deployed, the loop must be modified in certain ways to eliminate its designed-in bandwidth limitations. For example, some long loops are deployed with load coils that "tune" the loop to the voice band. They make the transmission of frequencies above the voice band impossible but allow the relatively low-frequency voice band components to be carried across a long local loop. High-frequency signals tend to deteriorate faster than low-frequency signal components, so the elimination of the high frequencies extends the transmission distance and reduces transmission impairments that would result from the uneven deterioration of a rich, multifrequency signal. These load coils, therefore, must be removed if digital services are to be deployed.

A local loop is only incapable of transporting high-frequency signal components if it is *designed* not to carry them. The capacity is still there; the network design, through load coil deployment, simply makes that additional bandwidth unavailable. DSL services, especially ADSL, take advantage of this "disguised" bandwidth.

DSL Technology

In spite of the name, DSL—*Digital* Subscriber Line—is an analog technology. The devices installed on each end of the circuit are sophisticated high-speed modems that rely on complex encoding schemes to achieve the high bit rates that DSL offers. Furthermore, several of the DSL services, specifically ADSL, G.lite, VDSL, and RADSL—all discussed later in this section—are designed to operate in conjunction with voice across the same local loop. ADSL is the most commonly deployed service and offers a great deal to both business and residence subscribers.

DSL Services

DSL comes in a variety of flavors designed to provide flexible, efficient, high-speed service across the existing telephony infrastructure. From a consumer point-of-view, DSL, especially ADSL, offers a remarkable leap forward in terms of available bandwidth for broadband access to the Web. As content has steadily moved away from being largely text-based and has become more graphical, the demand for faster delivery services has grown for some time now. DSL may provide the solution at a reasonable cost to both the service provider and the consumer.

Businesses will also benefit from DSL. Remote workers, for example, can rely on DSL for LAN and Internet access. Furthermore, DSL provides a good solution for VPN access as well as for ISPs looking to grow the bandwidth available to their customers. It is available in a variety of both symmetric and asymmetric services and therefore offers a high-bandwidth access solution for a variety of applications. The most common DSL services are ADSL, ADSL2+, HDSL, HDSL-2, RADSL, and VDSL. The special case of G.lite, a form of ADSL, will also be discussed.

Asymmetric Digital Subscriber Line (ADSL)

When the World Wide Web and flat-rate access charges arrived, the typical consumer phone call went from roughly four minutes in duration to several times that. All the engineering that led to the overall design of the network based on an average four-minute hold time went out the window as the switches staggered under the added load. Never was the expression, "In its success lies the seeds of its own destruction," more true. When ADSL arrived, it provided the offload required to save the network.

The typical ADSL installation is shown in Figure 6-4. No change is required to the two-wire local loop; however, minor equipment changes are required. First, the customer must have an ADSL modem at their premises. This device allows both the telephone service and a data access device, such as a PC, to be connected to the line.

The ADSL modem is more than a simple modem, in that it also provides the frequency-division multiplexing process required to separate the voice and data traffic for transport across the loop. The device that actually does this is called a splitter, in that it splits the voice traffic away from the data. It is usually bundled as part of the ADSL modem although it can also be installed as a card in the PC, as a stand-alone device at the demarcation point, or on each phone at the premises. The most *common* implementation is to integrate the splitter as part of the DSL modem; this, however, is the least *desirable* implementation because this design can lead to crosstalk between the voice and data circuitry inside the

Figure 6-4
Layout of typical Digital Subscriber Line (DSL)

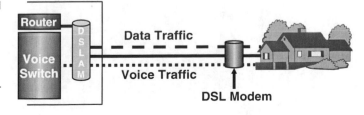

device. When voice traffic reaches the ADSL modem, it is immediately encoded in the traditional voice band and handed off to the local switch when it arrives at the central office. The modem is often referred to as an ADSL transmission unit for remote use, or ATU-R. Similarly, the device in the central office often called an ATU-C (for "central office").

When a PC wishes to transmit data across the local loop, the traffic is encoded in the higher frequency band reserved for data traffic. The ADSL modem knows to do this because the traffic is arriving on a port reserved for data devices. Upon arrival at the CO, the data traffic does not travel to the local switch; instead, it stops at the ADSL modem that has been installed at the CO end of the circuit. In this case, the device is actually a bank of modems that serves a large number of subscribers, and is known as a Digital Subscriber Line Access Multiplexer, or DSLAM (pronounced "dee-slam"). A rack of DSLAMs is shown in Figure 6-5.

Instead of traveling on to the local switch, the data traffic is now passed around the switch to a router, which in turn is connected to the Internet. This process is known as a *line-side redirect*.

The advantages of this architecture are fairly obvious. First, the redirect offloads the data traffic from the local switch so that it can go back to doing what it does best—switching voice traffic. Second, it creates a new line of business for the service provider. As a result of adding the router and connecting the router to the Internet, the service provider instantly becomes an ISP. This is a near-ideal combination, because it allows the service provider to become a *true service provider* by offering much more than simple access and transport.

Figure 6-5
A bay of
DSLAMs

As the name implies, ADSL provides two-wire asymmetric service—that is, the upstream bandwidth is different from the downstream. In the upstream direction, data rates vary from 16 to 640 Kbps, while the downstream bandwidth varies from 1.5 to 8 Mbps. Because most applications today are asymmetric in nature, this disparity poses no problem for the average consumer of the service.

A Word about the DSLAM

This device has received a significant amount of attention recently because of the central role that it plays in the deployment of broadband access services. Obviously, the DSLAM must interface with the local switch so that it can pass voice calls on to the PSTN. However, it often interfaces with a number of other devices as well. For example, on the customer side, the DSLAM may connect to a standard ATU-C, directly to a PC with a built-in NIC, to a variety of DSL services, or to an integrated access device of some kind. On the trunk side (facing the switch), the DSLAM may connect to IP routers as described before, to an ATM switch, or to some other broadband service provider. It therefore becomes the focal point for the provisioning of a wide variety of access methods and service types.

ADSL2

ADSL2 is a newer standard that has superseded preexisting ADSL standards. It is interoperable with preexisting ADSL deployments and is critical for service providers and customer alike; it means that they can continue to use their installed base of equipment with minimal disruption.

ADSL2+

ADSL2+, on the other hand, is an extension of ADSL2. It is capable of doubling the transmission speed of typical ADSL installations to over 20 Mbps downstream on loops shorter than 8,000 feet.

One major proponent of so-called Next-Generation DSL is BellSouth, which has the largest installed base of fiber among all the ILECs. The company believes that the best path to better residential broadband delivery is through faster DSL. Its strategy, which is a good one, is focused on scaling up existing DSL facilities with ADSL2+ and using DSL bonding techniques to scale DSL bandwidth to as high as 24 Mbps —high enough to support HDTV-quality video. Its plan is to implement ADSL2+, which will boost its available DSL bandwidth to 12 Mbps. Early next year the carrier also will test DSL bonding, which allows multiple DSL connections to be aggregated into a single virtual 24 Mbps channel.

As far as applications are concerned, Bellsouth plans to test VoIP and will include a trial of Microsoft's IPTV platform.

High Bit Rate Digital Subscriber Line (HDSL)

The greatest promise of HDSL is that it provides a mechanism for the deployment of four-wire T1 and E1 circuits without the need for span repeaters, which can add significantly to the cost of deploying data services. It also means that service can be deployed in a matter of days rather than weeks, something customers certainly applaud.

DSL technologies in general allow for repeaterless facilities as far as 12,000 feet, while traditional four-wire data circuits—such as T1 and E1 —require repeaters every 6,000 feet. Consequently, many telephone companies are now using HDSL "behind the scenes" as a way to deploy these traditional services. Customers do not realize that the T1 facility they are plugging their equipment into is being delivered using HDSL technology. The important thing is that they don't *need* to know. All the customer should have to care about is that there is now a SmartJack installed in the basement, and through that jack they have access to 1.544 Mbps or 2.048 Mbps of bandwidth—period.

HDSL2

HDSL2 offers the same service that HDSL offers, with one added (and significant) advantage: It does so over a single pair of wire, rather than two. It also provides other advantages. First, it was designed to improve vendor interoperability by requiring less equipment at either end of the span (transceivers, repeaters). Second, it was designed to work within the confines of standard telephone company Carrier Serving Area (CSA) guidelines by offering a 12,000-foot, wire-run capability that matches the requirements of CSA deployment strategies. (See the discussion on CSA guidelines later in this section.)

A number of companies have deployed T1 access over HDSL2 at rates 40 percent lower than typical T-Carrier prices. Furthermore, a number of vendors including 3Com, Lucent, Nortel Networks, and Alcatel have announced their intent to work together to achieve interoperability among DSL modems.

Rate-Adaptive Digital Subscriber Line (RADSL)

Rate-Adaptive Digital Subscriber Line, or RADSL (pronounced "Radzel"), is a variation of ADSL designed to accommodate changing line conditions that can affect the overall performance of the circuit. Like ADSL, it relies on DMT encoding, which selectively "populates" subcarriers with transported data, thus allowing for granular rate-setting.

Very High-Speed Digital Subscriber Line (VDSL)

Very High-Speed Digital Subscriber Line (VDSL) is the newest DSL entrant in the bandwidth game, and shows promise as a provider of extremely high levels of access bandwidth—as much as 52 Mbps over a short local loop. VDSL requires *fiber-to-the-curb* (FTTC) architecture and recommends ATM as a switching protocol; from a fiber hub, copper tail circuits deliver the signal to the business or residential premises. Bandwidth available through VDSL ranges from 1.5 to 6 Mbps on the upstream side, and from 13 to 52 Mbps on the downstream side. Obviously, the service is distance sensitive, and actual achievable bandwidth drops as a function of distance. Nevertheless, even a short loop is respectable when such high-bandwidth levels can be achieved. With VDSL, 52 Mbps can be reached over a loop length of up to 1,000 feet—a not unreasonable distance by any means.

G.lite

Because the installation of splitters has proven to be a contentious and problematic issue, the need arose for a version of ADSL that did not require them. That version is known as either "ADSL Lite" or "G.lite" (after the ITU-T G-Series standards that govern much of the ADSL technology). In 1997, Microsoft, Compaq, and Intel created the Universal ADSL Working Group (UAWG),[1] an organization that grew to nearly 50 members dedicated to the task of simplifying the rollout of ADSL. In effect, the organization had four stated goals:

- To ensure that analog telephone service will work over the G.lite deployment without remote splitters, in spite of the fact that the quality of the voice may suffer slightly due to the potential for impedance mismatch.

- To maximize the length of deployed local loops by limiting the maximum bandwidth provided—Research indicates that customers are far more likely to notice a performance improvement when migrating from 64 Kbps to 1.5 Mbps than when going from 1.5 Mbps to higher speeds. Perception is clearly important in the marketplace, so the UADSL Working Group chose 1.5 Mbps as their downstream speed.

- To simplify the installation and use of ADSL technology by making the process as plug-and-play as possible.

- To reduce the cost of the service to a perceived reasonable level.

Of course, G.lite is not without its detractors. A number of vendors have pointed out that if G.lite requires the installation of microfilters at the premises on a regular basis, then true splitterless DSL is a myth, since microfilters are, in effect, a form of splitter. They contend that if the filters are required anyway, then they might as well be used in full-service ADSL deployments to guarantee high-quality service delivery. Unfortunately, this flies in the face of one of the key tenets of G.lite, which is to simplify and reduce the cost of DSL deployment by eliminating the need for an installation dispatch (a "truck roll" in the industry's parlance). The key to G.lite's success in the eyes of the implementers was to eliminate the dispatch, minimize the impact on traditional POTS telephones, reduce costs, and extend the achievable drop length. Unfortu-

---

[1]The group "self-dissolved" in the summer of 1999 after completing what they believed their charter to be.

nately, customers still have to be burdened with the installation of micro-filters, and coupled noise on POTS is higher than expected. Many vendors argue that these problems largely disappear with full-feature ADSL using splitters; a truck dispatch is still required, but again, it is often required to install the microfilters anyway, so there is no net loss. Furthermore, a number of major semiconductor manufacturers support both G.lite and ADSL on the same chipset, so the decision to migrate from one to the other is a simple one that does not necessarily involve a major replacement of internal electronics.

DSL Market Issues

DSL technology offers advantages to both the service provider and the customer. The service provider benefits from successful DSL deployment because it serves not only as a cost-effective technique for satisfying the bandwidth demands of customers in a timely fashion, but also because it provides a Trojan horse approach to the delivery of certain preexisting services. As we noted earlier, many providers today implement T1 and E1 services over HDSL because it offers a cost-effective way to do so. Customers are blissfully unaware of the fact; in this case, it is the service provider rather than the customer who benefits most from the deployment of the technology. From a customer point-of-view, DSL provides a cost-effective way to buy medium-to-high levels of bandwidth and, in some cases, embedded access to content.

Provider Challenges

The greatest challenges facing those companies looking to deploy DSL are competition from cable and wireless, pent-up customer demand, installation issues, and plant quality.

Competition from cable and wireless companies represents a threat to wireline service providers on several fronts. First, cable modems enjoy a significant amount of press and, therefore, are gaining well-deserved market share; in fact, more than 65 percent of the broadband installed base today is delivered via cable modems, *not* via DSL. The service they provide, for the most part, is well received and offers high-quality, high-speed access. Wireless, on the other hand, is a slumbering beast. It has largely been ignored as a serious competitor for data transport, but the wild success of Wi-Fi, EV-DO, and CDMA-1x solutions have pushed wireless into the limelight.

The second challenge is unmet customer demand. If DSL is to satisfy the broadband access requirements of the marketplace, it must be made available throughout ILEC service areas. This means that incumbent providers must equip their central offices with DSLAMs that will provide the line-side redirect required at the initial stage of DSL deployment. The law of primacy is evident here: The ILECs must get to market first with broadband offerings if they are to achieve and keep a place in the burgeoning broadband access marketplace.

The third and fourth challenges to rapid and ubiquitous DSL deployment are installation issues and plant quality. A significant number of impairments have proven to be rather vexing for would-be deployers of widespread DSL. These challenges fall into two categories: electrical disturbances and physical impairments. While solutions have been crafted for most of these, they still pop up occasionally as vexing problems.

Electrical Disturbances

The primary cause of electrical disturbance in DSL is crosstalk, caused when the electrical energy carried on one pair of wires "bleeds" over to another pair and causes interference (noise) there. Crosstalk exists in several flavors. *Near-end crosstalk* (NEXT) occurs when the transmitter at one end of the link interferes with the signal received by the receiver at the same end of the link, while *far-end crosstalk* (FEXT) occurs when the transmitter at one end of the circuit causes problems for the signal received by a receiver at the far end of the circuit. Similarly, problems can occur when multiple DSL services of the same type exist in the same cable and interfere with one another. This is referred to as self-NEXT or self-FEXT. When different flavors of DSL interfere with one another, the phenomenon is called foreign-NEXT or foreign-FEXT. Other problems that can cause errors in DSL, and therefore a limitation in the maximum achievable bandwidth of the system, are simple *radio frequency interference* (RFI) that can find its way into the system; and impulse, Gaussian, and random noise that exist in the background but can affect signal quality even at extremely low levels.

Physical Impairments

The physical impairments that can have an impact on the performance of a newly deployed DSL circuit tend to be characteristics of the voice network that typically have minimal effect on simple voice and low-speed data services. These include load coils, bridged taps, splices, mixed gauge loops, and weather conditions.

Load Coils and Bridged Taps

We have already mentioned the problems that load coils can cause for digital transmission; because they limit the frequency range that is allowable across a local loop they can seriously impair transmission. Bridged taps are equally problematic. When a multipair telephone cable is installed in a newly built-up area, it is generally some time before the assignment of each pair to a home or business is actually made. To simplify the process of installation when the time comes, the cable's wire pairs are periodically terminated in terminal boxes (sometimes called B-Boxes) installed every block or so. As we discussed in Chapter 3, there may be multiple appearances of each pair on a city street, waiting for assignment to a customer. When the time comes to assign a particular pair, the installation technician will simply go to the terminal box where that customer's drop appears and cross-connect the loop to the appearance of the cable pair (a set of lugs) to which that customer has been assigned. This eliminates the need for the time-consuming process of splicing the customer directly into the actual cable.

Unfortunately, this efficient process also creates a problem. While the customer's service has been installed in record time, there are now unterminated appearances of the customer's cable pair in each of the terminal boxes along the street. While these so-called bridged taps present no problem for analog voice, they can be a catastrophic source of noise due to signal reflections that occur at the copper-air interface of each bridged tap. If DSL is to be deployed over the loop, the bridged taps must be removed. And while the specifications indicate that bridged taps do not cause problems for DSL, actual deployment says otherwise. To achieve the bandwidth that DSL promises, the taps must be terminated.

Splices and Gauge Changes

Splices can result in service impairments due to cold solder joints, corrosion, or weakening effects caused by repeated bending of wind-driven aerial cable.

Gauge changes tend to have the same effects that plague circuits with unterminated bridged taps: When a signal traveling down a wire of one gauge jumps to a piece of wire of another gauge, the signal is reflected, resulting in an impairment known as intersymbol interference. The use of multiple gauges is common in loop deployment strategy because it allows outside plant engineers to use lower cost small-gauge wire where appropriate, cross-connecting it to larger gauge, lower resistance wire where necessary.

Weather

"Weather" is perhaps a bit of a misnomer: The real issue is moisture. One of the greatest challenges facing DSL deployment is the age of the out-side plant. Much of the older distribution cable uses inadequate insulation between the conductors (paper in some cases). In some cases the outer sheath has cracked, allowing moisture to seep into the cable itself. The moisture causes crosstalk between wire pairs that can last until the water evaporates. Unfortunately, this can take a considerable amount of time and result in extended outages.

Solutions

All of these factors have solutions. The real question is whether they can be remedied at a reasonable cost. Given the growing demand for broad-band access, there seems to be little doubt that the elimination of these factors would be worthwhile at all reasonable costs, particularly consid-ering how competitive the market for the local loop customer has become.

The electrical effects, largely caused by various forms of crosstalk, can be reduced or eliminated in a variety of ways. Physical cable deployment standards are already in place that, when followed, help to control the amount of near- and far-end crosstalk that can occur within binder groups in a given cable. Furthermore, filters have been designed that eliminate the background noise that can creep into a DSL circuit.

Physical impairments can be controlled to a point, although to a cer-tain extend the service provider is at the mercy of their installed network plant. Obviously, older cable will present more physical impairments than newer cable, but there are steps that the service provider can take to maximize their success rate when entering the DSL marketplace. The first step that they can take is to prequalify local loops for DSL service to the extent possible. This means running a series of tests using mechanized loop testing (MLT) to determine whether each loop's transmission performance falls within the bounds established by the service provider and existing standards and industry support organizations.

For DSL prequalification, MLT tests the following performance indicators (and others as required):

- Cable architecture
- Loop length
- Crosstalk and background noise

56K modems, ISDN, and DSL represent the primary access technologies that the telephone companies offer. Recently, however, a new contender has entered the game: the cable provider. The section that follows describes the technology they offer and the role that they play in the evolving access marketplace.

Cable-Based Access Technologies

In 1950, Ed Parsons placed an antenna on a hillside above his home in Washington State, attached it to a coaxial cable distribution network, and began to offer television service to his friends and neighbors. Prior to his efforts, the residents of his town were unable to pick up broadcast channels because of the blocking effects of the surrounding mountains. Thanks to Parsons, community antenna television (CATV) was born; from its roots came cable television.

Since that time the cable industry has grown into a massive industry. In the United States alone 15,000 headends have the ability to deliver content to more than 150 million homes in more than 50,000 communities over more than 1 million miles of coaxial and fiber-optic cable. As the industry's network has grown, so too have the aspirations of those deploying it. Their goal is to make it much more than a one-way medium for the delivery of television and pay-per-view; they want to provide a broad spectrum of interactive, two-way services that will allow them to compete head-to-head with the telephony industry. To a large degree, they are succeeding. The challenges they face, however, are daunting.

According to the Leichtman Research Group, the 20 largest cable and DSL providers in the United States, which represent approximately 95 percent of the market, enjoyed record net subscriber additions in the third quarter of 2004. Combined net adds for the quarter totaled over 2.3 million subscribers—a not insignificant figure. The largest broadband providers now account for over 30.9 million high-speed Internet subscribers, with cable delivering service to 18.8 million broadband subs and DSL providing service to just over 12 million.

The top cable broadband providers retain a 6.6 million subscriber advantage over DSL, with a 61 percent share of the total market. In 2004, from the end of third quarter 2003 to the end of third quarter 2004, cable and DSL added 8.3 million net subscribers.

Playing in the Broadband Game

Unlike the telephone industry that began its colorful life under the scrutiny of a small number of like-minded individuals (Alexander Graham Bell and Theodore Vail, among others), the cable industry came about thanks to the combined efforts of hundreds of innovators, each building on Parson's original concept. As a consequence, the industry, while enormous, is in many ways fragmented. Powerful industry leaders like John Malone and Gerald Levine were able to exert "Tito-like" powers to unite the many companies, turning a loosely cobbled-together collection of players into cohesive, powerful corporations with a shared vision of what they were capable of accomplishing.

Today, the cable industry is a force to be reckoned with and upgrades to the original network are underway. This is a crucial activity that will ensure the success of the industry's ambitious business plan and provide a competitive balance for the traditional telcos.

The Cable Network

The traditional cable network is an analog system based on a tree-like architecture. The headend, which serves as the signal origination point, serves as the signal aggregation facility. It collects programming infor-

Figure 6-6
Satellite receive
antennas at
cable headend
facility

Figure 6-7
Layout of
typical cable
distribution
network

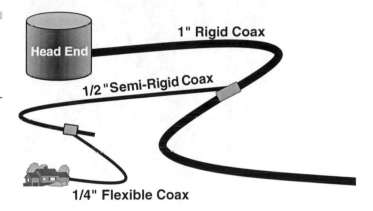

mation from a variety of sources including satellite and terrestrial feeds.
Headend facilities often look like a mushroom farm; they are typically
surrounded by a variety of satellite dishes (see Figure 6-6).

The headend is connected to the downstream distribution network by
one-inch diameter, rigid coaxial cable, as shown in Figure 6-7. That cable
delivers the signal—usually a 450 MHz collection of 6 MHz channels—to
a neighborhood, where splitters divide the signal and send it down half-
inch diameter, semi-rigid coax that typically runs down a residential
street. At each house, another splitter (see Figure 6-8) pulls off the signal
and feeds it to the set-top box in the house over the drop wire, a local loop
of flexible quarter-inch coaxial cable.

Figure 6-8
Signal splitter
in residential
cable
installation

While this architecture is perfectly adequate for the delivery of one-way television signals, its shortcomings for other services should be fairly obvious to the reader. First of all, it is, by design, a broadcast system. It does not typically have the ability to support upstream traffic (from the customer toward the headend) and, therefore, is not suited for interactive applications. Second, because of its design, the network is prone to significant failures that have the potential to affect large numbers of customers. The tree structure, for example, means that if a failure occurs along any "branch" in the tree, every customer from that point downward loses service. Contrast this with the telephone network where customers have a dedicated local loop over which their service is delivered. Second, because the system is analog, it relies on amplifiers to keep the signal strong, as it is propagated downstream. These amplifiers are powered locally—they do not have access to central office power as devices in the telephone network do. Consequently, a local power failure can bring down the network's ability to distribute service in that area.

The third issue is one of customer perception. For any number of reasons, there is a general perception that the cable network is not as capable or as reliable as the telephone network. As a consequence of this perception, the cable industry is faced with the daunting challenge of convincing potential voice and data customers that they are in fact capable of delivering high-quality service. Some of the concerns are justified. In the first place, the telephone network has been in existence for almost 125 years, during which time its operators have learned how to optimally design, manage, and operate it in order to provide the best possible service. The cable industry, on the other hand, came about 50 years ago and didn't benefit from the rigorously administered, centralized management philosophy that characterized the telephone industry. Additionally, the typical 450 MHz cable system did not have adequate bandwidth to support the bi-directional transport requirements of new services.

Furthermore, the architecture of the legacy cable network, with its distributed power delivery and tree-like distribution design, does not lend itself to the same high degree of redundancy and survivability that the telephone network offers. Consequently, cable providers have been hard-pressed to convert customers who are vigorously protective of their telecommunications services.

The Ever-Changing Cable Network

Faced with these harsh realities and the realization that the existing cable plant could not compete with the telephone network in its original analog incarnation, cable engineers began a major rework of the network in the early 1990s. Beginning with the headend and working their way outward, they progressively redesigned the network to the extent that, in many areas of the country, their coaxial "local loop" is capable of competing on equal footing with the telco's twisted pair—and in some cases beating it.

The process they have used in their evolution consists of four phases. In the first phase, they converted the headend from analog to digital.

This allowed them to digitally compress the content, resulting in far more efficient utilization of available bandwidth. Second, they undertook an ambitious physical upgrade of the coaxial plant, replacing the one-inch trunk and half-inch distribution cable with optical fiber. This brought about several desirable results. First, by using a fiber feeder, network designers were able to eliminate a significant number of the amplifiers responsible for the failures the network experienced due to power problems in the field. Second, the fiber makes it possible to provision significantly more bandwidth than coaxial systems allow. Third, because the system is digital, it suffers less from noise-related errors than its analog predecessor did. Finally, an upstream return channel was provisioned (see Figure 6-9), which makes possible the delivery of true interactive services such as voice, Web surfing, and videoconferencing.

The third phase of the conversion had to do with the equipment provisioned at the user's premises. The analog set-top box has now been replaced with a digital device that has the ability to take advantage of the capabilities of the network, including access to the upstream channel. It decompresses digital content, performs content ("stuff") separation, and provides the network interface point for data and voice devices.

The final phase is business conversion. Cable providers look forward to the day when their networks will compete on equal footing with the twisted-pair networks of the telephone company and customers will see them as viable competitors. In order for this to happen, they must demonstrate that their network is capable of delivering a broad variety of competitive services, that the network is robust, that they have oper-

Figure 6-9
Upstream and downstream channels in modern cable systems

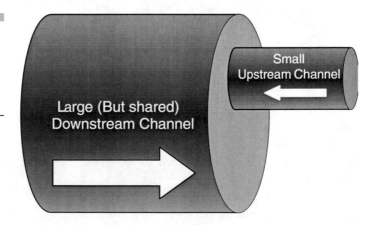

ations support systems (OSSs) that will guarantee the robustness of the infrastructure, and that they are cost competitive with incumbent providers. They must also create a presence for themselves in the business centers of the world. Today they are almost exclusively residential service providers; if they are to break into the potentially lucrative business market, they must have a presence there.

Cable Modems

As cable providers have progressively upgraded their networks to include more fiber in the backbone, their plan to offer two-way access to the Internet has become a reality. Cable modems offer access speeds of up to 10 Mbps, and so far the market uptake has been spectacular.

Cable modems provide an affordable option to achieve high-speed access to the Web, with current monthly subscription rates in the neighborhood of $40. They offer asymmetric access, that is, a much higher downstream speed than upstream, but for the majority of users this does not represent a problem, since the bulk of their use will be for Web surfing—during which the bulk of the traffic travels in the downstream direction anyway.

While cable modems do speed up access to the Web and other online services by several orders of magnitude, there are a number of downsides that must be considered. The greatest concern that has been voiced about cable modems is security. Because cable modems are "always on," they represent an easy entry point for hackers looking to break into machines. It is, therefore, *critical* that cable subscribers use some form of firewall software or a router that has the ability to perform filtering.

Data over Cable Standards

As interest grew in the late 1990s for broadband access to data services over cable television networks, CableLabs®, working closely with the ITU and major hardware vendors, crafted a standard known as the Data Over Cable Service Interface Specification, or DOCSIS®. The standard is designed to ensure interoperability among cable modems as well as to assuage concerns about data security over shared cable systems, DOCSIS has done a great deal to resolve marketplace issues.

Under the standards, CableLabs crafted an original cable modem certification standard called DOCSIS 1.0 which guarantees that modems carrying the certification will interoperate with any headend equipment, are ready to be sold in the retail market, and will interoperate with other certified cable modems. Engineers from Askey, Broadcom, Cisco Systems, Ericsson, General Instrument, Motorola, Philips, 3Com, Panasonic, Digital Furnace, Thomson, Terayon, Toshiba, and Com21 participated in the development effort.

The DOCSIS 1.1 specification was released in April 1999, and included two additional functional descriptions, which began to be implemented in 2000. The first specification details procedures for guaranteed bandwidth, as well as a specification for quality of service guarantees. The second specification is called Baseline Privacy Interface Plus (BPI+); it enhances the current security capability of the DOCSIS standards through the addition of digital certificate-based authentication and support for multicast services to customers.

DOCSIS™ 2.0

In December 2002 the ITU-T announced the approval of a standard defining the second-generation data over cable system, known as DOCSIS 2.0.

ITU recommendation J.122 gives cable operators the ability to offer speeds up to 600 times faster than a standard dial-up telephone modem provides. The new standard can be used as the foundation upon which IP-based telephony services can be offered.

The enhancements that J.122 provides over its predecessor are primarily focused on the upstream transmission path from the customer to the network. Changes include increased capacity and improved robustness over the upstream path.

While the DOCSIS name is in widespread use, CableLabs now refers to the overall effort as the CableLabs Certified Cable Modem Project.

Wireless Access Technologies

To understand wireless communications it is necessary to examine both radio and telephone technologies, because the two are inextricably intertwined. In 1876, Alexander Graham Bell, a part-time inventor and a teacher of hearing-impaired students, invented the telephone while attempting to resolve the challenge of transmitting multiple telegraph signals over a shared pair of wires. His invention changed the world forever.

In 1896, a mere 20 years later, Italian engineer and inventor Guglielmo Marconi developed the spark gap radio transmitter, which eventually allowed him to transmit long-wave radio signals across the Atlantic Ocean as early as 1901. Like Bell, his invention changed the world; for his contributions, he was awarded the Nobel Prize in 1909.

It wasn't until the 1920s, though, when these two technologies began to dovetail, that their true promise was realized. Telephony provided interpersonal, two-way, high-quality voice communications, but required the user to be stationary. Radio, on the other hand, provided mobile communications, but was limited by distance, environmentally induced signal degradation, and spectrum availability. Whereas telephony was advertised as a universally available service, radio was more of a catch-as-catch-can offering that was subject to severe blocking. If a system could be developed that combined the signal quality and ubiquity of telephony with the mobility of radio, however, a truly promising new service offering could be made available.

Today, cellular telephony (and other services like it) provides high-quality, *almost* ubiquitous, wireless telephone service. Thanks to advances in digital technology, wireless telephony also offers services identical to those provided by the wired network. And pricing for wired and wireless services are now reaching parity: Flat-rate nationwide pricing models are commonplace today that have no roaming or long-distance restrictions.

A Bit of History

The road to the modern wireless communications network was fraught with technical, bureaucratic, and social speed bumps. In 1905, listeners in New York City, accustomed to the staccato beat of Morse Code broadcasts spewing from their radios, were astounded to hear music and voice as Reginald Fessenden, a local radio commentator, broadcast the results of the annual New York City yacht races. Listeners could hear the results immediately, instead of waiting for the late edition of the paper. Public perception of the possibilities of radio suddenly began to grow.

At roughly the same time, the United States military, recognizing the potentially strategic value of radio communications in the armed forces, helped to create the Radio Corporation of America (RCA), ushering wireless technology into the technological forefront.

The first wireless applications were developed in 1915 and 1916 for ship-to-ship and ship-to-shore communications. In 1921, the Detroit Police Department inaugurated the first land-based mobile application for radio communications, when they installed one-way radios in patrol cars—the original "calling all cars" application. The system required police officers to stop at a phone box to place a call to their precinct when a call came in, but the ability to quickly and easily communicate with roving officers represented a giant step forward for the police and their ability to respond quickly to calls.

Unfortunately, the system was not without its problems. Even though the transmitter was powerful, Detroit's deep concrete canyons prevented the signal from reaching many areas. Equally vexing was the problem of motion: The delicate vacuum tubes, subjected to the endless vibrations stemming from life in the trunk of a moving vehicle, caused both physical and electronic problems that were so bad that in 1927 the Detroit police department shut down the station out of sheer frustration. It wasn't until 1928, when Purdue University student Robert Batts developed a super heterodyne radio that was somewhat resistant to buffeting and vibration, that Detroit reinaugurated their radio system.

Other cities followed in Detroit's footsteps. By 1934, there were nearly 200 radio-equipped city police departments communicating with nearly 5,000 roving vehicles. Radio had become commonplace, and in 1937 the FCC stepped in and allocated 19 additional radio channels to the 11 that they had already made available for police department use. No one realized it, but the spectrum battles had begun.

Early radio systems were based on *amplitude modulation* (AM) technology. One drawback of AM radio is that the strength of the transmitted signal diminishes according to the square of the distance. A signal measured at 20 watts, one mile from the transmitter will be *one quarter* that strength, or 5 watts, *two* miles from the transmitter. When Edwin Armstrong (see Chapter 1) announced and demonstrated his invention of *frequency-modulated* (FM) radio in 1935, he turned the broadcast industry on its ear. He offered a technology that not only eliminated the random noise and static that plagued AM radio, but also lowered transmitter power requirements and improved the ability of receivers to lock onto a weak signal.

World War II provided an ideal testing ground for FM. AM radio, used universally by the Axis powers during the war, was easily jammed by powerful Allied transmitters. FM, on the other hand, used only by Allied forces, was unaffected by AM jamming techniques. As we mentioned earlier, many historians believe that FM radio was crucially important to the Allies' success in World War II.

Radio's Evolution

Radio is often viewed as having evolved in two principal phases. The first was the pioneer phase, during which the fundamental technologies were developed, initial technical bugs were ironed out, applications were created (or at least conceived), and bureaucratic and governmental haranguing over the roles of various agencies initiated. This phase was characterized by the development and acceptance of FM transmission, the dominance of military and police usage of radio, and the challenge of building a radio that could withstand the rigors of mobile life. The pioneer phase lasted until the early 1940s, when the commercial phase began.

During the 1940s, radio technology advanced to the point that FM-based car phones were commercially available. Initially, AT&T provided mobile telephone service under the umbrella of its nationwide monopoly. In 1949, however, the Justice Department filed suit against AT&T, settling the case with the Consent Decree of 1956. Under this Consent Decree, AT&T kept its telephone monopoly, but was forced to give up a number of other lines of business, including the manufacture of mobile radiotelephones. This allowed companies like Motorola to enter the fray and paved the way for the creation of radio common carriers, or RCCs, much to AT&T's chagrin. These RCCs were typically small businesses that offered wireless service to several hundred people in a fixed geographical area.

A Touch of Technology

Before proceeding further, let's introduce some fundamentals of radio transmission.

For radio waves to carry information such as voice, data, music, or images, certain characteristics of the waves are changed or *modulated*.

Those characteristics include the amplitude, or loudness, of the wave; the frequency, or pitch, of the wave; and the phase, or angle of inflection, of the wave. By modulating these characteristics, a *carrier wave* can be made to transport information, whether it is a radio broadcast, a data stream, or a telephone conversation. (Please review this topic in Chapter 1 for more information.)

Transmission "space" is allocated based on ranges of available frequency. This range is known as a spectrum, a word that means "a range of values." The radio spectrum represents a broad span of electromagnetic frequencies between one kilohertz (1,000 cycles per second) and ten *quintillion* (that's ten followed by 19 zeroes) cycles per second. Most radio-based applications operate at frequencies between 1 kHz and 300 GHz.

One concern that has surfaced repeatedly over the years since the inception of radio is spectrum availability. Different services require differing amounts of spectrum for proper operation. For example, a modern FM radio channel requires 30 kHz of bandwidth, while an AM channel only requires 3 kHz. Television, on the other hand, requires 6 MHz. As you might imagine, battles between the providers of these disparate services over spectrum allocation became quite heated in the years that followed.

Improving the Model

The first FM radiotelephone systems were massively inefficient. They required 120 kHz of channel bandwidth to transport a 3 kHz voice signal. As the technology advanced, though, this requirement was reduced. In the 1950s the FCC mandated that channels be halved to 60 kHz, and by the 1960s things had advanced to the point that the spectrum could be further reduced to 30 kHz, where it remains today in analog systems. Realize that this is *still* a 10:1 ratio.

AT&T introduced the first commercial mobile telephone service in St. Louis in 1946. It relied on a single, high-power transmitter placed atop one of the city's tallest buildings and had an effective service range of about 50 miles. This FM system was fully interconnected with AT&T's wireline telephone network.

The service was not cheap. There was a basic $15 monthly charge, and usage was billed at a flat $0.15 per minute. In spite of the cost, the service was quite popular; a fact that led to its undoing.

System engineers based their design on existing radio systems, which were primarily designed for dispatch applications. The traffic patterns of radio dispatches are different than telephony: A dispatch occupies several seconds of airtime, while a telephone call typically lasts several minutes. Almost immediately, blocking became a serious problem. In fact, in New York City, where the service quickly became popular, AT&T had 543 active subscribers in 1976 and a waiting list of nearly 4,000 anxious-to-be-connected customers that the system's limited capacity could not accommodate.

Radio design played a role in improving spectrum utilization. The first systems—such as the one in St. Louis—used a scheme called *nontrunked radio*. In nontrunked radio systems, the available spectrum was divided into channels and groups of users were assigned to those channels. The advantage of this technique is that nontrunked radios were relatively inexpensive. The downside was that certain channels could become severely overloaded, while others remained virtually unused.

The invention of *trunked radio* relieved this problem immensely. Trunked radios were *frequency agile,* meaning that all radios could access all channels. When a user placed a call, the radio unit would search for an available channel and use it. With the arrival of this capability, blocking became a nonissue. The downside, of course, was that frequency-agile radios, because of their more complex circuitry, were significantly more expensive than their nontrunked predecessors.

Most of this work was conducted during the turbulent 1960s and led to Bell Labs' introduction of Improved Mobile Telephone Service (IMTS). IMTS used two 30 kHz, narrowband FM channels for each conversation, and provided full-duplex talk paths, direct dialing, and automatic trunking. Introduced commercially in 1965, IMTS is considered to be the predecessor of modern cellular telephony.

The Spectrum Battles Heat Up

Starting in the mid-1940s, mobile telephony went head-to-head against the television industry in a pitched battle for spectrum. In 1947, the Bell System proposed to the FCC a plan for a large-scale mobile telephone system, asking them to allocate 150 two-way, 100 kHz channels. The proposal was not accepted.

In 1949, while the FCC wrestled with the assignment of spectrum in the 470 to 890 MHz range that would become UHF television, the telephone industry argued that they should be granted a piece of the electromagnetic pie. Their issue was that the 6 MHz of bandwidth required for a *single* TV channel was more than had *ever* been allocated for mobile telephony. Unfortunately, television had penetrated the American household—Captain Kangaroo, Roy Rogers, and Winky-Dink captivated viewers, and consumers were hungry for additional programming. Wireless telephony wasn't even on their radar screens; it would have to wait.

In the 20 years that followed, mobile telephony continued to take a backseat to television. UHF usage grew slowly; in 1957 the Bell System petitioned the FCC to give them a piece of the earmarked spectrum in the 800 MHz range that was virtually unused. But television was still the darling of the nation, and the FCC was ferociously dedicated to expanding the deployment of UHF television. In 1962, the government signed into law the All Channels Receiver Act that mandated that all new televisions must have both VHF and UHF receivers. Remember when televisions had two tuner knobs on them? TV won again.

Between 1967 and 1968, the FCC and the House of Representatives, under pressure from the telephone industry, studied the issue of spectrum allocation once again. In 1968, the FCC convened the "Cellular Docket," a contentious and highly visible collection of lawmakers who eventually ruled that mobile telephony's concerns could only be solved by giving them spectrum. In 1970, the FCC reallocated UHF channels 70 to 83 from television to mobile services. From the resulting 115 MHz electromagnetic chunk, 75 MHz was allocated to mobile telephony, with 40 MHz available immediately and 35 MHz held in reserve.

Once spectrum was allocated, the political games began. It was roundly assumed that AT&T would design, install, and operate the wireless network as an extension of its own universal wireline network. After all, the spectrum allocation was "deeded" to the wireline telephone companies; since AT&T provided service to roughly 85 percent of the market, this assumption was somewhat natural. Their Advanced Mobile Phone System (AMPS) architecture—based on the same design philosophy as the wireline network—relied on a massively expensive switching infrastructure, hundreds of cells, and centralized control of all functions.

AT&T's only competitors in this market were the RCCs, mentioned earlier. Among them was Motorola, even then a large player in the industry. Initially, Motorola sided with AT&T, but when AT&T chose other vendors to manufacture the equipment required to establish the network, Motorola changed its spots and sided with the other 500-some-odd RCCs in the country to sue the FCC for unfair treatment. Their suit called for the Commission to deny AT&T's application for development of AMPS and to reexamine the spectrum allocation. In 1970, Motorola, in partnership with another large RCC, applied for permission to offer wireless service in Washington, D.C. After examining the petition, the courts decided that the answer to the industry's woes lay in a competitive market. In 1980 they began another cellular rulemaking effort to determine the regulatory structure of the market they were attempting to create.

Three options emerged from their discussions. The first was preservation of the monopolistic, single-operator concept; the second advocated an open market, in which competition was opened to all comers and the market would sort itself out; and the third involved a duopoly approach, in which two systems would be allowed in each major market area. After long debate, regulators and lawmakers decided on the duopoly concept, with two, 20 MHz systems ("A" frequencies and "B" frequencies) allocated in each market. The "A" frequencies were allocated to the nonwireline company, while the "B" frequencies went to the wireline carrier.

The first commercial cellular telephone system became operational in October of 1983. By 1985 there were 340,000 cellular subscribers; today there are over 16 million, with annual revenues of nearly $11 billion. It's interesting to note that in 1982, AT&T proudly predicted that there would be over 100,000 cellular subscribers by 1990.

Cellular Telephony

As modern as cellular telephony is considered to be, it was originally conceived in the 1940s at Bell Labs, as part of a plan to overcome the congestion and poor service issues associated with mobile telephone systems.

There are four key design principles of cellular systems (see Figure 6-10), which are the same today as they were in the 1940s. They are the use of many of low-power, small coverage-area transmitters instead of a single, powerful, monolithic transmitter to cover a wide area; frequency reuse; the concept of cell splitting; and central control and cell-to-cell handoff of calls. These concepts are fairly straight forward. The first relies on the philosophy that the whole is greater than the sum of its parts. By using a large number of low-power transmitters scattered across a broad coverage zone, each capable of handling multiple simultaneous calls, more users can be supported than with a single, monolithic transmitter.

The second, *frequency reuse*, takes into account the fact that cellular telephony, like all radio-based services, has been allocated a limited number of frequencies by the FCC. By using geographically small, low-power cells, frequencies can be reused by nonadjacent cells.

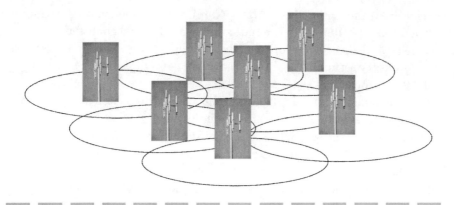

Figure 6-10
An array of low-power cells in a cellular network: Notice how the coverage areas overlap, allowing handoff of a call to occur from one cell to another. Also, because of low power and distance, frequencies can be reused.

When usage areas become saturated by excessive usage, cells can be split. When traffic engineers observe that the number of callers refused service because of congestion reaches a critical level, they can split the original cell into two cells, by installing a second site within the same geographical area that uses a different set of nonadjacent frequencies. This has an extended impact; as the cells become smaller, the total number of calls that the aggregate can support climbs dramatically. Because of cellular geometry, if the radius of the cell is halved, the number of supported calls is quadrupled. The smaller the cell, therefore, the larger the total number of simultaneous callers that can be accommodated. Of course, this also causes the cost of deployment to climb dramatically, and while the architectural goal of most cellular providers is to create a mosaic of thousands of very small cells—called microcells or picocells—that's an expensive proposition and will not happen immediately.

Finally, cellular systems rely on a technique called *cell-to-cell handoff*, which simply means that the cellular network has the ability to track a call (using relative signal strength as an indicator) as the user moves across a cell. When the signal strength detected by the current cell is perceived by the system to be weaker than that detected by the cell the user is approaching, the call is handed off to the second cell. Ideally, the user hears nothing. Cell handoff and other cellular system capabilities are under the central control of a mobile telephone switching office, or MTSO (sometimes pronounced "Mitso").

OK, But How Does It Work?

When a user turns a cellular phone on, several things happen. First, the phone identifies itself to anyone willing to listen (hopefully, a local cell) by transmitting a unique identification code on a frequency that is designated as a *control channel*. The control channel is used for the transmission of operations and maintenance messages between cellular telephones and cell sites. If the phone is within the operating area of a cell site, the site registers the presence of the phone within its operating area, notifies the MTSO to which all the cells in an area are connected, and tracks its position—based on signal strength—as it moves around the cell.

When the user wants to place a call, he or she simply pushes the right buttons, which create simulated touch-tone sounds. Once dialing is complete, the user pushes the Send button, which causes the handset to transmit the buffered digits across the control channel to the local cell. The local cell hands the call off to the MTSO. The MTSO analyzes the digits, instructs the handset and the cell to use a particular set of frequencies for the call, and routes the call to the appropriate destination. MTSOs are interconnected to the wireline network and, therefore, can terminate calls at any location, including to another cellular user.

If driving while talking, the user may approach a cell boundary. The MTSO, which tracks the relative signal strength of each user as they move among the various cells within its domain, will effect handoff of a call from one cell to another if the user's movement (based on signal strength) indicates that they are approaching a cell boundary.

Access Methods

The original analog cellular telephony systems relied on a technique called *frequency-division multiple access* (FDMA) as the access and frequency sharing scheme between mobile users and the cellular network. In FDMA systems, the available spectrum is divided into channels that are assigned to users on demand. One or more of the channels are reserved and set aside as control channels, used to transmit maintenance and operations information between the mobile phone and the network.

Each conversation requires two 30 kHz channels—one for the *forward,* or base-to-mobile direction, and one for the *reverse,* or mobile-to-base station direction. This pairing of channels permits true, full-duplex telephony. Most analog systems have since gone by the wayside in favor of the far more popular, efficient, and battery-friendly digital systems. During the developmental battles, two techniques emerged.

The first, *time-division multiple access,* or TDMA, resembles FDMA in that it divides the available frequency spectrum into channels. That, however, is where the resemblance ends. In TDMA each of the analog channels carries telephone calls that are time-division multiplexed— that is, they share access to the channel. As in FDMA, a control channel is reserved for communication between the network and mobile users.

The biggest advantage that TDMA systems have over FDMA systems is that they support significantly more users. If each channel is divided into four time slots, then the system capacity is quadrupled. And while TDMA electronics are significantly more complex, the fact that they are digital means that they can easily evolve as technology advances.

The second digital access technique is called *code-division multiple access,* or CDMA. CDMA systems are dramatically different from FDMA and TDMA systems in that they do not channelize the available bandwidth—instead, they allow all users to access and use the available spectrum simultaneously. This technique is called *spread spectrum transmission.* Spread-spectrum techniques are described in detail later in this chapter.

Not only are digital systems more secure than narrowband technologies, they also support significantly larger numbers of simultaneous users. In fact, whereas FDMA systems support a single user per frequency slot, CDMA systems can support hundreds.

Radio-based telephony has enjoyed a wild, tumultuous ride along the way to its position today as a mainstream, foundation-level technology. Starting in the late 1800s with the parallel work of Marconi and Bell, radio and telephony wandered down different paths until fairly recently, when they converged and joined forces, leading to the development of cellular telephony.

The story doesn't end with cellular telephony, however. Today, mobile users are clamoring for the ability to extend the reach of LANs, videoconferencing systems, medical image devices, and database access, without having to deal with the restrictions of a copper tether. Developing nations have realized that with cellular technology, telephony infrastructures can be installed in countries in a fraction of the time it takes to install a wired network. In fact, my work in Africa has shown me the degree to which humans demonstrate their ability to innovate in the face of technological challenge. In August 2004, I traveled to Johannesburg, South Africa, to work on a project with one of my clients there. When the

Figure 6-11
Metal shipping container converted to a central office and calling center in Africa

work was complete we took a trip into the bush to do some wildlife photography, and on the way to the game area we passed the most remarkable sight.

The one word that comes to mind when I think of South Africa—or Africa in general, for that matter—is *vibrant*. I have never seen, in any of the countries I travel to, the degree of entrepreneurial vibrancy that I regularly see in Africa. As we approached the intersection of two dirt roads in the middle of the bush northwest of town, I saw the usual gaggle of people standing on all four corners of the intersection, selling the usual collection of fresh fruit, roasted corn, cell phone chargers, toys, and newspapers. On my left, a row of baboons sat on a highway guardrail like old men watching the world go by from the front porch of a general store.

As I looked beyond the cacophony at the intersection, I saw what looked like a metal shipping container sitting in the dirt. And sure enough, that's what it was (see Figure 6-11)—a shipping container that had been converted into a mobile telephony central office and phone services store. Vents on the roof hid a microwave link that connected the switching equipment inside the box to a remote office, and hanging on the wall inside were phones that could be used for a fee by those who did not own their own mobiles.

This innovation was the result of a challenge put before the wireless players in South Africa. The government handed out spectrum with the following proviso: "We give you this with the expectation that you will cover 70 percent (or so) of your served area within 5 years. If you don't we will come after you with stiff penalties." The licensed providers took this challenge to heart. Many of them farmed out the services deployment to contractors, who set up little offices like the one in the picture. Within 2 years they covered 90 percent of the served area. Imagine. In Africa, as in so many regions of the world, wireless is the dominant connectivity solution, even though wireline is far more prevalent and far better in terms of quality than it has ever been before in these regions. Yet because the capital and expense costs to deploy wireless are so much lower than with wireline (because there is no need to dig trenches or put up poles), and because wireless now supports satisfactory levels of data applications, mobility has become the access method of choice in many parts of the world. And well it should. Furthermore, because of the high cost of telephone calls in many countries, many subscribers use their mobile devices for chat—they rely on IM/SMS to stay in touch with their friends. The cost to send an SMS message is far cheaper than a phone call; so, in many cases, people use their mobile handsets exclusively to chat!

This is what I love about telecommunications: When technology is applied to real human problems and challenges, magic happens.

▉ Wireless Today

Perhaps no area of the greater telecommunications industry has been more fraught with misunderstanding and misdirection than wireless. Because of poor market positioning, blind reliance by service providers on "technology" as a synonym for "marketplace solution," capability promises that could not be delivered, and occasionally less than stellar customer service, wireless has become a symbol for much of what is perceived to be wrong with the telecommunications industry. The financials bear this out: The average wireless customer in the United States talked 50 percent more in 2001 than in 2000, but paid only 5 percent more—and 2002 through 2004 numbers reflect a continued trend as usage climbs and revenue declines. Usage is dramatically up, but average revenue per user (ARPU) is dramatically down—not a healthy picture. In the last 5 years, average ARPU has dropped from $65 to roughly $45. In 2001, the subscriber base for wireless grew 14 percent in the United States to roughly 177 million users. This amounted to a 63 percent market penetration rate, which will most likely grow to exceed 90 percent penetration by 2006. According to Pyramid Research, 13 million new subscribers joined the service ranks in 2003, a drop from 17 million in 2001 and 2002. Meanwhile, prices have dropped 10–15 percent, which has obviously led to difficulty in achieving revenue growth and managing profitability. Furthermore, subscriber acquisition costs have climbed dramatically; according to industry reports, Sprint saw acquisition costs go from $285 in the third quarter of 2001 to $395 a year later, and later numbers bear out this continued increase.

Like most segments of the industry, wireless faces further consolidation. As I write this section of the book, Sprint and Nextel confirmed this very morning their intent to merge, and others will undoubtedly follow in the wake of the collapsed telecom bubble. Today, six national operators serve 75 percent of the overall market. These players struggle to uniquely address the needs of a wider array of identified market segments including high-volume business users, corporations with large numbers of users looking for deeply discounted pricing plans, extended families looking for deals, kids, and of course, new users. They are also looking for ways to satisfy the demands of prepay customers who are credit-challenged or have not yet established credit. Only recently, Virgin Mobile entered the U.S. market and is already making significant inroads. Virgin is a *mobile virtual network operator (MVNO)*, which means that they own no infrastructure. They buy huge volumes of minutes of use from other carriers at deep discounts and then resell them in prepaid mode, which means that they also have no billing infrastructure to support. The only organization they *do* support is—you guessed it, customer service. What a concept.

Access Evolution

Wireless access technologies have undergone an evolution comprising three generations. First generation systems, which originated in the late 1970s and continued to be developed and deployed throughout the 1980s, were analog and designed to support voice. They were characterized by the use of FDMA technology. In FDMA systems, users are assigned analog frequency pairs (and access to a control channel) over which they send and receive. One frequency serves as a voice transmit channel, the other as a receive channel. The control channel is used for signaling.

Second generation systems (2G), which came about in the early 1990s, were all-digital and still primarily voice oriented, although rudimentary data transport was part of the service package. In 2G systems, digital access became the norm through such technologies as TDMA. In TDMA systems, the available frequency is broken into channels as it is in

FDMA, but here the similarity between the two ends. Each channel is shared among a group of users based on a time-division technique that ensures fair and equal access to the available channel pairs. This increases the so-called packing ratio, that is, the number of simultaneous users that the system can support.

An offshoot of 2G TDMA systems is 2.5G, the most common of which is *Global System for Mobile Communications* (GSM). GSM is widely deployed in Europe and other parts of the world (well over 750 million subscribers globally) and goes beyond 2G by offering a suite of user services that enhances the value of the technology. These services include not only voice but also two-way radio, paging, *short messaging service* (SMS, similar to Instant Messaging), and the incorporation of Smart Card technology that permits the customization of the handset and its features. These services are both popular and lucrative: In Europe, where mobile penetration is nearly 85 percent, more people send and receive SMS/IM text messages via their mobile phones than use the Internet for similar functions. The domestic U.S. market is a challenge, however: In the United States, only 5 percent of the 170 million cellular users send or receive text messages—as a result of this and other factors, only 1 percent of the $72.8 billion mobile industry comes from data services, although the number is growing as wireless data applications evolve and penetrate the market. In Europe, however, the story is quite different. Because mobile operators in Europe charge relatively high prices for text messages, the service is enormously lucrative. In response to the game show "Big Brother," for example, which invited the viewing audience to actively take part in the outcome of each episode, 5.4 million text message-based votes were cast, generating £1.35 million in revenue. The UK-based service provider mmO2 enjoyed dramatically higher revenues in July 2002 because of the "Big Brother" message flood.

GSM is now present in the United States through Cingular—the joint venture wireless provider owned by SBC (60 percent) and Bellsouth (40 percent); AT&T Wireless, which recently sold its wireless holdings to Cingular; and T-Mobile, the only all-GSM network provider in the country. T-Mobile's activity in the Wi-Fi space is equally impressive as they "innervate" coffee shops all over the world with broadband wireless access. Their activities have an unintended positive consequence: Many of the Wi-Fi service providers use DSL as the backhaul technology to carry customer traffic back to the CO, resulting in increased sales of DSL.

Related to GSM is an enhancement called *Enhanced Data Rate for Global Evolution* (EDGE). Originally developed by Ericsson and called

GSM384, EDGE allows a service provider to offer 384 Kbps service when eight GSM timeslots are used—a less-than-efficient use of available bandwidth. The conversion to EDGE requires that an additional transceiver be added to base stations and naturally requires the use of handsets that can handle the additional protocol. The initial devices that are planned will offer high-speed downstream only; later devices will support high-speed, two-way service.

EDGE is one technology that paves the way for the implementation of the *Universal Mobile Telephony System* (UMTS). According to the ETSI,

> UMTS will be a mobile communications system that can offer user benefits including high-quality wireless multimedia services to a convergent network of fixed, cellular and satellite components. It will deliver information directly to users and provide them with access to new and innovative services and applications. It will offer mobile personalized communications to the mass market regardless of location, network or terminal used.

Native-mode GSM offers relatively low-speed data access. In response to demands for higher mobile bandwidth, a number of new technological add-ons have been created including the *Generalized Packet Radio Service* (GPRS), which can conceivably achieve higher data rates by clustering as many as eight 14.4 Kbps channels to offer as much as 115 Kbps to the mobile user. There are, of course, downsides to this model including the ability to support fewer simultaneous users because of channel clustering, and the need to build an overlay on the existing network to support packet-mode data transport. Nevertheless, GPRS deployment is proceeding apace.

The third generation of wireless systems (3G) offers broadband access to wireless users over a high user count, digital network. One access protocol that will be deployed is CDMA. In CDMA networks, there is no channelization per se.

The Reality

There are effectively two evolutionary 3G paths underway in the United States, one rooted in GSM, the other in CDMA. Both have merit, both will undoubtedly survive for some time to come, but they are different and must be explained.

The GSM path begins with traditional 2.5G access. Over time, GPRS is added for data service, and in many cases EDGE is later added to the mix. This is often referred to in the industry as *Wideband CDMA* and leads to the ultimate end state, the UMTS.

The CDMA path, on the other hand, begins with CDMA2000 and adds high-speed data through an overlay service known as *1x Evolution-Data Only* (EV-DO). 1xEV-DO technology, sometimes called "high data rate" (HDR), offers high-bandwidth wireless connectivity that is optimized for packet data over CDMA networks. EV-DO supports wireless Web access at download speeds ranging from about 384 Kbps to 2.4 Mbps, sometimes eclipsing the speed of DSL and cable modems.

EV-DO is being rolled out by both Verizon Wireless and Sprint, and trials are sprouting up all over the country. Ultimately, if the technology is successful, a second generation will be added called *Evolution-Data/Video* (EV-DV). Watch this technology closely—it may be the tipping point technology for CDMA-based 3G networks.

In the United States, carriers are all over the map with regard to the wireless access technologies they have chosen to base their networks upon, one reason for the relatively slow deployment of a seamless wireless broadband infrastructure. Verizon and Sprint PCS rely on CDMA2000; Cingular and T-Mobile are mostly GSM-based (there is some legacy TDMA technology scattered about as well); and Nextel, soon to be part of Sprint, uses a proprietary standard based on Motorola's iDEN technology.

These technological disparities (and similarities) could be indicators of ongoing consolidation activity within the industry where it makes sense. Alternatively (or additionally), handset manufacturers will have to manufacture multiprotocol handsets capable of roaming and operating in all accepted protocol networks. Qualcomm offers a dual-mode chipset that allows international roamers to operate seamlessly in both GSM and CDMA environments; Samsung offers a dual-mode five-band phone for both CDMA and GSM environments, as does Motorola.

3G Challenges

When I talk about 3G wireless I often refer to it as a colossal failure— "a technology looking for a problem to fix." The American telecom market suffers from an ongoing love affair with the technologies that underlie the functionality of the network rather than with the capabilities and applications that the technologies facilitate. In my office I have a folder of 3G ads from 1999 to 2000 that herald the service providers' collective abilities to deliver 2 Mbps of bandwidth to the cell phone! Yet, nowhere in those advertising blasts does *anyone* talk about the applications that a 2 Mbps cell phone might be able to take advantage of.

3G deployment has been plagued with issues. In July 2002, Finland-based Sonera abandoned its $9 billion 3G joint venture with Spain's Telefónica in Germany. At roughly the same time, Orange postponed its Sweden rollout of 3G until 2006. Similarly, Sweden's Tele2 threatened to abandon its planned deployment of 3G in Norway unless the license requirements were changed. In Spain, operators demanded a deposit refund. And why? Because W-CDMA, which was mandated by the EU commission as the de facto standard for 3G systems and upon which the enormously expensive European spectrum auctions were based, did not work as advertised. It had significant hardware interoperability issues. On the other hand, CDMA2000 worked well, as evidenced by the presence of some 17 million users in Korea, Japan, and the United States. Don't lose sight of GSM, however: Its users still dominate the planet in terms of sheer numbers—more than 750 million of them.

The Quest for Broadband

At the time of this writing there are about 80 million homes in the United States with broadband access, a number that is expected to grow to well over 150 million by the end of 2005. According to a report from the highly influential Gartner Group, in order for 3G to succeed, 50 percent of the population must have access to 75 percent of the offered services, 5 percent of the population must have the most recent devices, and an obvious and widely lauded "killer application" must be evident. Reality rears its ugly head, however: Today, only about 25 percent of United States and Canadian subscribers have *access* to broadband, while another 20 percent simply cannot get it. This lack of broadband access cannot be allowed to continue: Studies show that widespread broadband access will add hundreds of billions of dollars to the national economy and will add three billion work hours annually—numbers that tend to make regulators and legislators sit up and take notice.

The Wireless Data Conundrum

As voice revenues decline through competitive commoditization, wireless carriers are looking for alternative revenue sources that they can turn to. One is wireless data.

The first protocol that was announced for data transport over mobile environments was the *Wireless Application Protocol* (WAP). Originally developed by Phone.com for mobile Internet access, it was largely a disappointment among users. Early on in its deployment, Germany's D2 network administrators announced that customers were using it less than a minute a day. The WAP acronym was soon redefined as the "wrong approach to portability" because of the complexity involved in its use (one article reported that it took 32 clicks and scrolls to access a simple stock quote). Many, however, believe that WAP failed because service providers that implemented it were unwilling to share revenues with the content providers that they supposedly partnered with. Historically, given the choice of protecting their existing service margins or allowing new markets to emerge based on new models, service providers choose what they are most comfortable with—protecting existing service margins. Now they are changing the way they view this because of the success of text messaging and other applications, but they still have a long way to go. For all intents and purposes WAP is dead.

Wireless data faces a number of significant challenges, each of which could become a showstopper. First, most wireless data networks are designed as overlay IP networks, reflecting the ongoing evolution from circuit to packet architectures in the public network. Unfortunately, these designs don't integrate seamlessly with the preexisting PSTN.

Second, spectrum limitations severely limit the number of voice channels that can be provisioned as well as the amount of bandwidth that is available for data. Low bandwidth applications such as SMS and IM serve the market adequately for the time being, but over time—and not all that *much* time—higher bandwidth applications such as multimedia messaging (including video and still images), gaming, and multiuser collaboration applications will drive the demand upward. Mobile gaming alone is expected to grow from 7.9 percent of the user base actively playing games wirelessly to 35 percent by mid-year 2008.

Third, broadband users (cable and DSL) have come to expect always-on connectivity. Because of current spectrum limitations, most wireless players will, because of cost and spectral scarcity, drop broadband connections after a period of inactivity. This will reduce the perceived effectiveness and value of the service in the minds of customers. It's a well-known fact that always-on broadband users generate more than ten times the traffic of a dial-up user—the implication being services revenue and an upward pressure on capital expansion.

Fourth, handsets and other mobile access devices such as PDAs must become more richly functional. They must support multiple access protocols so that they can connect via not only standard cellular protocols but Bluetooth and 802.11 as well. Most manufacturers already offer these devices; growth will continue.

Finally, what are the so-called killer applications? If truth be told, the demand in mobile environments is not really for a new killer app, but rather for new killer ways to access *existing* applications. These may include (and in some cases already do) Web access, IM, e-mail, collaborative gaming, location-based services, personalized event notification, content aggregation, and personal information portals. In 2004 alone, more than 600 million mobile phones were sold. Interestingly, 70 million of them had built-in cameras, adding some credence to the belief that the ability to transmit and receive digital photographs from one cell phone to another device is a desirable application. (Although I have yet to meet anyone who regularly transmits photographs from their cell phone.) Two-thirds of all cell phones shipped today have color screens, and multiple service providers have announced multimedia applications for their mobile users including downloadable games and music clips, as well as a wide variety of customizable ring tones. Don't believe that these are desirable applications? In 2004, mobile games generated $1.1 billion in revenues, while downloadable ring tones generated an almost unbelievable $3.5 billion. Even so, a study performed recently by Bear Stearns-Booz Allen & Hamilton predicted that wireless data services are unlikely to exceed ten percent of carrier revenues by 2007. But here's a question for readers: *Should they?*

3G's Bottom Line

3G, billed as the ideal mobile data solution, has failed to live up to expectations or to satisfy the promises its developers made. Frankly, it has been a poorly positioned solution—a "technology in search of a problem to solve." Originally positioned as the facilitator of the wireless Internet, customers have come to realize that the wireless Internet today is largely a dream. It is doomed to fail as long as it continues to be a deadly combination of immature, not-yet-ready-for-the-market technology, poorly established expectations, slow and spotty connections, useless device screens, and largely nonexistent content. Comparing the experiences of surfing the Web via a computer and surfing it with a cell phone is a laughable exercise—there *is* no comparison. Even in Japan, where uptake of innovative services is always extremely high, the rollout of 3G services has, to a point, stalled. Reasons include service availability in urban areas (spotty at best), the physical dimensions of the handsets (they're heavy), the price of the handsets (three times that of the one it replaced), the cost of service (the average monthly invoice in Japan for subscribers climbed dramatically), and very short battery life. As a clerk in an electronics store in Tokyo told me, "You don't need 3G to send e-mails, download data, or even watch movie clips. You can do all that with the other services so there is no reason to use 3G."

Key to the success of broadband, obviously, is spectrum availability. The spectrum battles that are underway between broadcasters and telecom players create enormous problems that cannot be overcome without a shift in regulatory policy. Television has always been sacred territory in the United States and Canada, and has always enjoyed the rights of first refusal for newly available spectra. As a result, TV broadcasters have amassed huge bunkers of bandwidth. That they only use a relatively small percentage of what they hold appears to be irrelevant. It's high time federal regulators redistribute spectrum as appropriate to level the services playing field. "Water, water everywhere and not a drop to drink," is the mantra of would-be wireless providers. TV has historically controlled 15 percent of the most desirable spectrum (30–30,000 MHz) yet wireless television (i.e., antenna-based) serves only 11 million U.S. households. The rest are served by cable and satellite providers who don't use the most desirable spectrum to begin with. Cellular providers have half the spectrum allocated to them that TV has; yet, they must serve hundreds of millions of subscribers in that space. The ITFS spectrum that was set aside in the 1960s (see section on MMDS, which follows) for education purposes still sits largely unused; yet, Wi-Fi providers must pack 20 million users into half the spectrum available to ITFS users. The good news is that recent regulatory activity indicates that the FCC may be considering a "gerrymandering of the available spectrum," so stay tuned—no pun intended.

Fixed Wireless Technologies

A number of fixed broadband wireless technologies have emerged that present a potential alternative to traditional wired access infrastructures. These include *Local Multipoint Distribution Service* (LMDS), *Multichannel Multipoint Distribution Service* (MMDS), 802.11 (Wi-Fi), and *Freespace Optics*.

Local Multipoint Distribution Service (LMDS)

LMDS is a bottleneck resolution technology, designed to alleviate the transmission restriction that occurs between high-speed LANs and wide area networks (WANs). Today, local networks routinely operate at speeds of 100 Mbps (Fast Ethernet), 1,000 Mbps (Gigabit Ethernet), and even 10 Gbps, which means that any local loop solution that operates slower than any of those poses a restrictive barrier to the overall performance of the system. LMDS offers a good alternative to wired options. Originally offered as CellularVision, it was seen by its inventor, Bernard Bossard, as a way to provide cellular television as an alternative to cable. It is, in effect, wireless DSL.

Operating in the 28 GHz range, LMDS offers data rates as high as 155 Mbps, the equivalent of SONET OC-3c. Because it is a wireless solution, it requires minimal infrastructure and can be deployed quickly and cost effectively as an alternative to the wired infrastructure provided by incumbent service providers. After all, the highest cost component when building networks is not the distribution facility, but rather the labor required to trench it into the ground or build aerial facilities. Thus, any access alternative that minimizes the cost of labor will garner significant attention.

LMDS relies on a cellular-like deployment strategy under which the cells are approximately three miles in diameter. Unlike cellular service, however, users are stationary. Consequently there is no need for LMDS cells to support roaming. Antenna/transceiver units are generally placed on rooftops as they need unobstructed line of sight to operate properly. In fact, this is one of the disadvantages of LMDS (and a number of other wireless technologies): Besides suffering from unexpected physical obstructions, the service suffers from "rain fade" caused by absorption and scattering of the transmitted microwave signal because of atmospheric moisture. Even some forms of foliage will cause interference for LMDS, so the transmission and reception equipment must be mounted high enough to avoid such obstacles—hence the tendency to mount the equipment on rooftops.

Because of its high bandwidth, many LMDS implementations interface directly with an ATM backbone to take advantage of ATM's bandwidth and its diverse quality of service capability. If ATM is indeed the transport fabric of choice, then the LMDS service becomes a broadband access alternative to a network capable of transporting a full range of services including voice, video, image, and data—the full suite of multimedia applications.

The LMDS Market

Even with the gloomy forecasts coming from the dark corners of the telecommunications industry, there is hope for broadband wireless solutions such as LMDS.

Three key applications for LMDS have emerged in markets where the technology has been deployed: voice, remote LAN access, and interconnection and interactive television. These three applications alone make LMDS a powerful contender in the broadband access market.

Of course, fixed wireless has been hit hard, as have other segments of the telecom space. The three major suppliers—Advanced Radio Telecom, Winstar, and Teligent—filed for bankruptcy. All three took on enormous debt to build their networks quickly, but when the markets began to disappear in late 2000, they were left with few customers. Most of their assets have been picked up by carriers, especially long-distance carriers looking for a viable entry technology for the local loop market. Many left the market over security and reliability concerns associated with wireless. Nevertheless, industry analysts predict that fixed wireless is real, will survive the turbulent market, and will see revenues exceed $14 billion by 2006. In fact, other technologies such as WiMAX (described later in this chapter) may provide viable alternatives.

Multichannel Multipoint Distribution System (MMDS)

MMDS got its start as a "wireless cable television" solution. In 1963, a spectrum allocation known as the *Instructional Television Fixed Service* (ITFS) was executed by the FCC as a way to distribute educational content to schools and universities. In the 1970s the FCC established a two-channel metropolitan distribution service called the *Multipoint Distribution Service* (MDS). It was to be used for the delivery of Pay-TV signals, but with the advent of inexpensive satellite access and the ubiquitous deployment of cable systems the need for MDS went away.

In 1983, the FCC rearranged the MDS and ITFS spectrum allocation, creating 20 ITFS education channels and 13 MDS channels. In order to qualify to use the ITFS channels, schools had to use a minimum of 20 hours of airtime, which meant that ITFS channels tended to be heavily, albeit randomly, utilized. As a result, MMDS providers that use all 33 MDS and ITFS channels must be able to dynamically map requests for service to available channels in a completely transparent fashion, which means that the bandwidth-management system must be reasonably sophisticated.

MMDS is, in effect, wireless DSL. In spite of its roots in television distribution, there are no franchise issues for its use (there are, of course, licensing requirements). However, the technology is also limited in terms of what it can do. Unlike LMDS, MMDS is designed as a one-way broadcast technology and, therefore, does not typically allow for upstream communication. Beginning in November 1997, Bellsouth deployed an MMDS-based home entertainment system in New Orleans, offering more than 160 channels of programming. MMDS allows for transmission distances up to 35 miles, farther than LMDS, and by December 1998 the company had a similar presence in Atlanta, Charleston, Birmingham, and Jacksonville, as well. Today, their coverage area has expanded somewhat, and in late May 2003 they made a $65 million bid on MCI's MMDS properties, putting a $3.9 million deposit down on the firm's holdings.

Clearly, companies like Sprint and MCI see the value of wireless access to their own networks and have taken steps to bring about the technological convergence of innovative technologies to meet the growing customer demand for high-speed Internet access. The manufacturing sector certainly sees advantages in the LMDS and MMDS markets; more than a dozen companies manufacture equipment for the two technologies.

Freespace Optics

Freespace optics involves the transmission of a coherent beam of light from a source to a destination, eliminating the costly requirement to build a fiber infrastructure or comply with FCC spectrum licensing requirements.

A number of companies are involved in this technology including AirFiber, TeraBeam, Infrared Communication Systems, LSA Photonics, and LightPointe Communications. Bandwidth ranges from 155 Mbps (OC-3) to 622 Mbps (OC-12), over distances as far as 14 miles, with availability ratings as high as 99.98 percent.

Freespace optics are limited in capability compared to the fiber alternative because the technology relies on line-of-sight transmission. As a result, distances are limited by 10 to 15-mile operational horizons due to the curvature of the Earth, building construction, atmospheric conditions, and limitations on laser power due to safety restrictions. They are, however, enjoying success in the marketplace because they require minimal infrastructure.

Bluetooth

Bluetooth has had a tumultuous ride in the last few years. First heralded as the next great connectivity option, its future is somewhat less assured today. The Bluetooth standard, named for tenth century Danish king, Harald Bluetooth, who united Denmark and part of Norway into a single kingdom, was developed by a group of manufacturers and allows any device, including computers, cell phones, and headphones, to automatically establish wireless device-to-device connections. The Bluetooth Special Interest Group, comprising more than 1,000 companies, works to support Bluetooth's role as a replacement technology for connectivity between peripherals, telephones, and computers. In effect it is a cable replacement solution and is often cited as the ideal way to simplify the rat's nest typically found behind most computer desks.

Bluetooth operates at 2.45 GHz, a spectrum allocation set aside for the use of *industrial, scientific, and medical devices* (ISM). The system is capable of transmitting data at 1 Mbps, although headers and handshaking information consume about 20 percent of this capacity. A number of other devices operate within this domain as well, including baby monitors, garage-door openers, and some cordless phones. The process of ensuring that Bluetooth and these other devices don't interfere with one another has been a crucial part of the design process and, today, is one of the major detractors to the technology. 2.5 GHz cordless phones have become so popular that interference between them and Bluetooth-equipped devices has become a significant nuisance in some cases.

Applications suggested for Bluetooth are interesting, and while they are distance limited because of the technology's low operating power level, plenty of opportunities have presented themselves. The standard specifies operational distances of up to 100 meters between Bluetooth-equipped devices, but because of power restrictions most devices will only be able to operate within a 10-meter radius. Bluetooth is designed for the creation of personal area networks (PANs) that rely on picocells (very small service areas) for connectivity. To date, the most successful applications for Bluetooth are indeed cable replacement solutions on mobile phones, PCs (keyboards and mice), and headphones. The technology is also enjoying some success in the sensor and monitoring markets and is experiencing uptake in the manufacturing, security, and health-care sectors.

The Wireless Future

So what does the future wireless environment look like? Most likely, some form of industry consolidation for both economic and technology reasons is in order. Today there are two primary wireless camps in the United States: the GSM companies, including Cingular and T-Mobile; and the CDMA companies, including Verizon Wireless and Sprint PCS (at the time of this writing the two companies are rumored to be considering a merger). The truth is that GSM's days as a widely deployed protocol are probably somewhat numbered in the long term, even with the imminent arrival and implementation of GPRS (and possibly EDGE). The saving grace is UMTS; it will extend the life of this service for GSM's approximately 750 million worldwide users.

CDMA, however, is rapidly becoming the preferred protocol for 3G implementations. North America, Russia, and China all have large CDMA networks, but GSM still has large numbers of subscribers globally. In reality, the answer may not be an either/or question, but rather a dual-mode solution. One answer might be to offer CDMA as an overlay in GSM areas that have access to the appropriate spectrum. And of course, UMTS is the ultimate goal that will "harmonize" the two access options.

One factor is that CDMA appears to be a more cost-effective solution than traditional GSM and its various evolutionary stages. Performance is another consideration: 1X offers third-generation service at speeds comparable to that of dial-up service. 1xEV-DO offers higher-speed, high-capacity wireless Internet connectivity that is optimized for packet transport. It provides a combination of high network performance and strong economic benefits that are unprecedented in mobile, portable, and fixed services networks. 1X has been available in Canada for a while, and 1X EV-DO has been available in South Korea for quite some time. And Verizon and Sprint now offer the service in the United States as well.

Cognitive Radio

One other technology innovation that should be mentioned as part of the future of the industry is *cognitive radio*. Cognitive radio describes radios that provide software-based control of signal modulation techniques, wide-band or narrow-band operation, communications security functions (such as frequency hopping), and waveform requirements of current and evolving standards over a broad frequency range.

Cognitive radio is applicable across the wireless industry and provides efficient, inexpensive solutions to several challenges found in current systems. For example, cognitive radio-based end-user devices and network equipment can be dynamically programmed for better performance, delivery of enhanced on-demand feature sets, and the provisioning of a wide variety of advanced services that provide new revenue streams for the service provider.

The cognitive radio concept is based on the use of a simple hardware platform that allows customers to dynamically modify both the network and the end-user device to perform different functions at different times. This allows end-users to have choices with pay-as-you-go features, device independence, and scalable hardware. It allows network operators to differentiate their services without having to support multiple handsets, and to move quickly into new markets and offer new, tiered services to increase revenue. Cognitive radio also allows infrastructure suppliers to lower their costs and protect themselves from price erosion due to a reduction in network elements because of a common hardware platform.

Finally, cognitive radio allows application developers to create applications that are platform independent and that can be augmented easily through dynamic application of features, patches, and capabilities.

Cognitive radio applications span multiple wireless devices including mobile phones, smart phones, PDAs, and computing devices. Today's digital cellular networks use a variety of technologies for the air interface and support multiple data standards. To add to the complexity, the wireless industry is introducing GPS, Bluetooth, and a variety of equally disruptive protocols to the mix. Cognitive radio offers a solution to accommodate this plethora of standards, frequency bands, and applications by offering end-user devices that can be programmed using over-the-air software. With cognitive radio, a service provider will implement a common hardware platform and accommodate the interworking of all these standards and technologies via instantly downloadable software modules and firmware.

Radio Frequency Identification (RFID)

Radio frequency identification (RFID) has its origins in the 1940s, when the fundamental technology that underlies modern RFID systems was developed to discriminate between inbound friendly aircraft and inbound enemy aircraft. Allied aircraft carried transponders that broadcast a unique radio signal when interrogated by radar, identifying them as a "friendly." This *Identify: Friend or Foe* (IFF) system was the basis for the development of the technology set we refer to today as RFID. Modern aircraft still use transponders to automatically and uniquely identify themselves to ground controllers (the well-known "squawk" function). Many believe that RFID is an extension of the common barcode and, while the two share some application overlap, RFID is much more than a barcode.

RFID Functionality

A typical RFID system is a remarkably simple collection of technology components. It comprises a collection of transponder tags, which form the heart of the system; a reader, which energizes the tags, collects information from them, and delivers the information to backroom

Figure 6-12
RFID
transponder

analysis applications; and the application set that analyzes the data provided by the transponder tags to the reader. This application set is critically important, because RFID systems, by their very design, generate enormous amounts of data that must be analyzed and acted upon if the system is to have value.

Each transponder, an example of which is shown in Figure 6-12, has a unique "serial number" (a card ID, or CID) that identifies the tag and therefore whatever it is attached to—a pallet of products, an identification card, or a beef cow in a herd. When the tag is within the operational range of a reader, the reader's magnetic field energizes the tag, causing it to go through a series of functions that culminate in the transmission of whatever piece of data is stored in the tag's memory. This information is programmable and might contain detailed product information, product perishability data, routing information, a sheep's bloodline, and so on. For the most part, the tags are passive, meaning that they have no battery but are, in fact, powered inductively by the RF signal emitted by the reader. These tags necessarily have a relatively short read range—as much as a foot, no more—but active tags, which do have internal power, can broadcast up to 20 feet under the right conditions. Passive tags are often used in applications where proximity to a reader is assured, such as in a warehouse or transportation-based supply chain environment. Active devices are commonly seen in such applications as automated toll-taking systems on major freeways. The EZ-Pass system deployed in the northeastern United States is a good example.

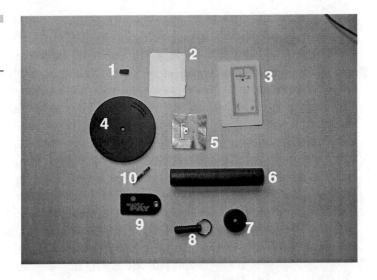

Readers are nothing more than RF emitters connected to back-end software analysis systems. The reader's role is to emit an RF signal that activates the tag or tags within its operational area, to transmit a series of simple commands to the tags, and to collect the data returned as a result of the transmitted commands. Because multiple tags may be activated simultaneously—as in a supply chain environment—most readers have anticollision capability that allows them to control a multitag environment, forcing each tag to take a turn so that simultaneously activated devices can successfully transmit their information without interference from adjacent tags.

Tags come in a variety of forms, as shown in Figure 6-13. The wedge-shaped device (1) is designed to be attached in a number of ways to track-able items. The wedge shape makes it possible to guarantee the orientation of the tag in environments where signal strength is weak or impeded by metal in the transmission area. The paper tag (2) is often seen in bookstores, inserted between the pages of books as a theft deterrent. The tag contains a single bit of information. If the bit is set by the "deactivator" at checkout, the book was properly purchased. If not, it sets off the alarm upon exiting the store.

The card (3) is a typical contactless smart card, used in corporate access control environments, hotel room keyless entries, etc. The large disk (4) can be attached to large pallets or other containers via the convenient hole in the center of the disk, and is reusable. The thin sheet tag (5) is affixed to a sticky backing and can be affixed to the inside of computers, library books, and other small items.

The cylindrical tag (6) is an interesting transponder tag. The one shown is designed for high heat and corrosive environments, often used, for example, in automobile painting lines that are subject to extremes of temperature and caustic substances. The tag can be attached to an automobile body, for example, and as it passes through the line broadcasts the color that is to be applied and the chassis style so that the robotic painters apply the paint correctly. The cylinder style also comes in a ceramic form factor and is designed to be embedded in the stomach of cattle. The tag contains bloodline history, shipping information, and other data that is invaluable to veterinarians tracking disease. And because the device is ceramic, it is impervious to corrosive stomach acids.

The disk (7) is similar to the larger disk described earlier; it can be attached to smaller packages and, like its larger cousin, is reusable. In fact, these devices often remain attached to their shipping containers.

The key fobs shown in (8) and (9) are RFID tags designed to facilitate the purchase of fuel. The Speedpass (8) is commonly seen, and replaces the need for a credit card.

The glass transponder (10) is intriguing. These devices are designed for subcutaneous use, often injected under the skin of livestock, fish, and other wildlife that veterinarians and wildlife biologists wish to track. Sturgeon are often tagged so that aquatic biologists can track the movement and survival rate of the fish.

Spectrum Considerations

RFID systems are generally divided into two categories: *passive (or near-field systems)* and *active (or far-field systems)*. Passive systems typically operate at very short distances between the reader and the transponder, and usually operate in the 13.56 MHz range of the electromagnetic spectrum. Active systems, which are more powerful and therefore capable of operating at significantly greater distances from the reader, operate in various regions of the spectrum between 800 MHz and 1 GHz, although there are a few other frequencies within which active RFID systems occasionally operate.

Applications for RFID

For the most part, RFID has been used as an extension of the well-known barcode system. It has numerous advantages over barcodes, however.

Barcode labels can fall off, be torn or smudged, and must be properly oriented so that the laser reader can see the printed label. RFID devices do not suffer from these limitations: They are not subject to tearing or smudging and for the most part do not require specific orientation—as long as they are within the operational range of the reader they can be activated and read. As a result, accuracy is increased and corporations see a reduction over time in both *operating expense* (OPEX) and *capital expense* (CAPEX) due to lowered personnel and equipment requirements.

Applications for RFID are wide ranging and include:

- Personnel identification
- Access control
- Security guard monitoring
- Duty evasion
- Food production control
- Blood analysis identification
- Water analysis
- Refuse collection identification
- Timber grade monitoring
- Road construction material identification
- Toxic waste monitoring
- Vehicle parking monitoring
- Valuable objects insurance identification
- Gas bottle inventory control
- Asset management
- Stolen vehicle identification
- Production line monitoring
- Car body production
- Parts identification
- Barrel stock control
- Machine tool management

This list represents a sample of possible applications; others emerge daily. One of the most intriguing new areas for RFID deployment lies within the realm of homeland security. There are clearly traditional applications for RFID—IFF, access control, package handling and identification, fire control, personnel movement, and production management—under the purview of defense and homeland security that are well understood and fully deployed. A new area that is enjoying a great deal of scrutiny today, however, is *port security*. Large ports are vulnerable environments because of the number of ships that come and go, and the far larger number of containers that enter the port system every hour of every day—over 90 percent of the world's volume of shipped goods travel by ship. Until recently it has been next-to-impossible to examine every inbound container because of sheer volumes: Long Beach, shown in Figure 6-14, handles the equivalent of 11 million containers a year, while Singapore and Hong Kong—arguably the largest *roll-on, roll-off* (RO-RO) ports in the world in terms of container volume—each move approximately 15 million units annually.

Shipboard container doors are closed and locked before the vessel leaves port. The doors are then sealed with antitamper protection to ensure that if they are opened in transit, the seal is broken and the fact that the container was opened will be evident to authorities upon inspection at the destination port. The concern, however, is that containers opened in transit may carry weapons or other destructive cargo and, by the time the tampered-with container is detected, the contents are already in-port.

A number of firms—including Hi-G-Tek, Savi Technology, and E. J. Brooks—build RFID-based electronic seals for containers that not only show that the container door has been opened, but also transmit the fact to a shipboard reader, which then notifies authorities so that the ship can be intercepted and searched before entering port.

Figure 6-14
The Long
Beach Harbor
Container Port

Department of Defense and RFID

The Department of Defense (DoD) is extremely interested in RFID applications. In a November 2003 summit on the technology, the DoD confirmed its commitment to RFID. The organization has mandated that all suppliers place passive RFID tags on products at the lowest possible level that is cost-effective by January 2005; placement may be at the individual product, pallet, or case level, and will vary somewhat by product type.

The DoD's primary interest in RFID is based on its move toward what it calls *knowledge-enabled logistics*. While the military is a unique "business," it still relies on effective supply-chain management to get the job done. Naturally, RFID lends itself to improved supply-chain processes and faster deployment of resources to a forward theater. Frankly, the similarities between military and civilian requirements are far greater than the differences; consider the quote in the following sidebar.

Network Impacts

Today, RFID's widespread implementation is somewhat limited by the cost of the tags—roughly $0.50 each, which means that no one is going to put them on a $0.30 toothbrush. However, it is widely believed that within a year or so the price per tag will drop to the sub-cent level, which means that they will be widespread. How widespread? Some analysts predict that by 2007 there will be more than *five trillion* of them deployed. Consider, now, the impact on the network. Each tag can hold as many as 50 bytes of data. If each tag is activated by a reader once a day (although most will be activated more often that), and a significant percentage of that traffic finds its way into the network . . . well, you do the math.

Final Thoughts

"In its ultimate form, the entire theater of operations will be networked. Sensors will reside on every piece of equipment and every person populating the field of operations, and information collected by those sensors will be processed in real time using artificial intelligence support to prioritize threats and challenges. In-charge personnel will be able to choose from a

portfolio of response options to identify and select targets.

"As a result, the time between sensing, processing, deciding and acting will fall dramatically, allowing forces to target the opposition before they can respond."

Retail executive, talking about his firm's current RFID
product-tracking initiative

RFID currently falls into that wonderful Mark Twain-like space that is characterized by the quote, "The only thing worse than people talking about you is *nobody* talking about you." RFID is certainly getting its share of negative attention, largely due to concerns voiced by privacy advocates. Organizations like the Committee Against Privacy Invasion and Numbering (CASPIAN) have risen up to fight the technology's widespread deployment because of concerns about the ability of various agencies to track an individual's movements and purchases without authorization. And while it is easy to assign these concerns with conspiracy theories, organizations like CASPIAN serve the same purpose in the technology world that Greenpeace serves in the oil industry: They force the industry to be at the top of its game by creating public awareness of potential hazards, real or not. RFID does need to be monitored, and while concerns over its abuse should be heard, the advantages that RFID brings in such applications as security, law enforcement, defense, health care, product manufacturing, veterinary medicine, food and water protection, and supply-chain management far outweigh the risks associated with the potential for misuse of the technology.

What's Next?

So what's coming among the wireless ranks? Within the next 18 months, expect significant segment buy-down and consolidation as the players close competitive ranks. I would expect to see a renewed focus on the applications that wireless technology can most effectively deliver to mobile users, particularly voice, e-mail, instant messaging, and perhaps some set of value-added services: games, downloadable ring-tones, photo sharing, multipoint IM communication, etc. I would also expect to see the emergence of location-based services that rely on GPS technology to enhance intelligent purchasing applications and guarantee carrier-grade requirements such as E911. RFID will undoubtedly play a growing role as well; already it is showing up in retail, health care, and passport covers.

Finally, I would expect the price wars to continue as wireless access seeks its natural and acceptable price level in the face of growing competition among the players.

Satellite Technology

In October 1945, Arthur C. Clarke published a paper in *Wireless World* entitled, "Extra-Terrestrial-Relays: Can Rocket Stations Give World-Wide Radio Coverage?" In his paper, Clarke proposed the concept of an orbiting platform that would serve as a relay facility for radio signals sent to it that could be turned around and retransmitted back to Earth with far greater coverage than was achievable through terrestrial transmission techniques. His platform would orbit at an altitude of 42,000 kilometers (25,200 miles) above the equator, where it would orbit at a speed identical to the rotation speed of the Earth. As a consequence, the satellite would appear to be stationary to Earth-bound users.

Satellite technology may prove to be the primary communications gateway for regions of the world that do not yet have a terrestrial wired infrastructure, particularly given the fact that they are now capable of delivering broadband services. In addition to the United States, the largest markets for satellite coverage are Latin America and Asia, and particularly Brazil and China.

Geosynchronous Satellites

Clarke's concept of a stationary platform in space forms the basis for today's geostationary or geosynchronous satellites. Ringing the equator like a string of pearls, these devices provide a variety of services including 64 Kbps voice, broadcast television, video-on-demand services, broadcast and interactive data, and point-of-sale applications, to name a few. And while satellites are viewed as technological marvels, the real magic lies more with what it takes to harden them for the environment in which they must operate and what it takes to get them there than it does their actual operational responsibilities. Satellites, in effect, are nothing more than a sophisticated collection of assignable, on-demand repeaters —in a sense, the world's longest local loop.

From a broadcast perspective, satellite technology has a number of advantages. First, its one-to-many capabilities are unequaled. Information from a central point can be transmitted to a satellite in geostationary orbit; the satellite can then rebroadcast the signal back to Earth, covering an enormous service footprint.

Because the satellites appear to be stationary, the Earth stations actually *can* be. One of the most common implementations of geosynchronous technology is seen in the *very small aperture terminal* (VSAT) dishes that have sprung up like mushrooms on a summer lawn. These dishes are used to provide both broadcast and interactive applications; the small DBS dishes used to receive TV signals are examples of broadcast applications, while the dishes seen on the roofs of large retail establishments, automobile dealerships, and convenience stores are typically (although not always) used for interactive applications such as credit card verification, inventory queries, e-mail, and other corporate communications. Some of these applications use a satellite downlink but rely on a telco return for the upstream traffic; that is, they must make a telephone call over a land line to offer two-way service.

One disadvantage of geosynchronous satellites has to do with their orbital altitude. On the one hand, because they are so high, their service footprint is extremely large. On the other hand, because of the distance from the Earth to the satellite, the typical transit time for the signal to propagate from the ground to the satellite (or back) is about half a second, which is a significant propagation delay for many services. Should an error occur in the transmission stream during transmission, the need to detect the error, ask for a retransmission, and wait for the second copy to arrive could be catastrophic for delay-sensitive services like voice and video. Consequently, many of these systems rely on forward error correction transmission techniques that allow the receiver to not only detect the error but correct it as well.

An interesting observation is that because the satellites orbit above the equator, dishes in the northern hemisphere always face south. The farther north a user's receiver dish is located, the lower it has to be oriented. Where I live in Vermont, the satellite dishes are practically lying on the ground—they almost look as of they are receiving signals from the depths of the mountains instead of a satellite orbiting 23,000 miles above the Earth.

Low/Medium Earth Orbit Satellites (LEO/MEO)

In addition to the geosynchronous satellite arrays, there are a variety of lower orbit constellations deployed known as *low and medium Earth orbit satellites* (LEO/MEO). Unlike the geosynchronous satellites, these orbit at lower altitudes—400-600 miles, far lower than the 23,000-mile altitude of the typical GEO bird. As a result of their lower altitude, the transit delay between an Earth station and a LEO satellite is virtually nonexistent. However, another problem exists with low Earth orbit technology. Because the satellites orbit pole-to-pole, they do not appear to be stationary, which means that if they are to provide uninterrupted service they must be able to hand off a transmission from one satellite to another before the first bird disappears below the horizon. This has resulted in the development of sophisticated satellite-based technology that emulates the functionality of a cellular telephone network. The difference is that in this case, the user does not appear to move; the cell does!

Iridium

Perhaps the best-known example of LEOs technology is Motorola's ill-fated Iridium deployment. Comprising 66 satellites[2] in a polar array, Iridium was designed to provide voice service to any user, anywhere on the face of the Earth. The satellites would provide global coverage and would hand off calls from one to another as the need arose. Unfortunately, Iridium's marketing strategy was flawed; their prices were high, their phones large and cumbersome (one newspaper article referred to them as "manly phones"), and their market significantly overestimated. Additionally, their system was only capable of supporting 64 Kbps voice services, a puny bandwidth allocation in these days of customers with broadband desires. In the last year the company failed but was pulled from the ashes. In June 2001 the company announced a resumption of service, but many still question the technology's viability: Iridium offers a maximum bandwidth level of 10 Kbps for $1.50 per minute. With the right kind of coaxing most traditional, terrestrial providers (AT&T, Sprint, WorldCom) offer comparable prices but significantly higher bandwidth.

Iridium is not alone. ICO Global Communications, another satellite-based global communications company, filed for bankruptcy in August

---

[2]The system was named "Iridium" because in the original design the system was to require 77 satellites, and 77 is the atomic number of that element. Shortly after naming it Iridium, however, the technologists in the company determined that they would only need 66 birds. They did not rename the system "Dysprosium."

1999. In November of that same year, Craig McCaw, Teledesic, and Eagle River offered salvation for the company with an investment package valued at as much as $1.2 billion.

Globalstar

Others have been successful, however. Globalstar's 48-satellite array offers voice, short messaging, roaming, global positioning, fax, and data transport up to 9,600 bps. And while the data rates are miniscule by comparison to other services, the converged collection of services they provide is attractive to customers who wish to reduce the number of devices they must carry with them in order to stay connected.

ORBCOMM

ORBCOMM is a partnership jointly owned by Orbital Sciences Corporation and Teleglobe Canada. Their satellites are nothing more than extraterrestrial routers which interconnect vehicles and Earth stations to facilitate deployment of such packet-based applications as two-way short messaging, e-mail, and vehicle tracking. Their constellation has a total of 35 satellites.

Angel Technologies

Angel Technologies uses a different technique, flying airplanes above metropolitan areas that provide 10 Mbps symmetrical Internet access. In the United States, Angel's services operate in the LMDS microwave ranges of 24 GHz, 28 GHz, and 38 GHz. In other countries, the 3 to 20GHz range will be employed. The airplane is similar to a corporate jet; it flies 10 miles above the target area in a 2.5-mile-diameter circle, offering a service footprint of up to 75 miles. The planes fly three shifts of eight hours each; the 40-kilowatt signal they generate is powered by the aircraft's engines and can penetrate weather conditions that are normally bad for microwaves.

Angel is working closely with Raytheon Corporation to develop the communication link, and with a variety of companies that have pioneered high-flying, long-duration aircraft.

Satellite Services: What's the Catch?

As a service provisioning technology, satellites may seem so far out (no pun intended) that they may not appear to pose a threat to more traditional telecommunications solutions. At one time, that is, before the advent of LEO technology, this was largely true. Geosynchronous satellites were extremely expensive, offered low bit rates, and suffered from serious latency that was unacceptable for many applications.

Today, this is no longer true. Today, GEO satellites offer high-quality, two-way transmission for certain applications. LEO technology has advanced to the point that it now offers low-latency, two-way communications at broadband speeds, is relatively inexpensive, and as a consequence poses a clear threat to terrestrial services. On the other hand, the best way to eliminate an enemy is to make the enemy a friend. Many traditional service providers have entered into alliances with satellite providers; consider the agreements that exist between DirecTV and SBC Corporation. By joining forces with satellite providers, service providers create a market block that will help them stave off the short-term incursion of cable. Between the minimal infrastructure required to receive satellite signals and the soon-to-be ubiquitous deployment of DSL over twisted pair, incumbent local telephone companies and their alliance partners are in a reasonably good position to counter the efforts of cable providers wishing to enter the local services marketplace. In the long term, however, wireless will win the access game.

Other Wireless Access Solutions

A number of wireless technologies have emerged in the last few years that are worth mentioning including 802.11, WiMAX, Ultra-Wideband, ZigBee, SF-OFCDM, Bluetooth, and the WAP.

A Goat Trail Technology

In keeping with the technologically schizophrenic nature of this book, we begin our discussion of Wi-Fi by talking about goat trails.

Go to any reasonably sized business park or college campus and you will see that the designers of the place have gone to great lengths to make the environment pleasing. They have tastefully positioned the buildings in such a way that they are optimally collocated, sometimes even exhibiting the sublime characteristics of *Feng Shui*. Interconnecting the buildings are carefully poured concrete sidewalks, angling up and over and up and over, providing walkways from one building to the next.

The careful observer, however, will also notice the warren of "goat trails" worn into the grass that run from building to building (see Figure 6-15), demonstrating that while the angular sidewalks are aesthetically pleasing, they violate the rule that observes correctly that the shortest distance between two points is a straight line. (In fact, I maintain that architects should dispense with the sidewalks altogether, wait six months beyond building occupancy, and then pave the goat trails. It makes a lot more sense and saves on concrete).

This is essentially how Wi-Fi came about and it's important because other "competing" standards did not follow this path—and are experiencing very different market acceptance profiles. The technology was introduced in 1992 when wireless LAN companies began to make use of the 2.4 GHz spectrum component where Wi-Fi operates today. It is important to note that the technology was released before the standards were complete and manufacturers, seeing the promise of a new technological innovation, jumped on the bandwagon and began building Wi-Fi devices. Standards development efforts followed and the result was what

Figure 6-15
Goat trail

you might expect: a collection of incompatible standards that created problems in the industry that were reminiscent of the "modem wars" of the 1990s, when two competing and incompatible 56 Kbps modem standards were released, creating implementation paralysis throughout the industry. Hence the goat trail comparison: Wi-Fi was created as the result of a "path of least resistance" progression model in which developers rushed to create products. And while this resulted in rapid introduction of Wi-Fi products and services, it created a variety of problems—including the release of dysfunctional standards.

The technology waddled around through varying degrees of commercial success for the intervening 5 years until June 1997, when the IEEE released the initial 802.11 standard for high-speed transmission using infrared and two forms of radio transmission within the unlicensed 2.4 GHz frequency band. The first of these was *frequency hopping spread spectrum* (FHSS), the second was *direct sequence spread spectrum* (DSSS). It is useful to understand the difference between these two techniques; they are described in the following sections. We begin with an explanation of spread spectrum.

Spread Spectrum

The idea behind spread-spectrum transmission is a simple concept: disguise a narrowband transmission within a broadband transmission by "spreading" or "smearing" the narrowband component within the broadband carrier. This is done using either of two basic techniques: *frequency hopping*, in which a control signal directs the two communicating devices to "hop" randomly from frequency to frequency to avoid eavesdropping; or *direct sequence*, in which the signal is combined with a random "noise" signal to disguise its contents under the command of a control signal that knows how to separate the signal "wheat" from the noise "chaff."

Frequency Hopping Spread Spectrum (FHSS)

The development of FHSS is one of those stories that is worth knowing about—simply because it is so much fun to tell and so remarkable in the telling. During World War II there was considerable angst among Allied forces over the Axis Powers' ability to thwart radio-controlled torpedoes launched from submarines by jamming the radio signals that guided

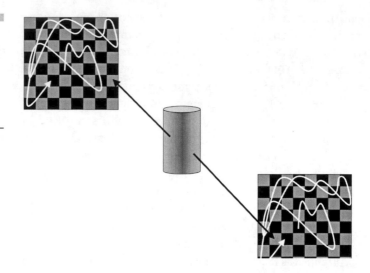

Figure 6-16
Central switch
"authority"
commands
end points to
"hop" as
instructed

them to their targets. To that end, a pair of industrious entrepreneurs filed patent number 2,292,387 titled, "Secret Communications System." The inventors were orchestra composer George Antheil and electrical engineer Hedy K. Markey, better known as early film star Hedy Lamarr. The technique described in their patent application is remarkably simple: A central "authority" (the base station, shown schematically in Figure 6-16) communicates with the two communicating endpoint devices, instructing them to "hop" randomly from frequency to frequency at randomly selected times and for random intervals, typically on the order of 5 ms. Or less. Only the base station (think of an orchestra conductor) and the two devices know when and where to jump—and for how long. To an outsider looking in—that is, any device wishing to eavesdrop on the conversation—the hopping process appears completely random. It isn't, however: The base station knows precisely what it is doing and when it is doing it, so the hopping behavior is actually pseudorandom, and the technique is often referred to as a "pseudorandom hopping code." This technique is most commonly used in CDMA cellular systems, and is (as you might expect) used extensively in secure military communications systems.

Direct Sequence Spread Spectrum (DSSS)

DSSS, sometimes called "noise-modulated spread spectrum," is a very different technique than its FHSS cousin. In frequency hopping, a "conversation" jumps from frequency to frequency on a seemingly random basis. In noise modulation, the actual signal is combined with a carefully crafted "noise" signal that disguises it. The otherwise narrowband signal, shown in Figure 6-17, is spread across a much wider channel—typically on the order of 1.25 MHz. The bits in the data stream are combined with "noise bits" to create a much broader signal, and as before, only the base station and the communicating devices know the code that must be used to extract the original signal from the noise. The code is sometimes called a chipping code, and the technique is referred to as producing a "chipped signal."

One way to think about this technique is as follows. Imagine that you have just been appointed Ambassador to South Africa (Congratulations!). As part of your predeparture indoctrination you have learned that the country has 11 official languages: Afrikaans, IsiNdebele, IsiXhosa, IsiZulu, Northern Sotho, Sesotho, Setswana, SiSwati, Tshivenda, Xitsonga, and English. Upon arriving in the country, you attend a reception in your honor, and in attendance are dignitaries from the many regions of South Africa, all speaking their local language over glasses of *mampoer* and South African wine. The room is filled with people and the cacophony of eleven distinct languages is simply noise to your ears. Suddenly, however, from across the room, emerging from the general din, you hear someone calling to you. From across the ballroom your newly appointed aide de camp is talking to you in English at a normal level, yet you understand her perfectly.

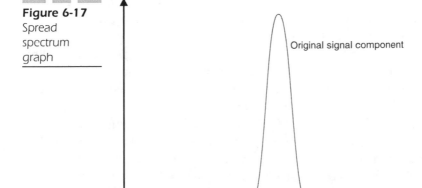

Figure 6-17
Spread
spectrum
graph

Original signal component

"Spread" signal component

If you understand this description, then you understand noise modulated spread spectrum. These two techniques, FHSS and DSSS, formed the functional basis of the original wireless LAN standards that ultimately coalesced into the phenomenon known as Wi-Fi.

Wi-Fi Today

Today several 802.11 standards exist. 802.11b and 802.11g devices transmit in the unlicensed 2.4 GHz range, while 802.11a devices transmit at 5 GHz. Needless to say this presents a radio compatibility problem that had to be overcome. As far as implementation history is concerned, 802.11b was the first version of the Wi-Fi standard to reach the market. It is the least costly and offers the lowest bandwidth of the three, transmitting in the 2.4 GHz spectral band at speeds up to 11 Mbps.

Next in the developmental lineup was 802.11a. Operating at 5 GHz, it can handle transmission speeds up to 54 Mbps. Its biggest challenge, however, was its incompatibility with the widely accepted 802.11b. To remedy this, 802.11g was introduced, which operates in the same 2.4 GHz band as 802.11b but offers transmission rates at speeds equivalent to 802.11a—upward of 54 Mbps. As a result, it is not only compatible with 802.11b but also benefits from the lower cost, unlicensed spectrum at 2.4 GHz.

To achieve the high-bandwidth levels that these systems offer, innovative data encoding techniques are used. One of the most common, employed in 802.11a and 802.11g, is called *orthogonal frequency division multiplexing* (OFDM). OFDM divides the radio signal into multiple subsignals, which are then transmitted simultaneously across a range of different frequencies to the receiver.

802.11b, on the other hand, uses a different encoding technique called *Complementary Code Keying* (CCK). As one knowledge base defines it, complementary codes are "binary complementary sequences with the mathematical property that their periodic auto-correlative vector sum is zero except at the zero shift." And while that definition may set the engineers in the audience all aquiver, it makes me feel like I'm reading the bridge column in the newspaper. So, in English, CCK is simply a technique that creates a set of 64, eight-bit "code words" that encode data for 11 Mbps data transmission in the 2.4 GHz band (802.11b). These code words are mathematically unique so that they can be distinguished from one another by the receiver, even when there is substantial noise and multipath interference. As a result, CCK-based systems can achieve substantially higher data rates by eliminating the impact of such effects that would otherwise limit its ability to achieve them.

It should also be noted that CCK only works with DSSS systems specified in the original 802.11 standard. It does not work with FHSS.

802.11 Physical Layer

All 802 standards address themselves to both the physical (PHY) and media access control (MAC) layers. At the PHY layer, IEEE 802.11 identifies three options for wireless LANs: diffused infrared, DSSS, and FHSS.

While the infrared PHY operates at a baseband level, the other two operate at 2.4 GHz, part of the ISM band. It can be used for operating wireless LAN devices and does not require an end-user license. All three PHYs specify support for the 1 Mbps and 2 Mbps data rate.

802.11 Media Access Control Layer

The 802.11 MAC layer, like CSMA/CD and token passing, presents the rules used to access the wireless medium. The primary services provided by the MAC layer are as follows:

- *Data transfer:* Based on a Carrier Sense, Multiple Access with Collision Avoidance (CSMA/CA) algorithm as the media access scheme

- *Association:* The establishment of wireless links between wireless clients and access points (APs)

- *Authentication:* The process of conclusively verifying a client's identity prior to a wireless client associating with an AP— 802.11 devices operate under an Open System in which any wireless client can associate with any AP without verifying credentials. True authentication is possible with the use of the Wired Equivalent Privacy Protocol or WEP, which uses a shared key validation protocol similar to that used in public key infrastructures (PKI). Only those devices with a valid shared key can be associated with an AP.

- *Privacy:* By default, data is transferred "in the clear;" any 802.11-compliant device can potentially eavesdrop, such as PHY 802.11 traffic that is within range. WEP encrypts the data before it is transmitted, using a 40-bit encryption algorithm known as RC4. The same shared key used in authentication is used to encrypt or decrypt the data; only clients with the correct shared key can decipher the data.

■ *Power management:* 802.11 defines an *active mode,* in which a wireless client is powered at a level adequate to transmit and receive, and a *power save mode,* under which a client is not able to transmit or receive, but consumes less power while in a "standby mode" of sorts.

802.11 has garnered a great deal of attention in recent months, particularly with the perceived competition from Bluetooth, another short-distance wireless protocol. Significantly more activity is underway in the 802.11 space, however, with daily product announcements throughout the industry. Various subcommittees have been created that address everything from security to voice transport to quality-of-service; it is a technology to watch.

Wi-Fi in Action

Wi-Fi is a remarkably simple technology. In fact, most new laptops come equipped with a built-in Wi-Fi card. The card, which is simply a radio, serves to connect the computer or PDA in which it is installed to a base station known as a hotspot. The hotspot is simply the central point of connection for a computer to the network; typically it is a small device that is connected to the Internet, often via a DSL facility, as shown in Figure 6-18. The hotspot typically contains an 802.11-based radio that has the ability to engage in simultaneous conversations with multiple

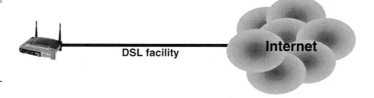

Figure 6-18
Wi-Fi hot spot
connected to
Internet via
DSL

DSL facility   Internet

802.11 cards. Today it is common to see them deployed in coffee shops, airports, hotels, and conference centers. The technology has also gained an enormous following in the residence and home office markets because of the degree to which it simplifies home or small office networking. It has been slower to catch on, however, in the enterprise space—read on.

Wi-Fi Security

Any time data is transmitted over a wireless facility there are concerns about the confidentiality of the information being transmitted. The broad appeal of Wi-Fi has resulted in the incorporation into Wi-Fi of an encryption technique known as WEP. WEP is an encryption technique with two variants: a 64-bit encryption technique and a 128-bit encryption scheme. 128-bit encryption is naturally more secure and is the most widely used technique.

Gaining access to a WEP-enabled system requires the use of a WEP key, which is a decryption code required for the system to recognize and allow a user access. One of the reasons Wi-Fi has been slow to deploy in the enterprise environment is its inherent vulnerability. Consider the following scenario: A well-meaning employee purchases a hotspot at a local office supply store. The employee installs it at the office as a favor to his or her colleagues, but fails to enable a WEP key. The hotspot is now wide open—and anyone driving by with a Wi-Fi-enabled laptop can gain access to the hotspot—and to the enterprise network that lies behind it. It is critical, therefore, that all hotspots be WEP-enabled.

Beyond Wi-Fi

Wi-Fi is a remarkable technology. It nominally offers 11 Mbps (or 54 Mbps) to the user over service radii of approximately 300 feet, and while the 11 Mbps is more typically 7 Mbps and the 54 more like 30, that's still a healthy chunk of access bandwidth. There are other technologies on the horizon, however, that may prove to be more desirable than Wi-Fi in the medium- to long-term future for certain applications. They include WiMAX, Ultra-Wideband, ZigBee, SF-OFCDM, and Bluetooth.

An Aside: Mobility vs. Ubiquity

As a telecom industry analyst and author of numerous books about the business and technology sides of the telecom marketplace, I tend to focus on the words used by the people and companies that comprise the industry, because the careless use of those words often leads to misinterpretation, which in turn leads to inappropriate investment and infrastructure decisions. To that end, I realized recently that two important words in our technology lexicon are used synonymously. The words "mobility" and "ubiquity" are *not* synonyms. However, applications, network designs, investment decisions, strategic planning efforts, and end-user device engineering are being done as if they were one and the same. At worst, this will lead to another spate of ill-placed investments and the requisite marketplace blowback; at best it will lead to confusion and the resulting slowdown of effective deployment. Hopefully, the former won't occur; the latter already has. So before this gets too far down a dead-end road, let me clarify the difference between the two, beginning with definitions.

According to the *American Heritage Dictionary*, "mobility" means "the quality or state of being mobile," while "ubiquity" is defined as "existence or apparent existence everywhere at the same time." From a communications perspective, mobility means being able to move freely while staying connected, as when engaging in the increasingly socially unacceptable practice of using a cell phone while driving. Ubiquity, on the other hand, means universal connectivity; that is, the ability to count on the presence of a connection of one kind or another from the bottom of the Grand Canyon to the top of Mount McKinley and everywhere in between.

From a development point-of-view, these two concepts are being used interchangeably, creating confusion and developmental paralysis. Consider, for example, Wi-Fi. Wi-Fi is a high-speed wireless technology that provides 11 Mbps (or more—or less) of access bandwidth to roaming laptops and PDAs. Its installation in restaurants and coffee shops by the likes of McDonald's and Starbucks is being advertised as a great step forward in mobility. But this isn't mobility: People aren't walking around Starbucks with a Venti Latte in one hand and a laptop in the other, surfing for MP3s. In fact, they are sitting in a booth or at a table enjoying the merits of *ubiquitous access.* Yes, the connection happens to be wireless, and these people are not in their homes or offices, but the presence of Wi-Fi in this case is not an addition to the pantheon of mobility applications. It is, however, an application of ubiquitous connectivity. So, in this particular situation, what value does the wireless loop add? It seems to me that Starbucks, McDonald's, and others could provide equal or better service to their customers who have a need to be connected by providing an RJ-45 cable or two at each table. It would be cheaper, more secure, and better in terms of service quality. Yet, there is a perceived sexiness associated with wireless that somehow precludes wired connectivity as an access option in ad hoc situations.

Wi-Fi, of course, provides a convenience factor that clearly has value to the user, however intangible that value may be. Being able to connect anywhere in the coffee shop without being physically tethered is an advantage, but how much of an advantage is it? Bluetooth, an alternative wireless technology, promised to eliminate dependence on wires between computer peripherals; yet, it remains a largely stillborn solution. Does it work? Yes. Does it provide a level of value that overcomes the price? Apparently not, because its usage levels are near zero.

Consider the following scenario: McDonald's, Starbucks, Barnes & Noble, Borders, major airports, and large department stores all purchase DSL lines. They terminate the lines on a low-cost router, which in turn connects to a multiport hub. They then install convenient RJ-45 connections at each table or at some easily recognized spot in the store. Laptop users can either provide their own cable or borrow/purchase one from the host store. The service is offered at no cost and provides differentiation for the business, or is available via subscription in much the same way that T-Mobile and Wayport already offer connectivity in hotels and airports. Note that the connectivity technology is wired, and while the businesses listed could also add a wireless access point to their network, providing connectivity for Wi-Fi users, they still offer near-ubiquitous connectivity using a wired option.

There is another issue that seems to have been lost in this discussion and that's the issue of power. Wireless connectivity sucks the life out of laptop batteries at a dizzying rate, which means that even with a *wireless* network loop a user will need a *wired* power loop in short order. In other words, they will have to be tethered anyway if they use the connection for more than ten or fifteen minutes, and this negates the sexiness of the wireless connection to a large extent.

So what is the point of this discussion? Both mobility and ubiquity are important in the emerging world of network usage, but they are not necessarily the same. Mobility implies the ability to connect to the network via a wireless local loop. Ideally it offers predictable high-bandwidth, easy, dependable connections, and secure transmission. Ubiquity, on the other hand, implies the ability to connect to the network anywhere, anytime, regardless of the characteristics of the loop. Ubiquitous access may include wireless as an option, but may also include wired solutions such as Ethernet, T1, and DSL.

Another issue is the usage of the two words in terms of parts of speech. Mobility, used as it is in our lexicon, is a noun, because in the minds of many it defines (incorrectly) an application. Mobility is no more an application than DSL is an application. It is a technology option—no more and no less.

Ubiquity, on the other hand, is used as an adjectival modifier to qualify the nature of an individual's access to the network. Ubiquitous access implies the delivery of something that is superior because it is—everywhere. But isn't wireline access universally available and, therefore, ubiquitous as well? And isn't it significantly more secure than a wireless connection, particularly in an enterprise application? After all, if a business decides to implement wireless (802.11) as its connectivity option of choice within a workplace, how can it possibly guarantee that (1) all users implement a secure wireless protocol over the local loop, and (2) signals do not leak out of the building? Wireless is a great technology, offering freedom and mobility to users, but there is a price. And that price can be steep if it is implemented without forethought.

I have observed in this book that customers are not really looking for the "next great killer application" but rather for a "killer" way to access existing applications, because those applications offer solutions to most of the challenges that users encounter. Consider the typical business user. As long as they are in their home office environment, wired connectivity is perfectly acceptable for both voice and data. When they leave the office and get in the car, *mobile telephony* becomes important. Mobile data has no application (thankfully!) in the car other than for those applications optimized for that environment—OnStar service, for example, or GPS-based guidance systems, or specific applications related to public safety. If, however, the user stops at Starbucks for coffee before going home and decides to check e-mail one last time, *ubiquitous* connectivity, whether wired or wireless, provides value to the user from a data perspective, while *mobile* telephony remains valuable for voice. An RJ-45 connection on the tabletop is just as serviceable as a Wi-Fi connection, not to mention far more secure and predictable.

The bottom line is this. *Mobility* defines the characteristics of a *lifestyle choice* that involves networking, whether personal or work-related, while *ubiquity* defines the *characteristics of the technology infrastructure* required to support the mobile lifestyle. "Anywhere, anytime connectivity" has become the mantra of the mobile user, and while wireless (Wi-Fi) is the most loudly proclaimed option, it is *not* the *only* option. This, I believe, is part of the reason that revenues associated with Wi-Fi remain elusive. It is sexy, cool, and functional. But of those three characteristics the only one that has revenue potentially associated with it is *functional*, and there are too many alternatives to wireless that offer lower cost, greater security, and more predictable connections. Until a service provider comes up with a compelling argument for Wi-Fi's performance superiority, the only companies that will make money on it will be those building wireless access points and routers.

Worldwide Interoperability for Microwave Access (WiMAX)

In April 2002, the IEEE published their 802.16 standard for *broadband wireless access* (BWA), also known as WiMAX. 802.16 specifies the details of the air interfaces for *wireless metropolitan area networks* (MANs). And while there are some similarities between Wi-Fi and WiMAX, in other respects they could not be more different. First of all, WiMAX was not created as a "goat trail" technology. Instead, developers first quietly created standards that were socialized through other standards bodies such as the ITU-TSS. As a result of this strategy, spectrum was allocated globally for 802.16 implementation through a 2-year, open-consensus procedure that involved hundreds of engineers from major operators and vendors around the world. Consequently, 802.16, while still a nascent technology, enjoys global acceptance and what will be a relatively trivial implementation phase once it becomes more widely deployed. Furthermore, the capabilities of the standard are impressive. Whereas Wi-Fi offers megabits of nominal bandwidth over service distances of 300 feet, *WiMAX offers 100 Mbps over a service radius of several miles*. And because it is orthogonal, it does not require line of sight for connectivity.

802.16 is initially targeted at the "first mile" challenge for metropolitan area networks. It operates between 10 and 66 GHz (the 2 to 11 GHz spectrum with point-to-multipoint and optional mesh topologies) and defines a MAC layer that supports multiple physical layer specifications specific to each frequency band. The 10 to 66 GHz standard supports two-way transmission options at a variety of frequencies including 10.5 GHz, 25 GHz, 26 GHz, 31 GHz, 38 GHz, and 39 GHz. It also supports device interoperability so that carriers can use multiple vendors' products. Furthermore, the standard for the 2 GHz to 11 GHz spectrum supports both unlicensed and licensed bands, a real boon for the entrepreneurial set that does so much to push the limits of any new innovation.

Over the last few years 802.16 has undergone a series of modifications, resulting in the existence of various flavors of the original standard including 802.16a and 802.16e. They are discussed in the sections that follow and shown in Table 6-2.

802.16a

IEEE 802.16a was ratified as an extension to the original 802.16 standard in January 2003. It enhances 802.16 and addresses radio systems that operate in the 2 GHz to 11 GHz frequency ranges. It addresses the requirements of both licensed and unlicensed implementations, and sup-

Table 6-2

WiMAX Evolution

Characteristic	802.16	802.16a	802.16e
Standard Date	December 2001	January 2003	EOY 2004
Frequency Range	10–66 GHz	Less than 11 GHz	Less than 6 GHz
Transmission Limits	Line of Sight	Non-Line of Sight	Non-Line of Sight
Bandwidth	32 to 134 Mbps in 28 MHz channels	Up to 75 Mbps in 20 MHz channels	Up to 15 Mbps in 5 MHz channels
Modulation Scheme	QPSK, 16 QAM, 64QAM	OFDM (256 subcarriers), QPSK, 16 QAM, 64QAM	OFDM (256 subcarriers), QPSK, 16 QAM, 64QAM
Mobility Options	Fixed	Fixed and "Portable"	Fixed and Mobile
Channel Bandwidth	20, 25, 28 MHz	Scalable from 1.5 to 20 MHz	Scalable from 1.5 to 20 MHz
Operating Radius	2–5 Km	7–10 Km	2–5 Km

ports point-to-multipoint networks as well as mesh topologies within the unlicensed region.

IEEE 802.16e

802.16e is an extension of 802.16a. Its potential impact is enormous in that it adds mobility to 802.16a systems. Given that the ultimate goal of 802.16 is to provide a technology solution that will bridge the gap between fixed and mobile wireless systems, the addition of mobility could support the end-to-end needs of a subscriber in both environments. Consider, however, what the addition of mobility means. I often hear people say that "WiMAX represents a real threat to Wi-Fi!" Given that spectrum has already been allocated around the globe, mobile WiMAX represents more than a potential threat to Wi-Fi—*it represents a REAL threat to 3G!* Consider the tens of billions of dollars that have been spent in the last few years on 3G spectrum, very little of which has actually been deployed. Then along comes WiMAX with its promise of wireless broadband connectivity with global reach, and suddenly 3G begins to appear a bit less attractive.

802.16e allows *wireless ISPs* (WISPs) to enter and take over a market with minimal investment in infrastructure, then offer a complete package of services to subscribers. Wireless broadband, in the form of WiMAX, could also compete favorably with such options as cable modem and DSL (in fact, some analysts refer to WiMAX as wireless DSL).

Several industry players are leading WiMAX's implementation. The first of these is Intel, which is making heavy investments into WiMAX as part of a strategy to take the lead in WiMAX the same way they did in Wi-Fi with Centrino. Their research shows that many people use their PDAs, broadband equipped mobile phones, and laptops to access data networks while mobile, a phenomenon that is causing a significant number of communities to build metro-based broadband access areas to serve them.

As a founding member of the industry-led, nonprofit WiMAX Forum, Intel is leading the charge to promote compatibility and interoperability among certified broadband wireless products. The forum's member companies support the industry-wide acceptance of 802.16 as well as the European ETSI HiperMAN wireless LAN standards, which will most likely be "signed into technological law" by spring 2005.

And what of the continuing perspective that WiMAX competes with Wi-Fi? Most informed players argue that WiMAX is really a complementary technology to Wi-Fi, particularly in the metro arena. They see it as a broadband wireless alternative to cable or DSL and believe that certified WiMAX products will begin to appear in the enterprise market in mid-2005, while residential WiMAX products will provide solutions that are more cost-effective than fixed solutions in rural or greenfield areas.

Ultra Wideband (UWB)

Ultra wideband (UWB) technology is designed for short-range, *wireless personal area networks* (WPANs). It is a short-range radio technology that will be used to transmit data between devices over distances of as much as 30 feet, and is defined as a radio technology that has a spectral operating range that occupies a bandwidth greater than 20 percent of the center frequency, or at least 500 MHz.

A UWB device transmits billions of electromagnetic pulses across a broad range of frequencies that are collectively several GHz wide. The receiver translates the received pulses into data by listening for a recognizable pulse sequence sent by the transmitter.

Like Wi-Fi, UWB systems use OFDM to operate over these extremely wide spectral components. The technology's combination of broad operating spectrum and lower power improves speed and reduces interference with other wireless systems. In the United States, the FCC has mandated that UWB radio transmissions be allowed to operate from 3.1 GHz up to 10.6 GHz, at limited transmit power. Consequently, UWB provides dramatic channel capacity at short range that limits interference.

So what are the applications for UWB? As we mentioned before, PANs are the primary intended application; imagine a home or small office network with the ability offer gigabits per second of bandwidth over distances that easily accommodate the entire office or home.

ZigBee

ZigBee, also known as IEEE 802.15.4, offers a cost-effective, standards-based wireless solution that specifically supports low data rates, low power consumption, security, and reliability. The name derives from a biological principal known as the *ZigBee Principle,* seen in honeybee hives. The colonial honeybee lives with a queen, a few male drones, and thousands of worker bees. The survival of the colony depends on a process of continuous communication of vital information among all members of the hive. The technique that honeybees use to communicate with each other is called the ZigBee Principle, which defines the zigzag dance pattern that the insects use to communicate critical information to each other. IEEE 802.15.4 was nicknamed ZigBee because it facilitates the ability of humans to emulate this survival behavior.

ZigBee is designed to support such applications as home and environmental controls including lighting, automatic meter reading, telemetry for smoke and carbon monoxide detectors, HVAC, heating, security, drapery and shade controls, set top boxes, and specialized applications such as medical sensing and monitoring. To date, ZigBee is the only standard that supports the unique requirements of remote monitoring and control networks as well as sensory network applications. It is designed to support the broad deployment of wireless networks that have low cost, low power requirements. In fact, ZigBee is designed to run for years on a single battery.

ZigBee products rely on the IEEE 802.15.4 physical radio standard, which operates globally in unlicensed bands at 2.4 GHz (global), 915 MHz (Americas) and 868 MHz (Europe). ZigBee supports data rates of 250 Kbs at 2.4 GHz (using 16 channels), 40 Kbs at 915 MHz (using ten channels) and 20 Kbs at 868 MHz (using a single channel). Transmission distances range from 10 to 100 meters, depending on power and environmental considerations. And while ZigBee is considered a nascent technology today, full-scale production is expected in 2005.

Late-breaking news: In mid-December 2004 the ZigBee standard was ratified. Companies such as Figure 8 Wireless, Freescale, CompXs, Eaton, and Atmel—long involved in the development of the technology—are ahead of the competitive pack because of their ongoing involvement in the ZigBee Alliance during the specification-writing phase. These companies already have ZigBee-based products. Table 6-3 compares ZigBee to the most common alternate technologies.

VSF-OFCDM

There are certain acronyms that are so *cool* that they are worth the time it takes to memorize what they expand into. *Variable spreading*

Table 6-3

Wireless Technology Characteristics

Characteristic	ZigBee	Wi-Fi	Bluetooth
Application	Monitoring, control	E-mail, Web access, some video	Cable replacement
Memory Requirements	4–32 KB	1 MB	256 KB
Battery Life (days)	100–1,000	0.5– 5	1–7
Network Nodes	Unlimited	32	7
Bandwidth (Kbps)	20–250	11–54	720
Range (meters)	1–100	1–100	1–10

factor orthogonal frequency and code-division multiplexing (VSF-OFCDM) is one of these. Currently being studied by DoCoMo in Japan, VSF-OFCDM can theoretically transmit at speeds as fast as 100 Mbps outside and as fast as 1 Gbps indoors. As a test bed, DoCoMo has built a mobile IP data network to stress the technology, and early results are quite favorable. And while this technology is nowhere near ready for prime time, it bears watching.

The Mobile Appliance

The mobile appliance concept is enjoying a significant amount of attention of late because it promises to herald in a whole new way of using network and computer resources—*if it works as promised.* The problem with so many of these new technologies is that they overpromise and underdeliver—precisely the opposite of what they're supposed to do for a successful rollout. 3G, for example, has been billed as "the wireless Internet." Largely as a result of that billing it has failed. It is *not* the Internet—far from it. The bandwidth isn't there, nor is a device that can even begin to offer the kind of image quality that Internet users have become accustomed to. Furthermore, the number of screens that a user must go through to reach a desired site (I have heard estimates as high as 22!) is far too high. Therefore, until the user interface, content, and bandwidth challenges are met and satisfied, the technology will remain exactly that—a technology. There is no application yet, and *that's* what people are willing to pay money for.

Chapter Summary

Access technologies—used to connect the customer to the network—come in a variety of forms and offer a broad variety of connectivity options and bandwidth levels. The key to success is to *not* be a bottleneck; access technologies that can evolve to meet the growing customer demands for bandwidth will be the winners in the game. DSL holds an advantage as long as it can overcome the availability challenge and the technology challenge of loop carrier restrictions. Wireless is hobbled by licensing and spectrum availability, both of which are regulatory and legal in nature, rather than technological limitations.

In the next chapter, we will discuss transport technologies including private-line, frame-relay, ATM, and optical networking.

Chapter 6 Questions

1. Is wireless an access or a transport technology?

2. "A 56 Kbps modem is not really a 56 Kbps modem." Explain.

3. Why is ISDN not a more popular technology in the United States?

4. Explain the differences between a BRI and a PRI.

5. "DSL is actually an analog technology." Explain.

6. Explain the purpose of a DSLAM.

7. What are the key differences between ADSL and ADSL2+?

8. Why is the cable industry a far greater threat to the ILECs than the CLECs?

9. Who uses LMDS and MMDS today?

10. What are the differences between FDMA, TDMA, and CDMA?

11.   Explain the differences between DSSS and FHSS.

12.   What is an MVNO?

13.   Who's going to win—CDMA or GSM? Why?

14.   Describe a possible application for cognitive radio.

15.   "RFID deployment represents an assault on privacy." Write a response that refutes this contention.

16.   Why are satellites falling out of favor as an access solution?

17.   "WiMAX is the next answer to 3G." Explain.

CHAPTER **7**

Transport
Technologies

We have now discussed the premises environment and access technologies. The next area we'll examine is *transport*.

Because businesses are rarely housed in a single building, and because their customers are typically scattered across a broad geographical area (particularly multinational customers), there is a growing need for high-speed, reliable, wide-area transport. "Wide-area" can take on a variety of meanings: For example, a company with multiple offices scattered across the metropolitan expanse of a large city requires inter-office connectivity in order to do business properly. On the other hand, a large multinational with offices and clients in Madrid, San Francisco, Hamburg, and Singapore requires connectivity to ensure that the offices can exchange information on a 24-hour basis.

These requirements are satisfied through the proper deployment of wide-area transport technologies. These can be as simple as a dedicated private line circuit or as complex as a virtual installation that relies on ATM for high-quality transport.

Dedicated facilities are excellent solutions because they are dedicated. They provide fixed bandwidth that never varies, and guarantee the quality of the transmission service. Because they are dedicated, however, they suffer from two disadvantages: First, they are expensive, and only cost effective when highly utilized. The pricing model for dedicated circuits includes two components: the mileage of the circuit and the bandwidth. The longer the circuit and the faster it is, the more it costs. Second, because they are not switched and are often not redundant because of cost, dedicated facilities pose the potential threat of a prolonged service outage should they fail. Nevertheless, dedicated circuits are popular for certain applications and widely deployed. They include such solutions as T1, which offers 1.544 Mbps of bandwidth; DS3, which offers 44.736 Mbps of bandwidth; and SONET, which offers a wide range of bandwidth from 51.84 Mbps to as much as 40 Gbps.

The alternative to a dedicated facility is a switched service, such as frame relay, ATM, or in some cases Gigabit Ethernet. These technologies provide "virtual circuits": Instead of dedicating physical facilities, they dedicate logical timeslots to each customer, who then shares access to physical network resources. In the case of frame relay, the service can provide bandwidth as high as DS3, thus providing an ideal replacement technology for lower-speed dedicated circuits. ATM, on the other hand, operates hand in glove with SONET and is Gigabit Ethernet, particularly when deployed in a switched configuration, offering speeds in excess of 1 Gbps. Finally, optical networking is carving out a large niche for itself as a bandwidth-rich solution with the potential for inherent quality of service (QoS).

We begin our discussion with dedicated private line technologies, otherwise known as point-to-point technologies.

Point-to-Point Technologies

Point-to-point technologies do exactly what their name implies: They connect one point directly with another. For example, it is common for two buildings in a downtown area to be connected by a point-to-point microwave or infrared circuit, because the cost of establishing it is far lower than the cost of putting in physical facilities in a crowded city. Many businesses rely on dedicated, point-to-point optical facilities to interconnect locations, especially businesses that require dedicated bandwidth for high-speed applications. Of course, point-to-point does not necessarily imply high bandwidth; many locations use 1.544 Mbps T1 or 2.048 Mbps E1 facilities for interconnection, and some rely on lower-speed circuits where higher bandwidth is not required.

Dedicated facilities provide bandwidth from as low as 2,400 bits per second to as high as multiple gigabits per second. Analog facilities running at 2,400 bps are not commonly seen but are often used for alarm circuits and telemetry, while circuits operating at 4,800 and 9,600 bps are still used to access interactive, host-based data applications.

Higher-speed facilities are usually digital and are often channelized by dedicated multiplexers and shared among a collection of users or by a variety of applications. For example, a high-bandwidth facility that interconnects two corporate locations might be dynamically subdivided into various-sized channels for use by a PBX for voice, a videoconferencing system, and data traffic, as shown in Figure 7-1.

Dedicated facilities have the advantage of always being available to the subscriber. They have the *disadvantage* of being there and accumulating charges whether they are being used or not. For the longest time, dedicated circuits represented the only solution that provided guaranteed bandwidth—switched solutions simply weren't designed for the heavy service requirements of graphical and data-intensive traffic. Over time, however, that has changed. A number of switched solutions have emerged in the last few years that provide guaranteed bandwidth and only accumulate charges when they are being used (although some of

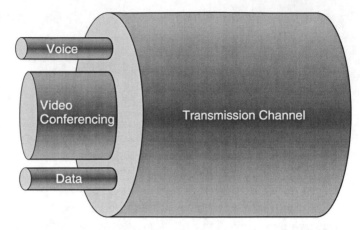

them offer very reasonable fixed-rate service). The two most common of these are frame relay and ATM; we begin with frame relay. Before we do, however, let's spend a few minutes discussing the hierarchy of switching. Part of this is review of prior material; most of it, though, is preparation for our discussion of frame relay and ATM, the so-called fast packet switching technologies.

The Switching Hierarchy

The switching hierarchy, shown in Figure 7-2, has two major subheadings—circuit switching and store-and-forward switching. Circuit switching is something of an evolutionary dead end in that it will not become something else: The only evolution for circuit switching is an evolution toward packet switching, at this point.

Packet switching evolved as one of the two descendents of store-and-forward technology. Message switching, the alternative and another evolutionary dead end, is inefficient and not suited to the bursty nature of most data services today. Packet switching continues to hold sway and, because of its many forms, is a valid solution for most data applications.

Packet switching has three major forms, two of which were discussed in earlier chapters. Connection-oriented packet switching manifests itself as virtual circuit service that offers the appearance and behavior of a dedicated circuit when, in fact, it is a switched service. It creates a path through the network that all packets from the same source (and going to the same destination) follow, the result of which is constant transmission

Figure 7-2
The switching
hierarchy

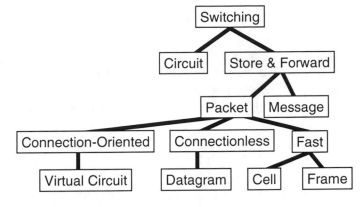

latency for all packets in the flow. Both frame relay and ATM, discussed later in this chapter, are virtual circuit technologies.

Connectionless packet switching does not establish a virtual dedicated path through the network; instead, it hands the packets to the ingress switch and allows the switch to determine optimum routing on a packet-by-packet basis. This is extremely efficient from a network point-of-view, but is less favorable for the client in that every packet is treated slightly differently, the results of which can be unpredictable delay and out-of-order delivery.

The third form of packet switching is called *fast packet*, two forms of which are frame relay and ATM. The technique is called "fast packet" because the processing efficiency is far better than traditional packet switching, due to reduced overhead.

Fast-packet technologies are characterized by low error rates, significantly lower processing overhead, high transport speed, minimal delay, and relatively low cost. The switches accomplish this by making several assumptions about the network. First, they assume (correctly) that the network is digital and based largely on a fiber infrastructure, the result of which is an inordinately low error rate. Second, they assume that, unlike their predecessors, the end-user devices are intelligent and therefore have the ability to detect and correct errors on an end-to-end basis (at a higher protocol layer), rather than stopping every packet at every node to check it for errors and correct them. These switches still check for errors, but if they find errored packets, they simply discard them, knowing that the end devices will realize that there is a problem and take corrective measures on an end-to-end basis. Think back to the protocol chapter for a moment: This is the difference between layer-two error detection and layer-four error detection.

Frame Relay

Frame relay came about as a private-line replacement technology and was originally intended as a data-only service. Today, it carries not only data but voice and video as well, and while it was originally crafted with a top speed of T1/E1, it now provides connectivity at much higher bandwidth levels.

In frame-relay networks, the incoming data stream is packaged as a series of variable length frames that can transport any kind of data — LAN traffic, IP packets, SNA frames, even voice and video. In fact, it has been recognized as a highly effective transport mechanism for voice, allowing frame-relay-capable PBXs to be connected to a frame-relay *permanent virtual circuit* (PVC), which can cost effectively replace private line circuits used for the same purpose. When voice is carried over frame relay, it is usually compressed for transport efficiency and packaged in small frames to minimize the processing delay of the frames. Several hundred voice channels can be encoded over a single PVC, although the number is usually smaller when actually implemented.

Frame relay is a virtual circuit service. When a customer wishes to connect two locations using frame relay, he or she contacts his or her service provider and tells the service representative the locations of the endpoints and the bandwidth they require. The service provider issues a service order to create the circuit. If at some point in the future the customer decides to change the circuit endpoints or upgrade the bandwidth, another service order must be issued. This service is a PVC and is the most commonly deployed frame-relay solution.

Frame relay is also capable of supporting *switched virtual circuit* (SVC) *service*, but SVCs, for the most part, are not available from service providers. With SVC service, customers can make their own modifications to the circuit by accessing the frame-relay switch in the central office and requesting changes. However, service providers do not currently offer SVC service because of billing and tracking concerns (customer activities are difficult to monitor). Instead, they allow customers to create a fully meshed network between all locations for a very reasonable price. Instead of making routing changes in the switch, the customer has a circuit between every possible combination of desired endpoints. As a result, customers get the functionality of a switched network, while the service provider avoids the difficulty of administering a network within which the customer is actively making changes.

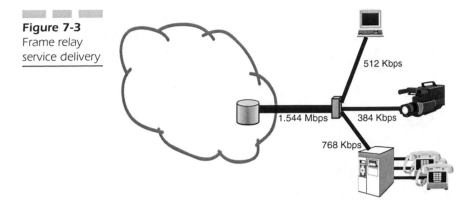

Figure 7-3
Frame relay
service delivery

In frame relay, PVCs are identified using an address called a *Data Link Connection Identifier*, or DLCI (pronounced "delsie"). At any given endpoint, the customer's router can support multiple DLCIs, and each DLCI can be assigned varying bandwidths based upon the requirements of the device/application on the router port associated with that DLCI. In Figure 7-3, the customer has purchased a T1 circuit to connect his router to the frame-relay network. The router is connected to a videoconferencing unit at 384 Kbps, a frame-relay-capable PBX at 768 Kbps, and a data circuit for Internet access at 512 Kbps. Note that the aggregate bandwidth assigned to these devices exceeds the actual bandwidth of the access line by 128 Kbps (1,664 – 1,536). Under normal circumstances this would not be possible, but frame relay assumes that the traffic that it will normally be transporting is bursty by nature. If the assumption is correct (and it usually is), there is very little likelihood that all three devices will burst at the same instant in time. As a consequence, the circuit's operating capacity can actually be "overbooked," a process known as *oversubscription*. Most service providers allow as much as 200 percent oversubscription, something customers clearly benefit from, provided the circuit is designed properly. This means that the salesperson must carefully assess the nature of the traffic that the customer will be sending over the link and ensure that enough bandwidth is allocated to support the requirements of the various devices that will be sharing access to the link. Failure to do so can result in an underengineered facility that will not meet the customer's throughput requirements. This is a critical component of the service delivery formula.

The throughput level, that is, the bandwidth that frame-relay service providers absolutely guarantee on a PVC-by-PVC basis, is called the *committed information rate* (CIR). In addition to CIR, service providers will often support an *excess information rate* (EIR), which is the rate above the CIR they will attempt to carry, assuming the capacity is available within the network. However, all frames above the CIR are marked as eligible for discard, which simply means that the network will do its best to deliver them but makes no guarantees. If push comes to shove, and the network finds itself to be congested, the frames marked discard eligible (DE) are immediately discarded at their point of ingress. This CIR/EIR relationship is poorly understood by many customers because the CIR is taken to be an indicator of the absolute bandwidth of the circuit. Whereas bandwidth is typically measured in bits per second, CIR is a measure of *bits in one second*. In other words, the CIR is a measure of the average throughput that the network will guarantee. The actual transmission volume of a given CIR may be higher or lower than the CIR at any point in time because of the bursty nature of the data being sent, but in aggregate the network will maintain an average, guaranteed flow volume for each PVC. This is a selling point for frame relay. In most cases, customers get more than they actually pay for, and as long as the switch loading levels are properly engineered, the switch (and therefore the frame-relay service offering) will not suffer adversely from this charitable bandwidth-allocation philosophy. The key to success when selling frame relay is to have a very clear understanding of the applications the customer intends to use across the link so that the access facility can be properly sized for the anticipated traffic load.

Managing Service in Frame-Relay Networks

Frame relay does not offer a great deal of granularity when it comes to QoS. The only inherent mechanism is the DE bit described earlier as a way to control network congestion. However, the DE bit is binary; it has two possible values, which means that a customer has two choices: The information being sent is either important, or it isn't—not particularly useful for establishing a variety of QoS levels. Consequently, a number of vendors have implemented proprietary solutions for QoS management. Within their routers (sometimes called frame-relay access devices, or FRADs), they have established queuing mechanisms that allow customers to create multiple priority levels for differing traffic flows. For example, voice and video—which don't tolerate delay well—could be assigned to a higher priority queue than the one to which asynchronous data traffic would be assigned. This allows frame relay to provide highly granular service. The downside is that this approach is proprietary, which means that the same vendor's equipment must be used on both ends of the circuit. Given the strong move toward interoperability, this is not an ideal solution because it locks the customer into a single-vendor situation.

Congestion Control in Frame Relay

Frame relay has two congestion control mechanisms. Embedded in the header of each frame-relay frame are two additional bits called the *forward explicit congestion notification bit* (FECN) and the *backward explicit congestion notification bit* (BECN). Both are used to notify devices in the network of congestion situations that could affect throughput.

Consider the following scenario. A frame-relay frame arrives at the second of three switches along the path to its intended destination, where it encounters severe local congestion (see Figure 7-4). The congested switch sets the FECN bit to indicate the presence of congestion and transmits the frame to the next switch in the chain. When the frame arrives, the receiving switch takes note of the FECN bit, which tells it the following: "I just came from that switch back there, and it's extremely congested. You can transmit stuff back there if you want to, but there's a good chance that anything you send will be discarded, so you might want

Figure 7-4
A frame encounters congestion in transit.

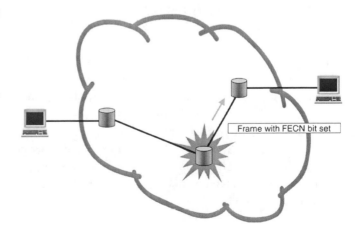

Frame with FECN bit set

Figure 7-5
BECN bit used
to warn a
station that is
transmitting
too much data
(in violation of
its agreement
with the
switch)

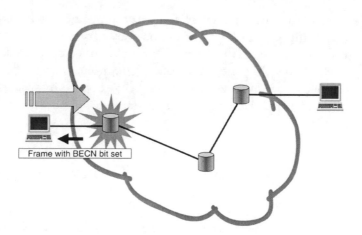

Frame with BECN bit set

to wait awhile before transmitting." In other words, the switch has been notified of a congestion condition, to which it may respond by throttling back its output to allow the affected switch time to recover.

On the other hand, the BECN bit is used to flow-control a device that is sending too much information into the network. Consider the situation shown in Figure 7-5, in which a particular device on the network is transmitting at a high volume, routinely violating the CIR and perhaps the EIR level established by mutual consent. The ingress switch—that is, the first switch the traffic touches—has the ability to set the BECN bit on frames going toward the offending device, which carries the implicit message, "Cut it out or I'm going to hurt you." In effect, the BECN bit notifies the offending device that it is violating protocol, and continuing to do so will result in every frame from that device being discarded— without warning or notification. If this happens, it gives the ingress switch the opportunity to recover. However, it doesn't fix the problem— it merely forestalls the inevitable, because sooner or later the intended recipient will realize that frames are missing and will initiate recovery procedures, which will cause resends to occur. However, it may give the affected devices time to recover before the onslaught begins anew.

The problem with FECN and BECN lies in the fact that many devices choose not to implement them. They do not necessarily have the inherent ability to throttle back upon receipt of a congestion indicator, although devices that can are becoming more common. Nevertheless, proprietary solutions are in widespread use and will continue to be for some time to come.

Frame Relay Summary

Frame relay is clearly a Cinderella technology, evolving quickly from a data-only transport scheme to a multiservice technology with diverse capabilities. For data and some voice and video applications, it shines as a WAN offering. In some areas, however, frame relay is lacking. Its bandwidth is limited to DS3, and its ability to offer standards-based QoS is limited. Given the focus on QoS that is so much a part of customers' mantra today, and the flexibility that a switched-solution permits, something else was required. That something was ATM.

Asynchronous Transfer Mode (ATM)

Network architectures often develop in concert with the corporate structures that they serve. Companies with centralized management authorities such as utilities, banks, and hospitals often have centralized and tightly controlled hierarchical data-processing architectures to protect their data. On the other hand, organizations that are distributed in nature—such as research and development facilities and universities—often have highly distributed data processing architectures. They tend to share information on a peer-to-peer basis and their corporate structures reflect this fact.

ATM came about not only because of the proliferation of diverse network architectures but also because of the evolution of traffic characteristics and transport requirements. To the well-known demands of voice, we now add various flavors of data, video, MP3, an exponentially large variety of IP traffic, interactive real-time gaming, and a variety of other content types that place increasing demands on the network. Further, we have seen a requirement arise for a mechanism that can transparently and correctly transport the *mix* of various traffic types over a single network infrastructure, while at the same time delivering granular, controllable, and measurable QoS levels for each service type. In its original form, ATM was designed to do exactly that, working with SONET or SDH to deliver high-speed transport and switching throughout the network—in the wide area, the metropolitan area, the campus environment, and the LAN, right down to the desktop—seamlessly, accurately, and fast.

Figure 7-6
The ATM cell

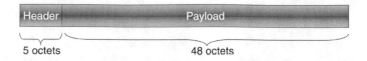

5 octets   48 octets

Today, because of competition from such technologies as QoS-aware IP transport, proprietary high-speed mesh networks, and Fast and Gigabit Ethernet, ATM, for the most part, has lost the race to the desktop. ATM is a cell-based technology, which simply means that the fundamental unit of transport—a frame of data, if you will—is of a fixed size; this allows switch designers to build faster, simpler devices, since they can always count on their switched payload being the same size at all times. That cell comprises a five-octet header and a 48-octet payload field, as shown in Figure 7-6. The payload contains user data; the header contains information that the network requires to both transport the payload correctly and ensure proper quality of service levels for the payload. ATM accomplishes this task well, but at a cost. The five-octet header makes up nearly 10 percent of the cell—a rather significant price to pay, particularly when other technologies such as IP and SONET add their own significant percentages of overhead to the overall payload. This reality is part of the problem: ATM's original claims to fame, and the reasons it rocketed to the top of the technology hit parade, were its ability to switch cells at tremendous speed through the fabric of the wide-area network and the ease with which the technology could be scaled to fit any network situation. Today, however, given the availability of high-speed IP routers that routinely route packets at terabit rates, ATM's advantages have begun to pale to a certain degree.

ATM Evolution

ATM has, however, emerged from the flames in other ways. Today, many service providers see ATM as an ideal aggregation technology for diverse traffic streams that need to be combined for transport across a WAN that will most likely be IP-based. ATM devices, then, will be placed at the edge of the network, where they will collect traffic for transport across the Internet or (more likely) a privately owned IP network. Furthermore, because it has the ability to be something of a chameleon by delivering diverse services across a common network fabric, it is further guaranteed a seat at the technology game.

It is interesting to note that the traditional, legacy telecommunications network comprises two principal "regions" that can be clearly distinguished from each other: the network itself, which provides switching, signaling, and transport for traffic generated by customer applications, and the access loop, which provides the connectivity between the customer's applications and the network. In this model, the network is considered to be a relatively intelligent medium, while the customer equipment is usually considered to be relatively "stupid."

Not only is the intelligence seen as being concentrated within the confines of the network, so too is the bulk of the bandwidth, because the legacy model indicates that traditional customer applications don't require much of it. Between central office switches, however, and between the offices themselves, enormous bandwidth is required.

Today, this model is changing. Customer equipment has become remarkably intelligent, and many of the functions previously done within the network cloud are now performed at the edge. PBXs, computers, and other devices are now capable of making discriminatory decisions about required service levels, eliminating any need for the massive intelligence embedded in the core.

At the same time, the bandwidth is migrating from the core of the network toward the customer as applications evolve to require it. There is still massive bandwidth within the cloud, but the margins of the cloud are expanding toward the customer.

The result of this evolution is a redefinition of the network's regions. Instead of a low-speed, low-intelligence access area and a high-speed, highly intelligent core, the intelligence has migrated outward to the margins of the network, and the bandwidth—once exclusively a core resource—is now equally distributed at the edge. Thus, we see something of a core and edge distinction evolving as customer requirements change.

One reason for this steady migration is the well-known fact within sales and marketing circles that products sell best when they are located close to the buying customer. They are also easier to customize for individual customers when they are physically closest to the situation for which the customer is buying them.

In "The Rise of the Stupid Network," David Isenberg makes the following observation:

> The Intelligent Network is a straight-line extension of four assumptions—scarcity, voice, circuit switching, and control. Its primary design impetus was not customer service. Rather, the Intelligent Network was a telephone

company attempt to engineer vendor independence, more automatic operation, and some "intelligent" new services into existing network architecture. However, even as it rolls out and matures, the Intelligent Network is being superseded by a Stupid Network, with nothing but dumb transport in the middle and intelligent user-controlled endpoints, whose design is guided by plenty, not scarcity, where transport is guided by the needs of the data, not the design assumptions of the network.

Isenberg continues:

A new network "philosophy and architecture," is replacing the vision of an Intelligent Network. The vision is one in which the public communications network would be engineered for "always-on" use, not intermittence and scarcity. It would be engineered for intelligence at the end-user's device, not in the network. And the network would be engineered simply to "Deliver the Bits, Stupid," not for fancy network routing or "smart" number translation.[1]

ATM Technology Overview

Because ATM plays such a major role in networks today, it is important to develop at least a rudimentary understanding of its functions, architectures, and offered services.

ATM Protocols

Like all modern technologies, ATM has a well-developed protocol stack (see Figure 7-7) that clearly delineates the functional breakdown of the service. The stack consists of four layers: the upper services layer, the ATM adaptation layer, the ATM layer, and the physical layer.

The *upper services layer* defines the nature of the actual services that ATM can provide. It identifies both constant and variable bit rate services. Voice is an example of a constant bit rate service, while signaling, IP, and frame relay are examples of both connectionless and connection-oriented variable bit rate services.

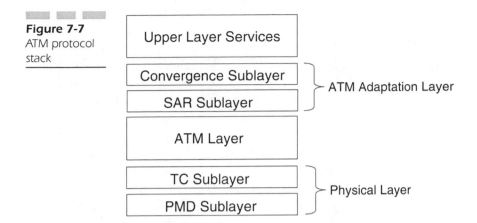

Figure 7-7
ATM protocol stack

The *ATM adaptation layer*, or AAL, has four general responsibilities:

- Synchronization and recovery from errors,
- Error detection and correction,
- Segmentation and reassembly of the data stream, and
- Multiplexing.

The AAL comprises two functional sublayers. The *convergence sublayer* provides service-specific functions to the services layer so that the services layer can make the most efficient use of the underlying cell-relay technology that ATM provides. Its functions include clock recovery for end-to-end timing management; a recovery mechanism for lost or out-of-order cells; and a timestamp capability for time-sensitive traffic such as voice and video.

The *segmentation and reassembly sublayer*, or SAR, converts the user's data from its original incoming form into the 48-octet payload "chunks" that will become cells. For example, if the user's data is arriving in the form of 64-kilobyte IP packets, SAR chops them into 48-octet payload pieces. It also has the responsibility to detect lost or out-of-order cells that the convergence sublayer will recover from, and to detect single-bit errors in the payload chunks.

The *ATM layer* has five general responsibilities:

- Cell multiplexing and demultiplexing,
- Virtual path and virtual channel switching,
- Creation of the cell header,

- Generic flow control, and
- Cell delineation.

Because the ATM layer creates the cell header, it is responsible for all of the functions that the header manages. The process, then, is fairly straightforward: The user's data passes from the services layer to the ATM adaptation layer, which segments the data stream into 48-octet pieces. The pieces are handed to the ATM layer, which creates the header and attaches it to the payload unit, thus creating a cell. The cells are then handed down to the physical layer.

The *physical layer* consists of two functional sublayers as well: the transmission convergence sublayer and the physical medium sublayer. The *transmission convergence sublayer* performs three primary functions. The first is called cell rate decoupling, which adapts the cell creation and transmission rate to the rate of the transmission facility by performing cell stuffing—similar to the bit stuffing process described earlier in the discussion of DS3 frame creation. The second responsibility is cell delineation, which allows the receiver to delineate between one cell and the next. Finally, it generates the transmission frame in which the cells are to be carried.

The *physical medium sublayer* takes care of issues that are specific to the medium being used for transmission, such as line codes, electrical and optical concerns, timing, and signaling.

The physical layer can use a wide variety of transport options, including the following:

- DS1/DS2/DS3

- E1/E3

- 25.6 Mbps UNI over UTP-3

- 51 Mbps UNI over UTP-5 (Transparent Asynchronous Transmitter/Receiver Interface, or TAXI)

- 100 Mbps UNI over UTP-5

- OC3/12/48c

Others, of course, will follow as transport technologies advance.

The ATM Cell Header

As we mentioned before, ATM is a cell-based technology that relies on a 48-octet payload field that contains actual user data, and a five-byte header that contains information needed by the network to route the cell and provide proper levels of service.

The ATM cell header, shown in Figure 7-8, is examined and updated by each switch it passes through, and comprises six distinct fields: the generic flow control field, the virtual path identifier, the virtual channel identifier, the payload type identifier, the cell loss priority field, and the header error control field.

- *Generic flow control* (GFC): This four-bit field is used across the UNI for network-to-user flow control. It has not yet been completely defined in the ATM standards, but some companies have chosen to use it for very specific purposes. For example, Australia's Telstra Corporation uses it for flow control in the network-to-user direction and as a traffic priority indicator in the user-to-network direction.

- *Virtual path identifier* (VPI): The eight-bit VPI identifies the virtual path over which the cells will be routed at the UNI. It should be noted that because of dedicated, internal flow control capabilities

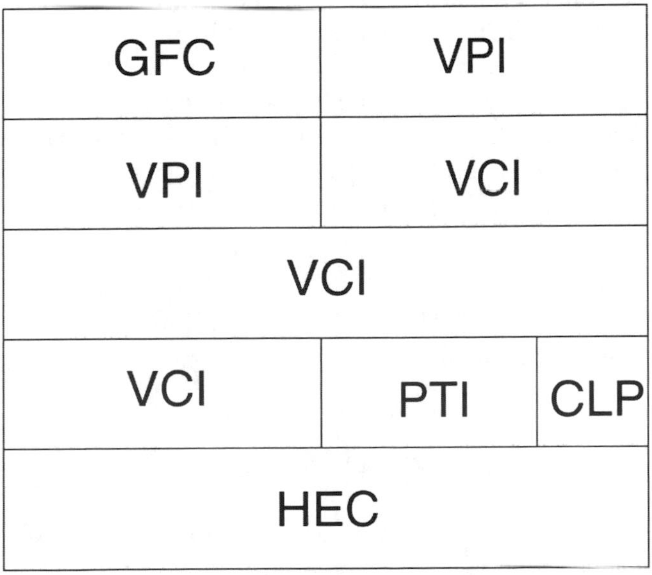

Figure 7-8
ATM cell
header details

within the network, the GFC field is not needed across the NNI. It therefore is redeployed: The four bits are converted to additional VPI bits, thus extending the size of the virtual path "field." This allows for the identification of more than 4,000 unique VPs. At the UNI, this number is excessive, but across the NNI it is necessary because of the number of potential paths that might exist between the switches that make up the fabric of the network.

■ *Virtual channel identifier* (VCI): As the name implies, the 16-bit VCI identifies the unidirectional virtual channel over which the current cells will be routed.

■ *Payload type identifier (PTI)*: The three-bit PTI field is used to indicate network congestion and cell type, in addition to a number of other functions. The first bit indicates whether the cell was generated by the user or by the network, while the second indicates the presence or absence of congestion in user-generated cells, or flow-related operations, administration, and maintenance (OA&M) information in cells generated by the network. The third bit is used for service-specific, higher-layer functions in the user-to-network direction, such as to indicate that a cell is the last in a *series* of cells. From the network to the user, the third bit is used with the second bit to indicate whether the OA&M information refers to segment or end-to-end-related information flow.

■ *Cell loss priority (CLP)*: The single-bit CLP field is a relatively primitive flow-control mechanism by which the user can indicate to the network which cells to discard in the event of a condition that demands that some cells be eliminated, similar to the DE bit in frame relay. It can also be set by the network to indicate to downstream switches that certain cells in the stream are eligible for discard, should that become necessary.

■ *Header error control (HEC)*: The eight-bit HEC field can be used for two purposes. First, it provides for the calculation of an eight-bit CRC that checks the integrity of the entire header. Second, it can be used for cell delineation.

Addressing in ATM

ATM is a connection-oriented, virtual circuit technology, meaning that communication paths are created through the network prior to actually

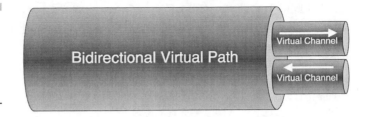

Figure 7-9
Addressing in
ATM with
virtual
channels,
paths

sending traffic. Once established, the ATM cells are routed based upon a virtual circuit address. A virtual circuit is simply a connection that gives the user the appearance of being dedicated to that user, when in point of fact the only thing that is actually dedicated is a time slot. This technique is generically known as *label-based switching*, and is accomplished through the use of routing tables in the ATM switches that designate input ports, output ports, input addresses, output addresses, and QoS parameters required for proper routing and service provisioning. As a result, cells do not contain explicit destination addresses, but rather contain timeslot identifiers.

There are two components to every virtual circuit address, as shown in Figure 7-9. The first is the virtual channel (VC), which is a unidirectional conduit for the transmission of cells between two endpoints. For example, if two parties are conducting a videoconference, they will each have a VC for the transmission of outgoing cells that make up the video portion of the conference.

The second "level" of the ATM addressing scheme is called a virtual path (VP). A VP is a bundle of VCs that have the same endpoints and that, when considered together, make up a bidirectional transport facility. The combination of unidirectional channels that we need in our two-way videoconferencing example makes up a VP.

ATM Services

The basic services that ATM provides are based on three general characteristics: the nature of the connection between the communicating stations (connection-oriented vs. connectionless), the timing relationship between the sender and the receiver, and the bit rate required to ensure proper levels of service quality. Based on those generic requirements, both the ITU-T and the ATM Forum have created service classes that address the varying requirements of the most common forms of transmitted data.

ITU-T Service Classes

The ITU-T assigns service classes based on three characteristics: connection mode, bit rate, and the end-to-end timing relationship between the end stations. They have created four distinct service classes based on the model shown in Figure 7-10. Class A service, for example, defines a connection-oriented, constant bit rate, timing-based service that is ideal for the stringent requirements of voice service. Class B, on the other hand, is ideal for such services as variable bit rate video, in that it defines a connection-oriented, variable bit rate, timing-based service.

Class C service was defined for such things as frame relay, in that it provides a connection-oriented, variable bit rate, timing-independent service. Finally, Class D delivers a connectionless, variable bit rate, timing-independent service that is ideal for IP traffic as well as SMDS.

In addition to service classes, the ITU-T has defined AAL service types, which align closely with the A, B, C, and D service types described previously. Whereas the service classes (A, B, C, D) describe the capabilities of the underlying network, the AAL types describe the cell format. They are AAL1, AAL2, AAL¾, and AAL 5. However, only two of them have really survived in a big way.

AAL1 is defined for Class A service, which is a constant bit rate environment ideally suited for voice and voice-like applications. In AAL1 cells, the first octet of the payload serves as a payload header that con-

Figure 7-10
ITU-T ATM
service
definitions

	Class A	Class B	Class C	Class D
AAL Type	1	2	5, 3/4	5, 3/4
Connection Mode	Connection-oriented	Connection-oriented	Connection-oriented	Connectionless
Bit Rate	Constant	Variable	Variable	Variable
Timing Relationship	Required	Required	Not required	Not required
Service Types	Voice, video	VBR voice, video	Frame relay	IP

tains cell sequence and synchronization information that is required to provision constant bit rate, fully sequenced service. AAL1 provides circuit emulation service without dedicating a physical circuit, which explains the need for an end-to-end timing relationship between the transmitter and the receiver.

AAL5, on the other hand, is designed to provide both Class C and D services, and while it was originally proposed as a transport scheme for connection-oriented data services, it turns out to be more efficient than AAL3/4 and accommodates connectionless services quite well.

To guard against the possibility of errors, AAL5 has an eight-octet trailer appended to the user data, which includes a variable-size *pad field* used to align the payload on 48-octet boundaries, a two-octet *control field* that is currently unused, a two-octet *length field* that indicates the number of octets in the user data, and, finally, a four-octet *CRC* that can check the integrity of the entire payload. AAL5 is often referred to as the "simple and easy adaptation layer," or SEAL, and it may find an ideal application for itself in the burgeoning Internet arena. Recent studies indicate that TCP/IP transmissions produce comparatively large numbers of small packets that tend to be around 48 octets long. That being the case, AAL5 could well transport the bulk of them in its user data field. Furthermore, the maximum size of the user data field is 65,536 octets; coincidentally, the same size as an IP packet.

ATM Forum Service Classes

The ATM Forum looks at service definitions slightly differently than the ITU-T does (see Figure 7-11). Instead of the A-B-C-D services, the ATM

Figure 7-11
ATM Forum
service
definitions

Service	Descriptors	Loss	Delay	Bandwidth	Feedback
CBR	PCR, CDVT	Yes	Yes	Yes	No
VBR-RT	PCR, CDVT, SCR, MBS	Yes	Yes	Yes	No
VBR-NRT	PCR, CDVT, SCR, MBS	Yes	Yes	Yes	No
UBR	PCR, CDVT	No	No	No	No
ABR	PCR, CDVT, MCR	Yes	No	Yes	Yes

Forum categorizes them as real-time and non-real-time services. Under the real-time category, they define constant bit rate services that demand fixed resources with guaranteed availability. They also define real-time, variable bit rate (VBR) service, which provides for statistical multi-plexed, variable-bandwidth service allocated on demand. A further subset of real-time VBR is peak-allocated VBR, which guarantees constant loss and delay characteristics for all cells in that flow.

Under the non-real-time service class, unspecified bit rate (UBR) is the first service category. UBR is often compared to IP in that it is a "best effort" delivery scheme in which the network provides whatever bandwidth it has available, with no guarantees made. All recovery functions from lost cells are the responsibility of the end user devices.

UBR has two subcategories of its own. The first, non-real-time UBR (NRT-UBR), improves the impacts of cell loss and delay by adding a network resource reservation capability. Available bit rate (ABR), UBR's other subcategory, makes use of feedback information from the far end to manage loss and ensure fair access to and transport across the network.

Each of the five classes makes certain guarantees with regard to cell loss, cell delay, and available bandwidth. Furthermore, each of them takes into account descriptors that are characteristic of each service described. These include peak cell rate (PCR), sustained cell rate (SCR), minimum cell rate (MCR), cell delay variation tolerance (CDVT), and burst tolerance (BT).

ATM Forum Specified Services

The ATM Forum has identified a collection of services for which ATM is a suitable, perhaps even desirable, network technology. These include cell relay service (CRS), circuit emulation service (CES), voice and telephony over ATM (VTOA), frame-relay bearer service (FRBS), LAN emulation (LANE), multiprotocol over ATM (MPOA), and a collection of others.

CRS is the most basic of the ATM services. It delivers precisely what its name implies: a "raw pipe" transport mechanism for cell-based data. As such it does not provide any ATM bells and whistles, such as quality of service discrimination; nevertheless, it is the most commonly implemented ATM offering because of its lack of implementation complexity.

CES gives service providers the ability to offer a selection of bandwidth levels by varying both the number of cells transmitted per second and the number of bytes contained in each cell.

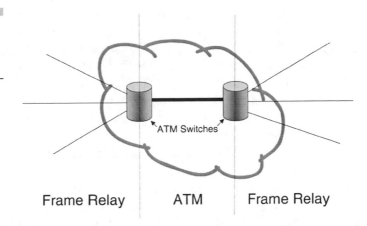

Figure 7-12
Frame-relay
bearer service
(FRBS) in ATM

VTOA is a service that has yet to be clearly defined. The ability to transport voice calls across an ATM network is a nonissue, given the availability of Class A service. What are not clearly defined, however, are corollary services such as 800/888 calls, 900 service, 911 call handling, enhanced services billing, SS7 signal interconnection, and so on. Until these issues are clearly resolved, ATM-based, feature-rich telephony will not become a mainstream service, but will instead be limited to simple voice—and there *is* a difference.

FRBS refers to the ability of ATM to interwork with frame relay. Conceptually, the service implies that an interface standard allows an ATM switch to exchange date with a frame-relay switch, thus allowing for interoperability between frame- and cell-based services. Many manufacturers are taking a slightly different tack, however: They are building switches with soft, chewy cell technology at the core, and surrounding the core with hard, crunchy interface cards to suit the needs of the customer.

For example, an ATM switch might have ATM cards on one side to interface with other ATM devices in the network but frame-relay cards on the other side to allow it to communicate with other frame-relay switches, as shown in Figure 7-12. Thus, a single piece of hardware can logically serve as both a cell- and frame-relay switch. This design is becoming more and more common, because it helps to avoid a future rich with forklift upgrades.

LANE allows an ATM network to move traffic transparently between two similar LANs but also allows ATM to transparently slip into the LAN arena. For example, two Ethernet LANs could communicate across the fabric of an ATM network, as could two token ring LANs. In effect, LANE allows ATM to provide a bridging function between similar LAN environments. In LANE implementations, the ATM network does not handle MAC functions such as collision detection, token passing, or beaconing; it merely provides the connectivity between the two communicating endpoints. The MAC frames are simply transported inside AAL5 cells.

One clear concern about LANE is that LANs are connectionless, while ATM is a virtual circuit-based, connection-oriented technology. LANs routinely broadcast messages to all stations, while ATM allows point-to-point or multipoint circuits only. Thus, ATM must look like a LAN if it is to behave like one. To make this happen, LANE uses a collection of specialized LAN emulation clients and servers to provide the connectionless behavior expected from the ATM network.

On the other hand, MPOA provides the ATM equivalent of *routing* in LAN environments. In MPOA installations, routers are referred to as MPOA servers. When one station wants to transmit to another station, it queries its local MPOA server for the remote station's ATM address. The local server then queries its neighbor devices for information about the remote station's location. When a server finally responds, the originating station uses the information to establish a connection with the remote station, while the other servers cache the information for further use.

MPOA promises a great deal, but it is complex to implement and requires other ATM "components," such as the Private NNI capability, to work properly. Furthermore, it's being challenged by at least one alternative technology, known as IP switching.

IP switching is far less overhead intensive than MPOA. Furthermore, it takes advantage of a known (but often ignored) reality in the LAN interconnection world: Most routers today use IP as their core protocol and the great majority of LANs are still Ethernet. This means that a great deal of simplification can be done by crafting networks to operate around these two technological bases. And in fact, this is precisely what IP switching does. By using existing, low-overhead protocols, the IP-switching software creates new ATM connections dynamically and quickly, updating switch tables on the fly. In IP-switching environments, IP resides on top of ATM within a device (see Figure 7-13) providing the best of both protocols. If two communicating devices wish to exchange information and they have done so before, an ATM mapping already exists and no layer three involvement (IP) is required—the ATM switch portion of the service simply creates the connection at high speed. If an

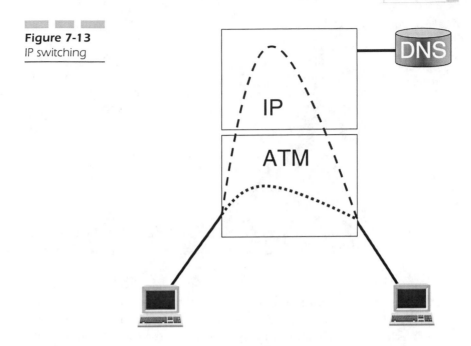

Figure 7-13
IP switching

address lookup is required, then the "call" is handed up to IP, which takes whatever steps are required to perform the lookup (a DNS request, for example). Once it has the information, it hands it down to ATM, which proceeds to set up the call. The next time the two need to communicate, ATM will be able to handle the connection.

There are other services looming on the horizon in which ATM plays a key role, including wireless ATM and video-on-demand for the delivery of interactive content such as videoconferencing and television. This leads to what I often refer to as "the great triumvirate": ATM, SONET or SDH, and broadband services. By combining the powerful switching and multiplexing fabric of ATM with the limitless transport capabilities of SONET or SDH, true broadband services can be achieved, and the idea of creating a network that can be all things to all services can finally be realized.

Optical Networking

Optical networking is often viewed as a point-to-point technology. It has achieved such a position of prominence in the last 2 years, however, that it qualifies as a transport option in its own right. Furthermore, optical

Figure 7-14
Bell
photophone
transmitter

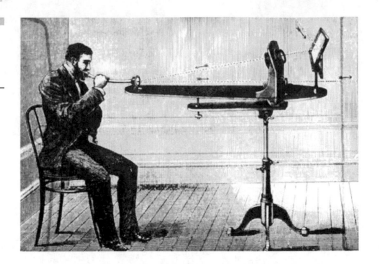

switching is fast becoming real and optical routing is not far behind. In this next section, we discuss the development of optical networking, the various technologies that it employs, and the direction it seems to be going in this fast-paced market.

Early Optical Technology Breakthroughs

In 1878, 2 years after perfecting his speaking telegraph (which became the telephone), Alexander Graham Bell created a device that transmitted the human voice through the air for distances up to 200 meters. The device, which he called the photophone, used carefully angled mirrors to reflect sunlight onto a diaphragm that was attached to a mouthpiece, as shown in Figure 7-14. At the receiving end (see Figure 7-15), the light was concentrated by a parabolic mirror onto a selenium resistor, which was connected to a battery and speaker. The diaphragm vibrated when struck by the human voice, which in turn caused the intensity of the light striking the resistor to vary. The selenium resistor, then caused the current flow to vary in concert with the varying sunlight, which caused the received sound to come out of the speaker with remarkable fidelity. This represented the birth of optical transmission.

Optical transmission in its early days was limited in terms of what it was capable of doing. Consider the following analogy: If you look through a 2-foot-square pane of window glass, it appears clear—if the glass is clean, it is virtually invisible. However, if you turn the pane on edge and look through it from edge to edge, the glass appears to be dark green.

Figure 7-15
Bell
photophone
receiver

Very little light passes from one edge to the other. In this example, you are looking through 2 feet of glass. Imagine trying to pass a high-bandwidth optical signal through 40 or more kilometers of that glass!

In 1966, Charles Kao and Charles Hockham at the UK's Standard Telecommunication Laboratory (now part of Nortel Networks) published their seminal work, demonstrating that optical fiber could be used to carry information provided its end-to-end signal loss could be kept below 20 dB per kilometer. Keeping in mind that the decibel scale is logarithmic, 20 dB of loss means that *99 percent of the light would be lost over each kilometer of distance.* Only 1 percent would actually reach the receiver—and that's a 1-kilometer run. Imagine the loss over today's fiber cables that are hundreds of kilometers long, if 20 dB was the modern performance criterion!

Kao and Hockham proved that metallic impurities in the glass—such as chromium, vanadium, iron, and copper—were the primary cause for such high levels of loss. In response, glass manufacturers rose to the challenge and began to research the creation of ultrapure products.

In 1970, Peter Schultz, Robert Maurer, and Donald Keck of Corning Glass Works (now Corning Corporation) announced the development of a glass fiber that offered better attenuation than the recognized 20 dB threshold. Today, fiber manufacturers offer fiber so incredibly pure that 10 percent of the light arrives at a receiver placed 50 kilometers away. Put another way, a fiber with 0.2 dB of measured loss delivers more than 60 percent of the transmitted light over a distance of 10 kilometers. Remember the windowpane example? Imagine glass so pure that you could see clearly through a window 10 kilometers thick.

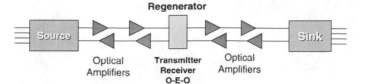

Figure 7-16
Components of
a typical optical
network

Fundamentals of Optical Networking

At their most basic level, optical networks require three fundamental components, as shown in Figure 7-16: a source of light, a medium over which to transport it, and a receiver for the light. Additionally, there may be regenerators, optical amplifiers, and other pieces of equipment in the circuit. We will examine each of these generic components in turn.

Optical Sources

Today the most common sources of light for optical systems are either light-emitting diodes or laser diodes. Both are commonly used, although laser diodes have become more common for high-speed data applications because of their coherent signal. While lasers have gone through several iterations over the years including ruby rod and helium-neon, semiconductor lasers became the norm shortly after their introduction in the early 1960s because of their low cost and high stability.

Light-Emitting Diodes (LEDs)

LEDs come in two varieties: *surface-emitting LEDs* and *edge-emitting LEDs*. Surface-emitting LEDs give off light at a wide angle and therefore do not lend themselves to the more coherent requirements of optical data systems because of the difficulty involved in focusing their emitted light into the core of the receiving fiber. Instead, they are often used as indicators and signaling devices. They are, however, quite inexpensive, and are therefore commonly found.

An alternative to the surface-emitting LED is the edge-emitting device. Edge emitters produce light at significantly narrower angles and have a smaller emitting area, which means that more of their emitted light can be focused into the core. They are typically faster devices than surface emitters, but do have a downside: They are temperature sensitive and, therefore, must be installed in environmentally controlled devices to ensure the stability of the transmitted signal.

Laser Diodes

Laser diodes represent the alternative to LEDs. A laser diode has a very small emitting surface, usually no larger than a few microns in diameter, which means that a great deal of the emitted light can be directed into the fiber. Because they represent a coherent source, the emission angle of a laser diode is extremely narrow. It is the fastest of the three devices.

Optical Fiber

When Peter Schultz, Donald Keck, and Robert Maurer began their work at Corning to create a low-loss optical fiber, they did so using a newly crafted process called *inside vapor deposition* (IVD). Whereas most glass is manufactured by melting and reshaping silica, IVD deposits various combinations of carefully selected compounds on the inside surface of a silica tube. The tube becomes the cladding of the fiber; the vapor-deposited compounds become the core. The compounds are typically silicon chloride ($SiCl_4$) and oxygen (O_2), which are reacted under heat to form a soft, sooty deposit of silicon dioxide (SiO_2), as shown in Figure 7-17. In some cases, impurities such as germanium are added at this time to cause various effects in the finished product. In practice, the $SiCl_4$ and O_2 are pumped into the fused silica tube as gases; the tube is heated in a high-temperature lathe, causing the sooty deposit to collect on the inside surface of the tube. The continued heating of the tube causes the soot to fuse into a glasslike substance.

This process can be repeated as many times as required to create a graded refractive index, if required. Ultimately, once the deposits are complete, the entire assembly is heated fiercely, which causes the tube to collapse, creating what is known in the optical fiber industry as a *preform*. An example of a preform is shown in Figure 7-18.

Figure 7-17
Creating a
multilayer
preform

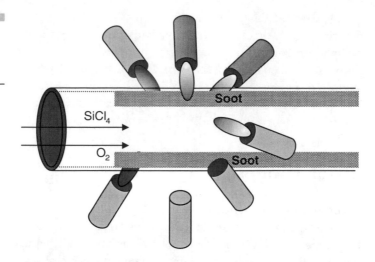

Figure 7-18
Preforms, ready
to be drawn

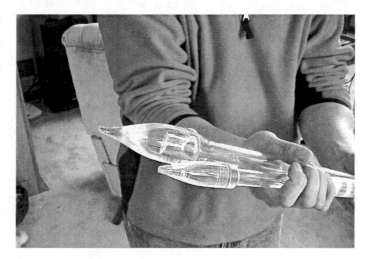

An alternative manufacturing process is called *outside vapor deposition* (OVD). In the OVD process, the soot is deposited on the surface of a rotating ceramic cylinder in two layers. The first layer is the soot that will become the core; the second layer becomes the cladding. Ultimately, the rod and soot are sintered to create a preform. The ceramic is then removed, leaving behind the fused silica that will become the fiber.

There are a number of other techniques for creating the preforms that are used to create fiber, but these are the principal techniques in use today.

The next step is to convert the preform into optical fiber.

Drawing the Fiber

To make fiber from a preform, the preform is mounted in a furnace at the top of a tall building called a *drawing tower*. The bottom of the preform is heated until it has the consistency of taffy, at which time the soft glass is drawn down to form a thin fiber. When it strikes the cooler air outside the furnace, the fiber solidifies. Needless to say, the process is carefully managed to ensure that the thickness of the fiber is precise; microscopes are used to verify the geometry of the fiber.

Other stages in the manufacturing process include monitoring processes to check the integrity of the product, a coating process that applies a protective layer, and a take-up stage where the fiber is wound onto reels for later assembly into cables of various types.

Optical Fibers

There are dozens of different types of fiber. Some of them are holdovers from previous generations of optical technology that are still in use and represented the best efforts of technology available at the time; others represent improvements on the general theme or specialized solutions to specific optical transmission challenges.

Generally speaking, there are two major types of fiber: *multimode*, which is the earliest form of optical fiber and is characterized by a large diameter central core, short distance capability, and low bandwidth; and *single mode*, which has a narrow core and is capable of greater distance and higher bandwidth. There are varieties of each that will be discussed in detail, later in the book.

To understand the reason for and philosophy behind the various forms of fiber, it is first necessary to understand the issues that confront transmission engineers who design optical networks.

Optical fiber has a number of advantages over copper: It is lightweight, has enormous bandwidth potential, has significantly higher tensile strength, can support many simultaneous channels, and is immune to electromagnetic interference. It does, however, suffer from several disruptive problems that cannot be discounted. The first of these is *loss* or *attenuation*, the inevitable weakening of the transmitted signal over distance that has a direct analog in the copper world. Attenuation is typically the result of two subproperties, *scattering* and *absorption*, both of which have cumulative effects. The second is *dispersion*, which is the spreading of the transmitted signal and is analogous to noise.

Scattering

Scattering occurs because of impurities or irregularities in the physical makeup of the fiber itself. The best known form of scattering is called *Rayleigh Scattering*; it is caused by metal ions in the silica matrix and results in light rays being scattered in various directions.

Rayleigh Scattering occurs most commonly around wavelengths of 1,000 nm, and is responsible for as much as 90 percent of the total attenuation that occurs in modern optical systems. It occurs when the wavelengths of the light being transmitted are roughly the same size as the physical molecular structures within the silica matrix; thus, short wavelengths are affected by Rayleigh Scattering effects far more than long wavelengths. In fact, it is because of Rayleigh Scattering that the sky appears to be blue: The shorter (blue) wavelengths of light are scattered more than the longer wavelengths of light.

Absorption

Absorption results from three factors: hydroxyl (OH^-, water) ions in the silica, impurities in the silica, and incompletely diminished residue from the manufacturing process. These impurities tend to absorb the energy of the transmitted signal and convert it to heat, resulting in an overall weakening of the optical signal. Hydroxyl absorption occurs at 1.25 and 1.39 μ; at 1.7 μ, the silica itself starts to absorb energy because of the natural resonance of silicon dioxide.

Dispersion

As mentioned earlier, dispersion is the optical term for the spreading of the transmitted light pulse as it transits the fiber. It is a bandwidth-limiting phenomenon and comes in two forms: *multimode dispersion*, and *chromatic dispersion*. Chromatic dispersion is further subdivided into *material dispersion* and *waveguide dispersion*.

Multimode Dispersion

To understand multimode dispersion, it is first important to understand the concept of a *mode*. Figure 7-19 shows a fiber with a relatively wide

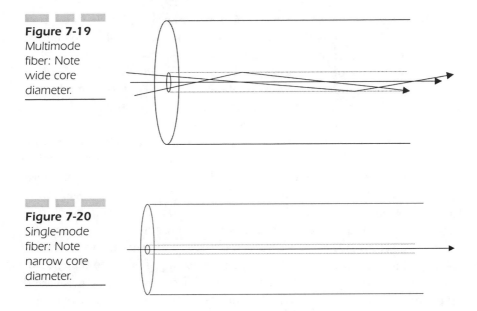

Figure 7-19
Multimode fiber: Note wide core diameter.

Figure 7-20
Single-mode fiber: Note narrow core diameter.

core. Because of the width of the core, it allows light rays arriving from the source at a variety of angles (three in this case) to enter the fiber and be transmitted to the receiver. Because of the different paths that each ray, or mode, will take, they will arrive at the receiver at different times, resulting in a dispersed signal.

Now consider the system shown in Figure 7-20. The core is much narrower and only allows a single ray, or mode, to be sent down the fiber. This results in less end-to-end energy loss and avoids the dispersion problem that occurs in multimode installations.

Chromatic Dispersion

The speed at which an optical signal travels down a fiber is absolutely dependent upon its wavelength. If the signal comprises multiple wavelengths, then the different wavelengths will travel at different speeds, resulting in an overall spreading or smearing of the signal. As discussed earlier, chromatic dispersion comprises two subcategories: material dispersion and waveguide dispersion.

Material Dispersion Simply put, material dispersion occurs because different wavelengths of light travel at different speeds through an optical fiber. To minimize this particular dispersion phenomenon, two factors must be managed. The first of these is the number of wavelengths that make up the transmitted signal. An LED, for example, emits a rather broad range of wavelengths—between 30 and 180 nm—whereas a laser emits a much narrower spectrum—typically less than 5 nm. Thus, a laser's output is far less prone to be seriously affected by material dispersion than is the signal from an LED.

The second factor that affects the degree of material dispersion is a characteristic called the *center operating wavelength of the source signal*. In the vicinity of 850 nm, red, longer wavelengths travel faster than their shorter, blue counterparts, but at 1,550 nm, the situation is the opposite: Blue wavelengths travel faster. There is, of course, a point at which the two meet and share a common minimum dispersion level; it is in the range of 1,310 nm, often referred to as the zero-dispersion wavelength. Clearly, this is an ideal place to transmit data signals, since dispersion effects are minimized here. As we will see later, however, other factors crop up that make this a less desirable transmission window than it appears. Material dispersion is a particularly vexing problem in single-mode fibers.

Waveguide Dispersion   Because the core and the cladding of a fiber have slightly different indices of refraction, the light that travels in the core moves slightly slower than the light that escapes into and travels in the cladding. This results in a dispersion effect that can be corrected by transmitting at specific wavelengths where material and waveguide dispersion actually occur at minimums.

Putting It All Together

So what does all of this have to do with the high-speed transmission of voice, video, and data? A lot, as it turns out. Understanding where attenuation and dispersion problems occur helps optical design engineers determine the best wavelengths at which to transmit information, taking into account distance, type of fiber, and other factors that can potentially affect the integrity of the transmitted signal. Consider the graph shown in Figure 7-21. It depicts the optical transmission domain, as well as the areas where problems arise. Attenuation (dB/km) is shown on the Y-axis; Wavelength (nm) is shown on the X-axis.

Figure 7-21
The optical
transmission
domain

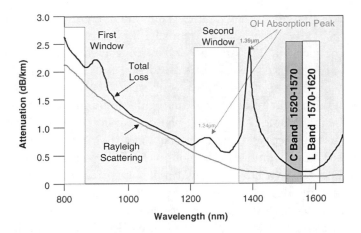

First of all, note that there are four *transmission windows* in the diagram. The first one is at approximately 850 nm, the second at 1,310 nm, the third at 1,550 nm, and the fourth at 1,625 nm; the last two are labeled "C" and "L" band, respectively. The 850 nm band was the first to be used because of its adherence to the wavelength at which the original LED technology operated. The second window at 1,310 nm enjoys low dispersion; this is where dispersion effects are minimized. At 1,550 nm, the so-called C-Band, emerges as the ideal wavelength at which to operate long-haul systems and systems upon which dense wavelength division multiplexing (DWDM) has been deployed because (1) loss is minimized in this region, and (2) dispersion minimums can be shifted here. The relatively new L-Band has enjoyed some early success as the next effective operating window.

Notice also that Rayleigh Scattering is shown to occur at or around 1,000 nm, while hydroxyl absorption by water occurs at 1,240 and 1,390 nm. Needless to say, network designers would be well-served to avoid transmitting at any of the points on the graph where Rayleigh Scattering, high degrees of loss, or hydroxyl absorption have the greatest degree of impact. Notice also that dispersion, shown by the lower line, is at a minimum point in the second window, while loss, shown by the upper line, drops to a minimum point in the third window. In fact, dispersion is minimized in traditional single-mode fiber at 1,310 nm, while loss is at minimums at 1,550 nm. So the obvious question becomes this: Which one do you want to minimize—loss or dispersion?

Luckily, this choice no longer has to be made. Today, *dispersion-shifted fibers* (DSF) have become common. By modifying the manufacturing process, engineers can shift the point at which minimum dispersion occurs from 1,310 nm to 1,550 nm, causing it to coincide with the minimum loss point such that loss and dispersion occur at the same wavelength.

Unfortunately, while this fixed one problem, it created a new and potentially serious alternative problem. DWDM has become a mainstay technology for multiplying the available bandwidth in optical systems. When DWDM is deployed over dispersion-shifted fiber, serious non-linearities occur at the zero dispersion point, which effectively destroy the DWDM signal. Think about it: DWDM relies on the ability to "chan-nelize" the available bandwidth of the optical infrastructure and main-tain some degree of separation between the channels. If dispersion is minimized in the 1,559 nm window, then the channels will effectively overlay each other in DWDM systems. Specifically, a problem called *four-wave mixing* creates "sidebands" that interfere with the DWDM chan-nels, destroying their integrity. In response, fiber manufacturers have created *nonzero dispersion-shifted fiber* (NZDSF) that lowers the disper-sion point to *near* zero and makes it occur just outside of the 1,550 nm window. This eliminates the nonlinear four-wave mixing problem.

Fiber Nonlinearities

A classic business quote, imminently applicable to the optical networking world, observes "in its success lie the seeds of its own destruction." As the marketplace clamors for longer transmission distances with minimal amplification, more wavelengths per fiber, higher bit rates, and increased signal power, a rather ugly collection of transmission impairments, known as *fiber nonlinearities*, rises to challenge attempts to make them happen. These impairments go far beyond the simple concerns brought about by loss and dispersion; they represent a significant performance barrier.

The "special relationship" that exists between transmission power and the refractive index of the medium gives rise to four service-affecting optical nonlinearities: *self-phase modulation* (SPM), *cross-phase modu-lation* (XPM), *four-wave mixing* (FWM), and *intermodulation*.

Self-Phase Modulation (SPM)

When SPM occurs, chromatic dispersion kicks in to create something of a technological double whammy. As the light pulse moves down the fiber, its leading edge increases the refractive index of the core, which causes a shift toward the longer wavelength, blue end of the spectrum. The trailing edge, on the other hand, decreases the refractive index of the core, causing a shift toward the shorter wavelength, red end of the spectrum. This causes an overall spreading or smearing of the transmitted signal—a phenomenon known as *chirp*. It occurs in fiber systems that transmit a single pulse down the fiber and is proportional to the amount of chromatic dispersion in the fiber: the more chromatic dispersion, the more SPM. It is counteracted with the use of large effective area fibers.

Cross-Phase Modulation (XPM)

When multiple optical signals travel down the same fiber core, they both change the refractive index in direct proportion to their individual power levels. If the signals happen to cross, they will distort each other. While XPM is similar to SPM, there is one significant difference: While self-phase modulation is directly affected by chromatic dispersion, cross-phase modulation is only minimally affected by it. Large effective area fibers can reduce the impact of XPM.

Four-Wave Mixing (FWM)

FWM mixing is the most serious of the power/refractive index-induced nonlinearities today because it has a catastrophic effect on DWDM-enhanced systems. Because the refractive index of fiber is nonlinear and because multiple optical signals travel down the fiber in DWDM systems, a phenomenon known as *third-order distortion* can occur that seriously affects multichannel transmission systems. Third-order distortion causes harmonics to be created in large numbers that have the annoying habit of occurring where the actual signals are, resulting in their obliteration.

Four-wave mixing is directly related to DWDM. In DWDM fiber systems, multiple simultaneous optical signals are transmitted across an optical span. They are separated on an ITU-blessed standard transmission grid by as much as 100 GHz (although most manufacturers today have reduced that to 50 GHz or less). This separation ensures that they do not interfere with each other.

Consider now the effect of dispersion-shifted fiber on DWDM systems. In DSF, signal transmission is moved to the 1,550 nm band to ensure that dispersion and loss are both minimized within the same window. However, minimal dispersion has a rather severe, unintended consequence when it occurs in concert with DWDM: Because it reduces dispersion to near zero, it also prevents multichannel systems from existing because it does not allow proper channel spacing. Four-wave mixing, then, becomes a serious problem.

Several things can reduce the impact of FWM. As the dispersion in the fiber drops, the degree of four-wave mixing increases dramatically. In fact, it is *worst* at the zero-dispersion point. Thus, the intentional inclusion of a small amount of chromatic dispersion actually helps to reduce the effects of FWM. For this reason, fiber manufacturers sell NZDSF, which moves the dispersion point to a point *near* the zero point, thus ensuring that a small amount of dispersion creeps in to protect against FWM problems.

Another factor that can minimize the impact of FWM is to widen the spacing between DWDM channels. This, of course, reduces the efficiency of the fiber by reducing the total number of available channels and, therefore, is not a popular solution, particularly since the trend in the industry is to move toward narrower channel spacing as a way to increase the total number of available channels. Already, several vendors have announced spacing as narrow as 5 GHz.

Finally, large effective area fibers tend to suffer less from the effects of FWM.

Intermodulation Effects

In the same way that cross-phase modulation results from interference between multiple simultaneous signals, intermodulation causes secondary frequencies to be created that are cross-products of the original signals being transmitted. Large effective area fibers can alleviate the symptoms of intermodulation.

Scattering Problems

Scattering within the silica matrix causes the second major impairment phenomenon. Two significant nonlinearities result: *Stimulated Brillouin Scattering* (SBS) and S*timulated Raman Scattering* (SRS).

Stimulated Brillouin Scattering (SBS)

SBS is a power-related phenomenon. As long as the power level of a transmitted optical signal remains below a certain threshold, usually on the order of three milliwatts, SBS is not a problem. The threshold is directly proportional to the fiber's effective area, and because dispersion-shifted fibers typically have smaller effective areas, they have lower thresholds. The threshold is also proportional to the width of the originating laser pulse: As the pulse gets wider, the threshold goes up. Thus, steps are often taken through a variety of techniques to artificially broaden the laser pulse. This can raise the threshold significantly, to as high as 40 milliwatts.

SBS is caused by the interaction of the optical signal moving down the fiber with the acoustic vibration of the silica matrix that makes up the fiber. As the silica matrix resonates, it causes some of the signal to be reflected back toward the source of the signal, resulting in noise, signal degradation, and a reduction of overall bit rate in the system. As the power of the signal increases beyond the threshold, more of the signal is reflected, resulting in a multiplication of the initial problem.

It is interesting to note that there are actually two forms of Brillouin Scattering. When (sorry, a little more physics) electric fields that oscillate in time within an optical fiber interact with the natural acoustic resonance of the fiber material itself, the result is a tendency to backscatter light as it passes through the material. This is called Brillouin Scattering. If, however, the electric fields are caused by the optical signal itself, the signal is seen to cause the phenomenon; this is called *Stimulated Brillouin Scattering*.

To summarize: Because of backscattering, SBS reduces the amount of light that actually reaches the receiver and causes noise impairments. The problem increases quickly above the threshold and has a more deleterious impact on longer wavelengths of light. One additional fact: In-line optical amplifiers such as erbium-doped fiber amplifiers (EDFAs) add to the problem significantly. If there are four optical amplifiers along an optical span, the threshold will drop by a factor of four.

Solutions to SBS include the use of wider-pulse lasers and larger effective area fibers.

Stimulated Raman Scattering (SRS)

SRS is something of a power-based crosstalk problem. In SRS, high-power, short-wavelength channels tend to bleed power into longer wave-length, lower power channels. It occurs when a light pulse moving down the fiber interacts with the crystalline matrix of the silica, causing the light to (1) be backscattered and (2) shift the wavelength of the pulse slightly. Whereas SBS is a backward-scattering phenomenon, SRS is a two-way phenomenon, causing both backscattering and a wavelength shift. The result is crosstalk between adjacent channels.

The good news is that SRS occurs at a much higher power level—close to a watt. Furthermore, it can be effectively reduced through the use of large effective area fibers.

▊ Optical Amplification

As long as we are on the subject of Raman Scattering, we should introduce the concept of optical amplification. This may seem like a bit of a non sequitur, but it really isn't; true optical amplification actually uses a form of Raman Scattering to amplify the transmitted signal!

Traditional Amplification and Regeneration Techniques

In a traditional metallic analog environment, transmitted signals tend to weaken over distance. To overcome this problem, amplifiers are placed in the circuit periodically to raise the power level of the signal. There is a problem with this technique, however: In addition to amplifying the signal, amplifiers also amplify whatever cumulative noise has been picked up by the signal during its trip across the network. Over time, it becomes difficult for a receiver to discriminate between the actual signal and the noise embedded in the signal. Extraordinarily complex recovery mechanisms are required to discriminate between optical wheat and noise chaff.

In digital systems, *regenerators* are used to not only amplify the signal, but to also remove any extraneous noise that has been picked up along the way. Thus, digital regeneration is a far more effective signal-recovery methodology than is simple amplification.

Even though signals propagate significantly farther in optical fiber than they do in copper facilities, they are still eventually attenuated to the point that they must be regenerated. In a traditional installation, the optical signal is received by a receiver circuit, converted to its electrical analog, regenerated, converted back to an optical signal, and transmitted onward over the next fiber segment. This optical-to-electrical-to-optical conversion process is costly, complex, and time consuming. However, it is proving to be far less necessary as an amplification technique than it used to be because of true optical amplification that has recently become commercially feasible. Please note that optical amplifiers *do not* regenerate signals; they merely amplify. Regenerators are still required, albeit far less frequently.

Optical amplifiers represent one of the technological leading edges of data networking. Instead of the O-E-O process, optical amplifiers receive the optical signal, amplify it as an optical signal, and then retransmit it as an optical signal—no electrical conversion is required. Like their electrical counterparts, however, they also amplify the noise; at some point, signal regeneration is required.

Optical Amplifiers: How They Work

It was only a matter of time before all-optical amplifiers became a reality. It makes intuitively clear sense that a solution that eliminates the electrical portion of the O-E-O process would be a good one; optical amplification is that solution.

You will recall that SRS is a fiber nonlinearity that is characterized by high-energy channels pumping power into low-energy channels. What if that phenomenon could be harnessed as a way to amplify optical signals that have weakened over distance?

Optical amplifiers are actually rather simple devices that, as a result, tend to be extremely reliable. The optical amplifier comprises the following: an input fiber, carrying the weakened signal that is to be amplified; a pair of optical isolators; a coil of doped fiber; a pump laser; and the output fiber that now carries the amplified signal. A functional diagram of an optical amplifier is shown in Figure 7-22.

The coil of doped fiber lies at the heart of the optical amplifier's functionality. Doping is simply the process of embedding some kind of functional impurity in the silica matrix of the fiber when it is manufactured.

Figure 7-22
Erbium-doped
fiber amplifier
(EDFA)

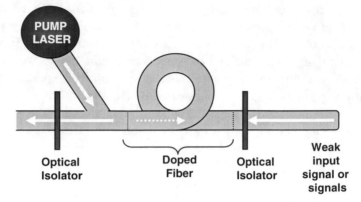

In optical amplifiers, this impurity is more often than not an element called *erbium*. Its role will become clear in just a moment.

The pump laser shown in the upper left corner of Figure 7-23 generates a light signal at a particular frequency—oftentimes 980 nm—in the *opposite direction* than the actual transmitted signal flows. As it turns out, erbium becomes atomically excited when it is struck by light at that wavelength. When an atom is excited by pumped energy, it jumps to a higher energy level (those of you who are recovering physicists will remember classroom discussions about orbital levels—$1S^1$, $1S^2$, $2S^1$, $2S^2$, $2P^6$, etc.), then falls back down, during which time it gives off a photon at a certain wavelength. When erbium is excited by light at 980 nm, it emits photons within the 1,550 nm region—coincidentally the wavelength at which multichannel optical systems operate. So, when the weak, transmitted signal reaches the coil of erbium-doped fiber, the erbium atoms, now excited by the energy from the pump laser, bleed power into the weak signal at precisely the right wavelength, causing a generalized amplification of the transmitted signal. The optical isolators serve to prevent errant light from backscattering into the system, creating noise.

EDFAs are highly proletariat in nature: They amplify anything, including the noise that the signal may have picked up. There will, therefore, still be a need at some point along the path of long-haul systems for regeneration, although far less frequently than in traditional copper systems. Most manufacturers of optical systems publish recommended span engineering specifications that help service providers and network designers take such concerns into account as they design each transmission facility.

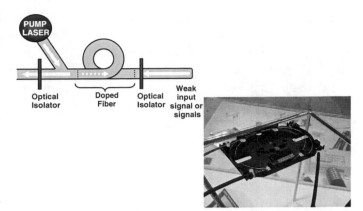

Figure 7-23
The pump laser in the upper left corner injects light at the opposite direction of the transmitted optical signal. Bottom right, an actual EDFA.

Other Amplification Options

There are at least two other amplification techniques in addition to EDFAs that have recently come into favor. The first of these is called *Raman amplification*, which is similar to EDFA in the sense that it relies on Raman effects to do its task, but different for other rather substantial reasons. In Raman amplification, the signal beam travels down the fiber alongside a rather powerful pump beam, which excites atoms in the silica matrix that in turn emit photons that amplify the signal. The advantage of Raman amplification is that it requires no special doping: Erbium is not necessary. Instead, the silica itself gives off the necessary amplification. In this case, the fiber itself becomes the amplifier!

Raman amplifiers require a significantly high-power pump beam (about a watt, although some systems have been able to reduce the required power to 750 mw or less) and even at high levels the power gain is relatively low. Their advantage, however, is that their induced gain is distributed across the entire optical span. Furthermore, it will operate within a relatively wide range of wavelengths, including 1,310 and 1,550 nm, currently the two most popular and effective transmission windows.

Semiconductor lasers have also been deployed as optical amplification devices in some installations. In semiconductor optical amplifiers, the weakened optical signal is pumped into the ingress edge of a semiconductor optical amplifier. The active layer of the semiconductor substrate amplifies the signal and regenerates it on the other side. The primary downside to these devices is their size: They are small, and their light-collecting ability is, therefore, somewhat limited. A typical single-mode fiber generates an intense spot of light that is roughly 10 microns in diameter; the point upon which that light impinges upon the semiconductor amplifier is less than a micron in diameter, meaning that a lot of the light is lost. Other problems also crop up including polarization issues, reflection, and variable gain. As a result, these devices are not in widespread use; EDFAs and Raman amplification techniques are far more common.

Optical Receivers

So far, we have discussed the sources of light, including LEDs and laser diodes; we have briefly described the various flavors of optical fiber and the problems they encounter as transmission media; now, we turn our attention to the devices that receive the transmitted signal.

The receive devices used in optical networks have a single responsibility: to capture the transmitted optical signal and convert it into an electrical signal that can then be processed by the end equipment. There may also be various stages of amplification to ensure that the signal is strong enough to be acted upon, and demodulation circuitry, which recreates the originally transmitted electronic signal.

Photodetector Types

While there are many different types of photosensitive devices, there are two used most commonly as photodetectors in modern networks: positive-intrinsic-negative (PIN) photodiodes and avalanche photodiodes (APDs).

Positive-Intrinsic-Negative (PIN) Photodiodes

PIN photodiodes are similar to the device described earlier in the general discussion of photosensitive semiconductors. Reverse biasing the junction region of the device prevents current flow until light at a specific wavelength strikes the substance, creating electron-hole pairs and allowing current to flow across the three-layer interface in proportion to the intensity of the incident light. While they are not the most sensitive devices available for the purpose of photodetection, they are perfectly adequate for the requirements of most optical systems. In cases where they are not considered sensitive enough for high-performance systems, they can be coupled with a preamplifier to increase the overall sensitivity.

Avalanche Photodiodes (APD)

APDs work as optical signal amplifiers. They use a strong electric field to perform what is known as *avalanche multiplication*. In an APD, the electric field causes current accelerations such that the atoms in the semiconductor matrix get excited and create, in effect, an avalanche of current to occur. The good news is that the amplification effect can be as much as 30 to 100 times the original signal; the bad news is that the effect is not altogether linear and can create noise. APDs are sensitive to temperature and require a significant voltage to operate them—30 to 300 volts depending on the device. However, they are popular for broadband systems and work well in the gigabit range.

We have now discussed transmitters, fiber media, and receivers. In the next section, we examine the fibers themselves, and how they have been carefully designed to serve as solutions for a wide variety of networking challenges and to forestall the impact of the nonlinearities described in this section.

Optical Fiber

As was mentioned briefly in a prior section, fiber has evolved over the years in a variety of ways to accommodate both the changing requirements of the customer community and the technological challenges that emerged as the demand for bandwidth climbed precipitously. These changes came in various forms of fiber that presented different behavior characteristics to the market.

Modes: An Analogy

The concept of "modes" is sometimes difficult to understand, so let me pass along an analogy that will help. Imagine a shopping mall that has a wide, open central area that all the shops open onto. An announcement comes over the PA system informing people that "The mall is now closed;

please make your way to the exit." Shoppers begin to make their way to the doors, but some wander from store to store, window-shopping along the way, while others take a relatively straight route to the exit. The result is that some shoppers take longer than others to exit the mall because there are different modes.

Now consider a mall that has a single, very narrow corridor that is only as wide as a person's shoulders. Now when the announcement comes, everyone heads for the exit, but they must form a single file line and head out in an orderly fashion. If you understand the difference between these two examples, you understand multimode vs. single-mode fiber. The first example represents multimode; the second represents single mode.

Multimode Fiber

The first of these examples was *multimode fiber*, which arrived in a variety of different forms. Multimode fiber bears this name because it allows more than a single mode or ray of light to be carried through the fiber simultaneously because of the relatively wide core diameter that characterizes the fiber. And while the dispersion that potentially results from this phenomenon can be a problem, there are advantages to the use of multimode fiber. For one thing, it is far easier to couple the relatively wide and forgiving end of a multimode fiber to a light source than that of the much narrower single-mode fiber. It is also significantly less expensive to manufacture (and purchase) and relies on LEDs and inexpensive receivers rather than the more expensive laser diodes and ultrasensitive receiver devices. However, advancements in technology have caused the use of multimode fiber to fall out of favor; single-mode fiber is far more commonly used today.

Multimode fiber is manufactured in two forms: *step-index fiber* and *graded-index fiber*. We will examine each in turn.

Multimode Step-Index Fiber

In step-index fiber, the index of refraction of the core is *slightly* higher than the index of refraction of the cladding. Remember that the higher the refractive index, the slower the signal travels through the medium. Thus, in step-index fiber, any light that escapes into the cladding because it enters the core at too oblique an angle will actually travel slightly faster in the cladding (assuming it does not escape altogether) than it would if it traveled in the core. Of course, any rays that are reflected repeatedly as they traverse the core also take longer to reach the receiver, resulting in a dispersed signal that causes problems for the receiver at the other end. Clearly, this phenomenon is undesirable; for that reason, graded-index fiber was developed.

Multimode Graded-Index Fiber

Because of the dispersion that is inherent in the use of step-index fiber, optical engineers created graded index fiber as a way to overcome the signal degradation that occurred.

In graded-index fiber, the refractive index of the core actually decreases from the center of the fiber outward. In other words, the refractive index at the center of the core is higher than the refractive index at the edge of the core. The result of this rather clever design is that, as light enters the core at multiple angles and travels from the center of the core outward, it is actually accelerated at the edge and slowed down near the center, causing most of the light to arrive at roughly the same time. Thus, graded-index fiber helps to overcome the dispersion problems associated with step-index multimode fiber. Light that enters this type of fiber does not travel in a straight line, but rather follows a parabolic path, with all rays arriving at the receiver at more or less the same time.

Graded-index fiber typically has a core diameter of 50 to 62.5 microns, with a cladding diameter of 125 microns. Some variations exist; there is at least one form of multimode graded index with a core diameter of 85 microns, somewhat larger than those described previously. Furthermore, the actual thickness of the cladding is important: If it is thinner than 20 microns, light begins to seep out, causing additional problems for signal propagation.

Graded-index fiber was commonly used in telecommunications applications until the late 1980s. Even though graded-index fiber is significantly better than step-index fiber, it is still multimode fiber and does not eliminate the problems inherent in being multimode. Thus was born the next generation of optical fiber: single-mode.

Single-Mode Fiber

There is an interesting mental conundrum that crops up with the introduction of single-mode fiber. The core of single-mode fiber is significantly narrower than the core of multimode fiber. Because it is narrower, it would seem that its ability to carry information would be reduced, due to limited light-gathering ability. This, of course, is not the case. As its name implies, it allows a single mode or ray of light to propagate down the fiber core, thus eliminating the intermodal dispersion problems that plague multimode fibers. In reality, single-mode fiber is a stepped-index design, because the core's refractive index is slightly higher than that of the cladding. It has become the de facto standard for optical transmission systems and takes on many forms depending on the specific application within which it will be used.

Most single-mode fiber has an extremely narrow core diameter, on the order of seven to nine microns, and a cladding diameter of 125 microns. The advantage of this design is that it only allows a single mode to propagate; the downside, however, is the difficulty involved in working with it. The core must be coupled directly to the light source and the receiver in order to make the system as effective as possible; given that the core is approximately one-sixth the diameter of a human hair, the mechanical process through which this coupling takes place becomes Herculean.

Single-Mode Fiber Designs

The reader will recall that we spent a considerable amount of time discussing the many different forms of transmission impairments (nonlinearities) that challenge optical systems. Loss and dispersion are the key contributing factors in most cases and do, in fact, cause serious problems in high-speed systems. The good news is that optical engineers have done yeoman's work creating a wide variety of single-mode fibers that address most of the nonlinearities.

Since its introduction in the early 1980s, single-mode fiber has undergone a series of evolutionary phases in concert with the changing demands of the bandwidth marketplace. The first variety of single-mode fiber to enter the market was called *nondispersion-shifted fiber* (NDSF). Designed to operate in the 1,310 nm second window, dispersion in these fibers was close to zero at that wavelength. As a result, it offered high bandwidth and low dispersion. Unfortunately, it was soon the victim of its own success. As demand for high-bandwidth transport grew, a third window was created at 1,550 nm for single-mode fiber transmission. It provided attenuation levels that were less than half those measured at 1,310 nm, but unfortunately was plagued with significant dispersion. Since the bulk of all installed fiber was NDSF, the only solution available to transmission designers was to narrow the linewidth of the lasers employed in these systems and to make them more powerful. Unfortunately, increasing the power and reducing the laser linewidth is expensive, so another solution emerged.

Dispersion-Shifted Fiber (DSF)

One solution that emerged was dispersion-shifted fiber. With DSF, the minimum dispersion point is mechanically shifted from 1,310 nm to 1,550 nm by modifying the design of the actual fiber so that waveguide dispersion is increased. The reader will recall that waveguide dispersion is a form of chromatic dispersion that occurs because the light travels at different speeds in the core and cladding.

One technique for building DSF (sometimes called "zero dispersion-shifted fiber") is to actually build a fiber of multiple layers. In this design, the core has the highest index of refraction and changes gradually from the center outward until it equals the refractive index of the outer cladding. The inner core is surrounded by an inner cladding layer, which in turn is surrounded by an outer core. This design works well for single-wavelength systems but experiences serious signal degradation when multiple wavelengths are transmitted, such as when it is used with DWDM systems. Four-wave mixing, described earlier, becomes a serious impediment to clean transmission in these systems. Given that multiple wavelength systems are fast becoming the norm today, the single-wavelength limit is a show stopper. The result was a relatively simple and elegant set of solutions.

The second technique was to eliminate or at least substantially reduce the absorption peaks in the fiber performance graph so that the second and third transmission windows merge into a single larger window, thus allowing for the creation of the fourth window described earlier which operates between 1,565 and 1,625 nm—the so-called L-Band.

Finally, the third solution came with the development of NZDSF. NZDSF shifts the minimum dispersion point so that it is *close* to the zero point, but not actually *at* it. This allows the nonlinear problems that occur at the zero point to be avoided, because it introduces a small amount of chromatic dispersion.

Why Does It Matter?

It is always good to go back and review why we care about such things as dispersion-shifting and absorption issues. Remember that the key to keeping the cost of the network down is to reduce maintenance and the need to add hardware or additional fiber when bandwidth gets tight. DWDM, discussed in detail later, offers an elegant and relatively simple solution to the problem of the cost of bandwidth. However, its use is not without cost. Multiwavelength systems will not operate effectively over dispersion-shifted fiber because of dramatic nonlinearities. So if DWDM is to be used, NZDSF must be deployed.

Optical Summary

In this section we have examined the history of optical technology and the technology itself, focusing on the three key components within an optical network: the light emitter, the transport medium, and the receiver. We also discussed the various forms of transmission impairment that can occur in optical systems and the steps that have been taken to overcome them.

The result of all this is that optical fiber, once heralded as a near-technological miracle because it only lost 99 percent of its signal strength *when transmitted over an entire kilometer*, has become the standard medium for transmission of high-bandwidth signals over great distances. Optical amplification now serves as an augmentation to traditional regenerated systems, allowing for the elimination of the optical-to-electrical conversion that must take place in copper systems. The result of all this is an extremely efficient transmission system that has the ability to play a role in virtually any network design in existence today.

Dense Wavelength Division Multiplexing (DWDM)

When high-speed transport systems such as SONET and SDH were first introduced, the bandwidth that they made possible was unheard of. The early systems that operated at OC-3/STM-1 levels (155.52 Mbps) provided volumes of bandwidth that were almost unimaginable. As the technology advanced to higher levels, the market followed Say's Law, creating demand for the ever more available volumes of bandwidth. There were limits, however; today, OC-48/STM-16 (2.5 Gbps) is extremely popular, but OC-192/STM-64 (10 Gbps) represents the practical upper limit of SONET's and SDH's transmission capabilities given the limitations of existing time-division multiplexing technology. The alternative is to simply multiply the channel count—and that's where WDM comes into play.

WDM is really nothing more than frequency-division multiplexing, albeit at very high frequencies. The ITU has standardized a channel separation grid that centers around 193.1 terahertz (THz), ranging from 191.1 THz to 196.5 THz. Channels on the grid are technically separated by 100 GHz, but many industry players today are using 50 GHz separation.

The majority of WDM systems operate in the C-Band (third window, 1,550 nm), which allows for close placement of channels and the reliance on EDFAs to improve signal strength. Older systems, which spaced the channels 200 GHz (1.6 nm) apart, were referred to simply as WDM systems; the newer systems are referred to as *Dense* WDM systems because of their tighter channel spacing. Modern systems routinely pack forty 10 Gbps channels across a single fiber, for an aggregate bit rate of 400 Gbps.

How DWDM Works

As Figure 7-24 illustrates, a WDM system consists of multiple input lasers, an ingress multiplexer, a transport fiber, an egress multiplexer, and of course, customer receiving devices. If the system has eight channels, such as the one shown in the diagram, it has eight lasers and eight receivers. The channels are separated by 100 GHz to avoid fiber non-linearities, or closer if the system supports the 50 GHz spacing. Each channel, sometimes referred to as a lambda (λ, the Greek letter and universal symbol used to represent wavelength), is individually modulated, and ideally, the signal strengths of the channels should be close to one another. Generally speaking, this is not a problem, because in DWDM

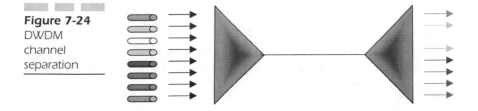

Figure 7-24
DWDM
channel
separation

systems the channels are closely spaced and therefore do not experience significant attenuation variation from channel to channel.

There is a significant maintenance issue that faces operators of DWDM-equipped networks. Consider a 16-channel DWDM system. This system has 16 lasers, one for each channel, which means that the service provider must maintain 16 spare lasers in case of a laser failure. The latest effort underway is the deployment of tunable lasers, which allow the laser to be tuned to any output wavelength, thus reducing the volume of spares that must be maintained and, by extension, the cost.

So what do we find in a typical WDM system? A variety of components, including multiplexers, which combine multiple optical signals for transport across a single fiber; demultiplexers, which disassemble the aggregate signal so that each signal component can be delivered to the appropriate optical receiver (PIN or APD); active or passive switches or routers, which direct each signal component in a variety of directions; filters, which serve to provide wavelength selection; and finally, optical add-drop multiplexers, which give the service provider the ability to pick up and drop off individual wavelength components at intermediate locations throughout the network. Together, these components make up the heart of the typical high-bandwidth optical network. And why is DWDM so important? Because of the cost differential that exists between a DWDM-enhanced network and a traditional network. To expand network capacity today by putting more fiber in the ground costs, on average, about $70,000 per mile. To add the same bandwidth using DWDM by changing out the endpoint electronics costs roughly one-sixth that amount. There is clearly a financial incentive to go with the WDM solution.

Optical Switching and Routing

DWDM facilitates the transport of massive volumes of data from a source to a destination. Once the data arrives at the destination, however, it must be terminated and redirected to its final destination on a lambda-by-lambda basis. This is done with switching and routing technologies.

Switching versus Routing: What's the Difference?

A review of these two fundamental technologies is probably in order. The two terms are often used interchangeably, and a never-ending argument is underway about the differences between the two.

The answer lies in the lower layers of the now-famous OSI Protocol Model. You will recall that OSI is a conceptual model used to study the step-by-step process of transmitting data through a network. It comprises seven layers, the lower three of which define the domain of the typical service provider. These layers, starting with the lowest in the seven-layer stack, are the physical layer (layer one), the data link layer (layer two), and the network layer (layer three). Layer one is responsible for defining the standards and protocols that govern the physical transmission of bits across a medium. SONET and SDH are both physical layer standards.

Switching, which lies at layer two (the data link layer) of OSI, is usually responsible for establishing connectivity within a single network. It is a relatively low-intelligence function and, therefore, is accomplished quite quickly. Such technologies as ATM, frame relay, and wireless access technologies such as FDMA, TDMA and CDMA, and LAN access control protocols (CSMA/CD, token passing) are found at this layer.

Routing, on the other hand, is a layer three (network layer) function. It operates at a higher, more complex level of functionality and is, therefore, more complex. Routing concerns itself with the movement of traffic between subnetworks and therefore complements the efforts of the switching layer. ATM, frame relay, LAN protocols, and the PSTN are switching protocols; IPRIP, OSPF, and IPX are routing protocols.

Switching in the Optical Domain

The principal form of optical switching is really nothing more than a very sophisticated digital cross-connect system. In the early days of data networking, dedicated facilities were created by manually patching the end points of a circuit at a patch panel, thus creating a complete four-wire circuit. Beginning in the 1980s, digital cross-connect devices such as AT&T's Digital Access and Cross-Connect (DACS) became common, replacing the time-consuming, expensive, and error-prone manual process. The digital cross-connect is really a simple switch, designed to establish "long-term temporary" circuits quickly, accurately, and inexpensively.

Enter the world of optical networking. Traditional cross-connect systems worked fine in the optical domain, provided there was no problem going through the O-E-O conversion process. This, however, was one of the aspects of optical networking that network designers wanted to eradicate from their functional requirements. Thus was born the optical cross-connect switch.

The first of these to arrive on the scene was Lucent Technologies' LambdaRouter. Based on a switching technology called Micro Electrical Mechanical System (MEMS), the LambdaRouter was the world's first all-optical cross-connect device. The product has since been discontinued, but the technology remains a good illustration of optical switching.

MEMS relies on micro-mirrors, an array of which is shown in Figure 7-25. The mirrors can be configured at various angles to ensure that an incoming lambda strikes one mirror, reflects off a fixed mirrored surface, strikes another movable mirror, and is then reflected out an egress fiber. The LambdaRouter and other devices like it offered switching speed, a relatively small footprint, bit rate and protocol transparency, nonblocking architecture, and highly developed database management. Fundamentally, these devices are very high-speed, high-capacity switches or

Figure 7-25
MEMS mirror
(Photo
courtesy
Lucent
Technologies)

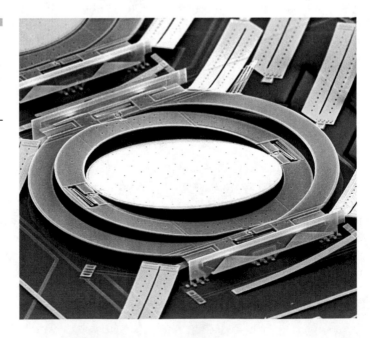

cross-connect devices. They are not routers, because they do not perform layer three functions. They will, however; now that the "nuclear winter" of the telecom industry has ended, all major manufacturers have announced plans to incorporate a layer three routing function into what will become their devices.

Optical networking, then, is a major component of the modern transport network. Together with switching technologies like frame relay and ATM, and physical layer standards like SONET, SDH, and copper-based transport technologies, the current network is able to handle the transport requirements of *any* traffic type while overlaying quality of service at the same time.

An Aside: A Trip to the Wave Venture

While working in Singapore recently, a friend approached me and said, "I just finished reading your *Optical Networking Crash Course* book. And I noticed that you didn't have much to say in it about submarine cables." "Yes," I replied, explaining that the book was really designed to explain the nature of terrestrial systems. I also added the observation that I had no experience in the submarine domain and didn't feel equipped to tackle that subject in detail in the book. "So you've never seen a cable laying ship?" he asked. No, I replied. "Would you like to?" He countered. "Absolutely!" I said, probably a little more enthusiastically than I should have. He explained that a good friend of his was the President of Asia Global Crossing, that they had a ship in port, and if I wanted he could probably set up a tour. So, he went away to make a phone call, and when he returned he told me that it was all set up—we would tour the ship on Friday.

Friday morning arrived, and we took a taxi down to the industrial pier where we boarded an ancient, scabby work boat that tooled us out to the *Wave Venture*, anchored a few miles offshore (see Figure 7-26). Approaching the vessel I was amazed at its size: a 142-meter cable-laying and maintenance ship, originally built as a passenger ferry in 1982 and converted to a cable ship in 2000. She can stay at sea for as long as 40 days without resupplying food and equipment for her 62 crew members.

We boarded the ship via a starboard boarding ladder and gathered on the fantail, where Martin Swaffield, master of the vessel, began the tour. We climbed up several flights of stairs to the driving bridge where he

Figure 7-26
Asia Global
Crossing's
Wave Venture

explained the principal responsibilities of the ship and its crew. "Our primary job is cable maintenance. We locate a cable, pull it up from the bottom, find the problem, repair it, and drop it back on the sea floor. We also
lay cable: We can put down as much as 15 kilometers a day depending on
the makeup of the bottom and obstacles we might have to go around, and
we do it all remotely. As you'll see in a few minutes we have a robot that
goes down and does all of the trenching and labor for us." I asked him
what the biggest problem was that they faced. "Unexploded ordinance,"
he responded. "We often pull up a cable that has been fouled in a fishing
net, and in an attempt to free the drag net, the fishing boat damages the
cable. Unfortunately, a lot of unexploded ordinance has been dropped
into the ocean over the years, sometimes intentionally, sometimes by
accident. It's often the case that when we drag up the cable we find an
unexploded mine or bomb in the net, dangling behind the stern of the
ship. At that point we stop all operations, call whatever Navy is closest,
and ask them to come deal with it."

From the driving bridge, we strolled over to the positioning bridge —
a separate control room that is used to hold the ship on station during
cable operations. Naturally it faces toward the stern of the vessel so that
Swaffield and his team can observe what is going on below. From the
windows of the positioning bridge we looked down on the entire working
area of the ship (see Figure 7-27). On the near right below us was the
control room for the robotic cable layer; on the left was the robot itself,
hanging from a giant crane. And far off in the distance was the stern of
the ship, where cables were retrieved for repair or dropped.

Figure 7-27
The aft work
deck

From the bridge we went back down to deck level where we walked into the covered area of the deck — looking for all the world like a giant garage (see Figure 7-28). On the left was a structure that resembled a restaurant walk-in refrigerator, which Martin confirmed was in fact a freezer where the repeaters were kept. "They cost a million dollars apiece; we keep them cold so that the electronics won't be shocked when we dump them overboard after installing them on cables."

Figure 7-28
Covered area
of deck

In the right was a small room where fiber splicing was done (see Figure 7-29). A long, narrow window ran the length of the room and cables to be spliced could be dragged through the window for repair.

This was also where I saw cross sections of submarine cable for the first time. The stuff is remarkable and comes in a variety of forms depending on where it will be deployed. For deep water deployment, where the danger of a ship dropping an anchor on it or the likelihood of a fishing net getting snagged on the cable is relatively low, a fairly thin unarmored cable, shown in Figure 7-30, is used. In shallower water, how-

Figure 7-29
Fiber splicing laboratory

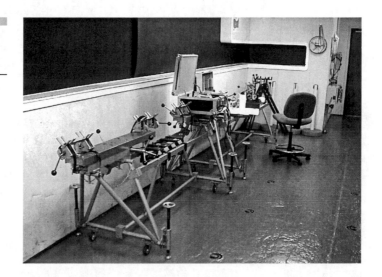

Figure 7-30
Deep water submarine cable: Note the lack of armor.

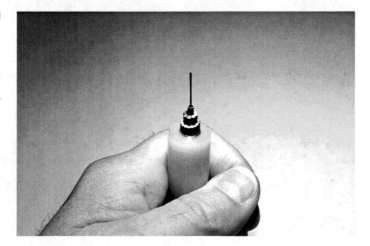

ever, where crushing incidents are more common, a heavier, armored cable is deployed, shown in Figure 7-31. Take a close look: The cable has four layers of armored strands, the largest of which is about the diameter of a pencil.

The robotic cable-laying device is a remarkable piece of machinery (see Figure 7-32). Controlled by an operator using a joystick like for a video game, it is dropped on a cable to the bottom, where it either achieves neutral buoyancy or crawls across the bottom on tractor treads. It has thrusters to control its position, digging tools (including a giant

Figure 7-31
Shallow water, heavily armored cable

Figure 7-32
Cable laying robot: Note the grappling arms on either side of the device.

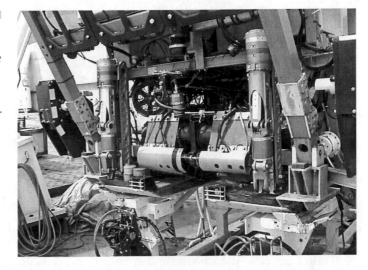

Figure 7-33
Cable tanks

circular rock saw that looks like something out of Roger Rabbit), and massive grappling arms with which it can gently pick up a fragile strand of unarmored cable or tear apart a rocky obstruction. It can operate in water as deep as 6,000 feet.

We then descended into the lower deck of the ship, where the cable tanks were located (see Figure 7-33). These tanks—three in all, about 50 feet in diameter and 11 feet deep—hold 1,200 miles of cable, and weigh about 1,700 tons. The cable is fed into the tanks from the main deck, where it is coiled by hand—a process that takes days. Cable is pulled from the tanks as needed and fed down the stern deck of the ship through the cable supports visible in Figure 7-34.

No question about it—this is one of the coolest tours I've ever had the honor to take.

Another Brief Aside: Broadband over Power Lines (BPL)

BPLs is the term created by the FCC to describe the new class of broadband modems that can be used to deliver IP-based broadband services over electric power lines. In 2003 the FCC adopted a Notice of Inquiry (NOI) that expressed enthusiasm for the potential of BPL to allow electric power lines to function as a third data connection to the home and to

Figure 7-34
Arrows show the path of cable as it is pulled from the cable tanks.

create competition for the naturally monopolistic copper telephone line and cable television coaxial cable line.

BPL modems transmit signals over electric power lines in much the same way that cable and DSL modems transmit over cable and telephone lines. Recent advances in processing capability enable BPL devices to overcome challenges that are inherent in transmitting over the electric power lines.

The FCC identifies two forms of BPL: in-house BPL, which uses transmission standards developed by the HomePlug Alliance, and Access BPL, a new technology to carry broadband Internet traffic over medium voltage power lines.

An *inductive coupler* connects the BPL modem to the power line. One key consideration is interference: There have been serious concerns expressed by the Amateur Radio Relay League (ARRL) about BPL's tendency to interfere with ham radio, a consideration that will be taken up by the FCC. When BPL devices are installed on underground power lines, the signal is shielded by the conduit and is unlikely to cause interference to other communications services.

IP and MPLS

It would be irresponsible to publish this book without detailed coverage of the next generation of transport network that is currently emerging. While frame relay and ATM continue to be important technologies for switched-broadband transport, they are expensive, complex, and cumbersome to operate. And given the degree to which IP has taken the lead position in networking discussions today, it makes sense that IP should have its space, particularly given the fact that it is now being deployed in conjunction with *Multiprotocol Label Switching* (MPLS) as a carrier-grade, layer three network infrastructure that offers truly granular QoS.

We begin with a discussion of the remarkable protocol suite called TCP/IP and how it works.

The Internet as we know it today began as a DoD project designed to interconnect DoD research sites. In December of 1968, the government research agency known as the Advanced Research Projects Agency (ARPA) awarded a contract to Bolt, Beranek, and Newman (BBN) to design and build the packet network that would ultimately become the Internet. It had a proposed transmission speed of 50 Kbps, and in September of 1969 the first node was installed at UCLA. Other nodes were installed on roughly a monthly basis at Stanford Research Institute (SRI), the University of California at Santa Barbara (UCSB), and the University of Utah. The ARPANET spanned the continental United States by 1971 and had connections to research facilities in Europe by 1973.

The original protocol selected for the ARPANET was called the Network Control Protocol (NCP). It was designed to handle the emergent requirements of the low-volume architecture of the ARPANET network. As traffic grew, however, it proved to be inadequate to handle the load, and in 1974 a more robust protocol suite was implemented, based on the TCP — an ironclad protocol designed for end-to-end network communications control. In 1978, a new design split the responsibilities for end-to-end vs. node-to-node transmission between two protocols: the newly crafted IP, designed to route packets from device to device and TCP, designed to offer reliable, end-to-end communications. Since TCP and IP were originally envisioned as a single protocol, they are now known as the *TCP/IP protocol suite*, a name that also incorporates a collection of protocols and applications that also handle routing, QoS, error control, and other functions.

One problem that occurred that the ARPANET planners didn't envision when they sited their nodes at college campuses was visibility. Naturally, they placed the switches in the raised floor facilities of the computer science department, and we know what is also found there: *Undergraduus Nerdus*, the dreaded computer science (or worse yet, engineering) student. In a matter of weeks the secret was out — ARPA's top secret network was top secret no longer. So, in 1983, the ARPANET was split into two networks. One half, still called ARPANET, continued to be used to interconnect research and academic sites; the other, called MIL-NET, was specifically used to carry military traffic and ultimately became part of the Defense Data Network.

That year was also a good year for TCP/IP. It was included as part of the communications kernel for the University of California's UNIX implementation, known as 4.2BSD (Berkeley Software Distribution) UNIX.

Extension of the original ARPANET continued. In 1986, the National Science Foundation (NSF) built a backbone network to interconnect four NSF supercomputing centers and the National Center for Atmospheric Research. This network, known as NSFNET, was originally intended to serve as a backbone for other networks, not as a stand-alone interconnection mechanism. Additionally, the NSF's Appropriate Use Policy limited transported traffic to noncommercial traffic *only*. NSFNET continued to expand, and eventually became what we know today as the Internet. And while the original NSFNET applications were multiprotocol implementations, TCP/IP was used for overall interconnectivity.

In 1994, a structure was put in place to reduce the NSF's overall role in the Internet. The new structure consists of three principal components. The first of these was a small number of *network access points* (NAPs), where ISPs would interconnect to the Internet backbone. The NSF originally funded four NAPs in Chicago (operated by Ameritech, now part of SBC), New York (really Pensauken, NJ, operated by Sprint), San Francisco (operated by Pacific Bell, now part of SBC), and Washington, D.C. (MAE-East, operated by MFS, now a division of MCI).

The second component was the *very High-Speed Backbone Network Service*, a network that interconnected the NAPs and was operated by MCI. It was installed in 1995 and originally operated at OC-3 (155.52 Mbps) but was upgraded to OC-12 (622.08 Mbps) in 1997.

The third component was the *routing arbiter*, designed to ensure that appropriate routing protocols for the Internet were available and properly deployed.

ISPs were given 5 years of diminishing funding to become commercially self-sustaining. The funding ended in 1998, and starting at roughly the same time, a significant number of additional NAPs have been launched. As a matter of control and management, three tiers of ISPs have been identified. *Tier 1* refers to national ISPs that have a national presence and connect to at least three of the original four NAPs. National ISPs include AT&T, Cable & Wireless, MCI, and Sprint. *Tier 2* refers to regional ISPs, which primarily have a regional presence and connect to less than three of the original four NAPs. Regional ISPs include Adelphia.net, Verizon.net, and BellSouth.net. Finally, *Tier 3* refers to local ISPs, or those that do not connect to a NAP but offer services via the connections of another ISP.

Managing the Internet

The Internet is really a network of networks, all interconnected by high-bandwidth circuits leased (for the most part) from various telephone companies. It is something of an erroneous conclusion to think of the Internet as a stand-alone network; in fact, it is made up of the same network components that make up corporate networks and public telephone networks. It's just a collection of leased lines that interconnect the routers that provide the sophisticated routing functions that make the Internet so powerful. As such, it is owned by no one and owned by everyone, but more importantly it is *managed* by no one and *managed* by everyone! There is no single authority that governs the Internet, because the Internet is not a single monolithic entity but rather a collection of entities. Yet, it runs well, perhaps better than other more centrally managed services!

That being said, there are certainly a number of organizations that provide oversight, guidance, and a certain degree of management for the Internet community. These organizations, described in some detail in this section, help to guide the developmental direction of the Internet with regard to such functions as communications standards, universal addressing systems, domain naming, protocol evolution, and so on.

One of the longest standing organizations with Internet oversight authority is the *Internet Activities Board* (IAB), which governs administrative and technical activities on the Internet. It works closely with the *Internet Society* (ISOC), a nongovernmental organization established in 1992 that coordinates such Internet functions as internetworking and applications development. The ISOC is also chartered to provide oversight and communications for the Internet Activities Board.

The *Internet Engineering Task Force* (IETF) is one of the best known Internet organizations and is part of the IAB. The IETF establishes working groups responsible for technical activities involving the Internet, such as writing technical specifications and protocols. Because of the organization's early, valuable commitment to the future of the Internet, the ISO accredited the organization as an "official" international standards body at the end of 1994.

On a similar front, the *World Wide Web Consortium* (W3C) has no officially recognized role but has taken a leading role in the development of standard protocols for the World Wide Wed to promote its evolution and ensure its interoperability. The organization has more than 400 member organizations and is currently leading the technical development of the Web.

Other smaller organizations play important roles as well. These include the *Internet Engineering Steering Group* (IESG), which provides strategic direction to the IETF and is part of the IAB; the *Internet Research Task Force* (IRTF), which engages in research that affects the evolution of the future Internet; the *Internet Engineering Planning Group* (IEPG), which coordinates worldwide Internet operations and helps Internet Service Providers (ISPs) achieve interoperability; and finally, the *Forum of Incident Response and Security Teams*, which coordinates CERTs in various countries.

Together these organizations ensure that the Internet operates as a single, coordinated "organism." Remarkable, isn't it? The least managed network in the world (and arguably the largest) is also the best run!

Naming Conventions on the Internet

The Internet is based on a system of domains, that is, a hierarchy of names that uniquely identify the "operating regions" on the Internet. For example, "IBM.com" is a domain, as are "Verizon.net," "fcc.gov," and "ShepardComm.com." The domains help to guide traffic from one user to another by following a hierarchical addressing scheme that includes a top-level domain, one or more subdomains, and a host name. The postal service relies on a hierarchical addressing scheme, and it serves as a good analogy to understand how Internet addressing works. Consider the following address:

> *John Hardy*
> *1237 Sunnyside Court*
> *Matanuska, Alaska*
> *USA*

In reality the address is upside down because a package addressed to John must be read from the bottom-up if it is to be delivered properly.

Let's assume that the package was mailed from Zemlya Graham Bell[2] in Franz Josef Land. To begin the routing process, the postal service will start at the bottom with USA, which narrows it down to a country. They will then go to Alaska, then on to Matanuska, then to the address in Matanuska, and finally to John. Messages routed through the Internet are handled in much the same way; routing arbiters deliver them first to a domain, then to a subdomain, then on to a hostname, where they are ultimately delivered to the proper account.

The assignment of Internet IP addresses was historically handled by the Internet Assigned Numbers Authority (IANA), while domain names were assigned by the Internet Network Information Center (InterNIC), which had overall responsibility for name dissemination, while regional NICs handled non-U.S. domains. The InterNIC was also responsible for the management of the Domain Name System (DNS), the massive, distributed database that reconciles host names and IP addresses throughout the Internet.

The InterNIC and its overall role have gone through a series of significant changes in the last decade. In 1993, Network Solutions, Inc. (NSI) was given the responsibility of operating the InterNIC registry by the NSF and had exclusive authority for the assignment of such domains as .com, .org, .net, and .edu. NSI's contract expired in April 1998 but was extended several times because no alternate agency existed to perform the task. In October 1998 NSI became the sole administrator for those domains, but a plan was created to allow users to register names in those domains with other companies. At roughly the same time, responsibility for IP address assignments was migrated to a newly created organization called the American Registry for Internet Numbers (ARIN). Shortly thereafter, in March 2000, NSI was acquired by VeriSign.

The most recent addition to the domain management process is the Internet Corporation for Assigned Names and Numbers (ICANN). Established in late 1998, ICANN was appointed by the U.S. National Telecommunications and Information Administration (NTIA) to manage the DNS.

Of course, it makes sense that sooner or later the most common top-level domains, which include such well-known suffixes as .com, .gov, .net, .org, and .net, would be deemed inadequate. And sure enough, in Novem-

---

[2]There is an island in Franz Josef Land, near the Arctic Circle, called Zemlya (Russian for "land") Alexander Graham Bell. I would love to know why—if anyone knows, I'd appreciate an e-mail!

ber 2000, the first new top-level domains, seven in all, were approved by ICANN. They are .aero, for the aviation industry; .biz, for businesses; .coop, for business cooperatives; .info, for general use; .museum, for (you guessed it); .name, for individuals; and .pro, for professionals.

The TCP/IP Protocol: What It Is and How It Works

TCP/IP has been around for a long time, and while we often tend to think of it as a single protocol that governs the Internet, it is in reality a fairly exhaustive collection of protocols that cover the functions of numerous layers of the protocol stack. And while OSI, discussed in an earlier chapter, has seven layers of protocol functionality, TCP/IP has only four, as shown in Figure 7-35. We will explore each layer in moderate detail in the sections that follow. For purposes of comparison, however, OSI and TCP/IP compare functionally as shown in Table 7-1.

The Network Interface Layer

TCP/IP was created for the Internet with the concept in mind that the Internet would not be a particularly well-behaved network. In other words, designers of the protocol made the assumption that the Internet would become precisely what it has become—a network of networks, using a plethora of unrelated and often conflicting protocols, transporting traffic with widely varying QoS requirements. The fundamental

Table 7-1	**TCP/IP Protocol Stack**	**OSI Reference Model**
TCP/IP compared to OSI	Network Interface Layer	Physical and Data Link Layers (L1-2)
	Internet Layer	Network Layer (L3)
	Transport Layer	Transport Layer (L4)
	Application Services Layer	Session, Presentation, and Application Layers (L5-7)

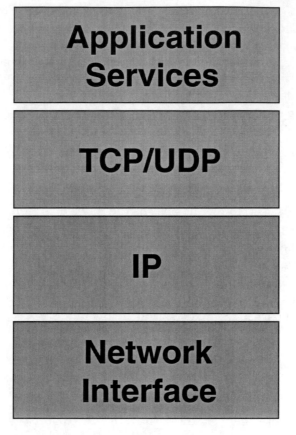

Application Services
TCP/UDP
IP
Network Interface

building block of the protocol, the IP packet, is designed to deal with all of these disparities, while TCP (and other related protocols, discussed later) take care of the QoS issues.

Two network interface protocols are particularly important to TCP/IP. The *Serial Line Internet Protocol* (SLIP) and *Point-to-Point Protocol* (PPP) are used to provide data link layer services in situations in which no other data link protocol is present, such as in leased-line or older dial-up environments. Most TCP/IP software packages for desktop applications include these two protocols, even though dial-up is rapidly fading into near-oblivion in the presence of growing levels of broadband access. With SLIP or PPP, a remote computer can attach directly to a host and connect to the Internet using IP rather than being limited to an asynchronous connection.

The Point-to-Point Protocol (PPP)

PPP, as its name implies, was created for the governance of point-to-point links. It has the ability to manage a variety of functions at the moment of connection and verification, including password verification, IP address resolution, compression (where required), and encryption for privacy or security. It can also support multiple protocols over a single connection, an important capability for dial-up users who rely on IP or some other network layer protocol for routing and congestion control. It also supports inverse multiplexing and dynamic bandwidth allocation via the *Multilink-PPP Protocol* (ML-PPP), commonly used in ISDN environments where bandwidth supersets are required over the connection.

The PPP frame (see Figure 7-36) is similar to a typical HDLC frame, with delimiting flags, an address field, a protocol identification field, information and pad fields, and a frame-check sequence for error control.

The Internet Layer

The Internet Protocol is the heart and soul of the TCP/IP protocol suite and Internet itself—and perhaps the most talked about protocol in history. IP provides a connectionless service across the network, which is sometimes referred to as an *unreliable* service because the network does not guarantee delivery or packet sequencing. IP packets typically contain an entire message or a piece (fragment) of a message that can be as large as 65,535 bytes in length. The protocol does *not* provide a flow-control mechanism.

IP packets, like all packets, have a header that contains routing and content information (see Figure 7-37). The bits in the packet are numbered from left to right starting at 0, and each row represents a single 32-bit word. An IP header must contain a *minimum* of five words.

Figure 7-36
PPP frame
format

Figure 7-37
IP header

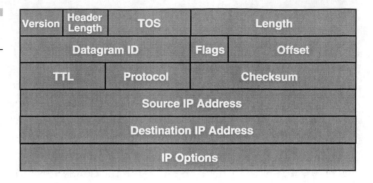

IP Header Fields

The IP header contains approximately 15 unique fields. The *Version field* identifies the version of IP that is being used to encode the packet (IPv4 vs. IP v6, for example). The *Internet Header Length* (IHL) *field* identifies the length of the header in 32-bit words. The maximum value of this field is 15, which means that the IP header has a maximum length of 60 octets.

The *Type of Service* (TOS) *field* gives the transmitting system the ability to request different classes of service for the packets that it transmits into the network. The TOS field is not typically supported in IPv4 but can be used to specify a service priority (0–7) or route optimization.

The *Total Length field* indicates the length (in octets) of the entire packet, including both the header and the data within the packet. The maximum size of an IP packet is 64 KB (OK, 65,535 bytes).

When a packet is broken into smaller "chunks" (a process called *fragmentation*) during transmission, the *Identification field* is used by the transmitting host to ensure that all of the fragments from a single message can be reassociated at the receiving end, to ensure message reassembly.

The *Flags* also play a role in fragmentation and reassembly. The first bit is referred to as the *More Fragments* (MF) bit and is used to indicate to the receiving host that the last fragment of a packet has been received, so that the receiver can reassemble the packet. The second bit is the *Don't Fragment* (DF) bit, which prevents packet fragmentation (for delay sensitive applications, for example). The third bit is unused and is always set to 0.

The *Fragment Offset field* indicates the relative position of this particular fragment in the original packet that was broken up for transmission. The first packet of a fragmented message will carry an offset value of 0, while subsequent fragments will indicate the offset in multiples of 8 bytes. Makes sense, no?

I love this field. A message is fragmented into a stream of packets and the packets are all sent on their merry way to the destination system. Somewhere along the way packet number 11 decides to take a detour and ends up in a routing loop on the far side of the world, trying in vain to reach its destination. To prevent this packet from living forever on the Internet, IP includes a *Time-to-Live* (TTL) *field*. This configurable field has a value between 0 and 255 and indicates the maximum number of hops that this packet is allowed to make before it is discarded by the network. Every time the packet enters a router, the router decrements the TTL value by one; when it reaches zero, the packet is discarded and the receiving device will ultimately invoke error control to ask for a resend.

The *Protocol field* indicates the nature of the higher layer protocol that is carried within the packet. Encoded options include values for ICMP (1), TCP (6), UDP (17), or OSPF (89).

The *Header Checksum field* is similar to a Frame-Check Sequence in HDLC and is used to ensure that the received IP header is free of errors. Keep in mind that IP is a connectionless protocol: This error check does not check the packet; it only checks the header.

When transmitting packets it is always a good idea to have a *source address* and a *destination address*. You can figure out what they are for.

▩ **Understanding IP Addresses**

As Figure 7-38 shows, IP addresses are 32-bits long. They are typically written as a sequence of four numbers, which represent the decimal value of each of the address bytes. These numbers are separated by periods ("dots" in telecom parlance), and the notation is referred to as *dotted decimal notation*. A typical address might be 168.152.20.10. These numbers are hierarchical; the hierarchy is described in the section that follows.

IP addresses are divided into two subfields. The *Network Identifier* (NET_ID) subfield identifies the subnetwork that is connected to the

Figure 7-38
32-bit IP
address: Each
"segment" of
the dotted
decimal
address (192,
168, 1, 1)
comprises
eight bits.

192.168.1.1

Internet. The NET_ID is most commonly used for routing traffic between networks. On the other hand, the *Host Identifier* (HOST_ID) subfield identifies the address of a system (host) within a subnetwork.

IP Address Classes

IP defines distinct *address classes*, which are used to discriminate between different sized networks. Classes A, B, and C are used for host addressing; the only difference between them is the length of the NET_ID subfield. A Class A address, for example, has an eight-bit NET_ID field and a 24-bit HOST_ID field. They are used for identification of very large networks and can identify as many as 16,777,214 hosts in each network. To date, only about 90 Class A addresses have been assigned.

Class B addresses have 16-bit NET_ID and 16-bit HOST_ID fields. They are used to address medium-sized-networks and can identify as many as 65,534 hosts within each network.

Class C addresses, which are far and away the most common, have a 24-bit NET_ID field and an eight-bit HOST_ID field. These addresses are used for smaller networks and can identify no more than 254 devices within any given network. There are 2,097,152 possible Class C NET_IDs which are commonly assigned to corporations that have fewer than 250 employees (or devices!).

There are two additional address types. Class D addresses are used for IP multicasting, such as transmitting a television signal to multiple recipients, whereas Class E addresses are reserved for experimental use.

Some addresses are reserved for specific purposes. A HOST_ID of 0 is reserved to identify an entire subnetwork. For example, the address 168.152.20.0 refers to a Class C address with a NET_ID of 168.152.20. A HOST_ID that consists of all ones (usually written "255" when referring to an all-ones byte, but also denoted as "-1") is reserved as a broadcast address and is used to transmit a message to all hosts on a particular network.

Subnet Masking

One of the most valuable but least understood tools in IP protocol management is called the *subnet mask*. Subnet masks are used to identify the portion of the address that specifies the network or the subnetwork for routing purposes. They also may be used to divide a large address into subnetworks or to combine multiple, smaller addresses to create a single large "domain." In the case of an organization subdividing its network, the address space is apportioned to identify multiple logical networks. This is accomplished by further dividing the HOST_ID subfield into a *Subnetwork Identifier* (SUBNET_ID) and a HOST_ID.

Adding to the Alphabet Soup: CIDR, DHCP, NAT, and PAT

As soon as the Internet became popular in the early 90s, concerns began to arise about the eventual exhaustion of available IP addresses. For example, consider what happens when a small corporation of 11 employees purchases a Class C address. They now control more than 250 addresses, of which they may only be using 25. Clearly, this is a waste of a scarce resource.

One technique that has been accepted for "address space conservation" is called *Classless Interdomain Routing* (CIDR). CIDR effectively limits the number of addresses assigned to a given organization, making the process of address assignment far more granular—and therefore efficient. Furthermore, CIDR has had a secondary, yet equally important, impact: It has dramatically reduced the size of the Internet routing tables because of the preallocation techniques used for address space management.

Other important protocols include *Network Address Translation* (NAT), which translates a private IP address that is being used to access the Web into a public IP address from an available pool of addresses, thus further conserving address space; *Port Address Translation* (PAT) and *Network Address Port Translation* (NAPT), which allow multiple systems to share a single IP address by using different "port numbers." Port numbers are used by transport layer protocols to identify specific higher layer applications.

Addressing in IP: The Domain Name System (DNS)

IP addresses are 32-bits long, and while not all that complicated, most Internet users don't bother to memorize the dotted decimal addresses of their systems. Instead, they use natural language host names. Most hosts, then, must maintain a comparative "table" of both numeric IP addresses and natural language names. From a host perspective, however, the names are worthless; they must use the numeric identifiers for routing purposes.

Because the Internet continues to grow at a rapid clip, a system was needed to manage the growing list of new Internet domains. That system is the DNS. It is a distributed database that stores host names and IP address information for all of the recognized domains found on the Internet. For every domain there is an *authoritative name server* that contains all DNS-related information about that domain, and every domain has at least one secondary name server that also contains the information. A total of thirteen *root servers*, in turn, maintain a list of all of the authoritative name servers.

How does the DNS actually work? When a system needs another system's IP address based upon its host name, the inquiring system issues a DNS request to a local name server. Depending on the contents of its local database, the local name server may be able to respond to the request. If not, it forwards the request to a root server. The root server, in turn, consults its own database and determines the most likely name server for the request and forwards it appropriately.

Early Address Resolution Schemes

When the Internet first came banging into the public psyche, most users were ultimately connected to the Internet via an Ethernet LAN. LANs use a local device address known as a *medium access control* (MAC) address, which is 48 bits long and nonhierarchical, which means that it cannot be used in IP networks for routing.

To get around this disparity, and to create a technique for relating MAC addresses to IP addresses, the *Address Resolution Protocol* (ARP) was created. ARP allows a host to determine a receiver's MAC address when it only knows the device's IP address. The process is simple: The host transmits an ARP request packet that contains the MAC broadcast address. The ARP request advertises the destination IP address and asks for the associated MAC address. Since every station on the LAN hears the broadcast, the station that recognizes its own IP address responds with an ARP message that contains its MAC address.

As ARP became popular, other address management protocols came into play. *Reverse ARP* (RARP) gives a diskless workstation (a dumb terminal, for all intents and purposes) the ability to determine its own IP address, knowing only its own MAC address. *Inverse ARP* (InARP) maps are used in frame-relay installations to map IP addresses to frame-relay virtual circuit identifiers. *ATMARP* and *ATMInARP* are used in ATM networks to map IP addresses to ATM virtual path/channel identifiers. And finally, *LAN Emulation ARP* (LEARP) maps a recipient's ATM address to its LAN Emulation (LE) address, which is typically a MAC address.

Routing in IP Networks

There are three routing protocols that are most commonly used in IP networks: the *Routing Information Protocol* (RIP), the *Open Shortest Path First* (OSPF), and the *Border Gateway Protocol* (BGP).

OSPF and RIP are used primarily for intradomain routing—within a company's dedicated network, for example. They are sometimes referred to as *interior gateways protocols*. RIP uses hop count as the measure of a particular network path's cost; in RIP, it is limited to 16 hops.

When RIP broadcasts information to other routers about the current state of the network, it broadcasts its entire routing table, resulting in a flood of what may be unnecessary traffic. As the Internet has grown, RIP has become relatively inefficient because it does not scale as well as it should in a large network. As a consequence the OSPF protocol was introduced. OSPF is known as a *link state protocol*. It converges (spreads important information across the network) faster than RIP, requires less overall bandwidth, and scales well in larger networks. OSPF-based routers only broadcast changes in status rather than the entire routing table.

The BGP is referred to as an *exterior gateway protocol* because it is used for routing traffic *between* Internet domains. Like RIP, BGP is a distance vector protocol, but unlike other distance vector protocols, BGP stores the actual route to the destination.

The Not-So-Distant Future: IP Version 6 (IPv6)

Because of the largely unanticipated growth of the Internet since 1993 or so (when the general public became aware of its existence), it was roundly concluded that IP version 4 was inadequate for the emerging and burgeoning needs of Internet applications. In 1995, IP version 6 (IPv6) was introduced, designed to deal with the shortcomings of version 4. Changes included increased IP address space from 32 to 128 bits, improved support for differentiable QoS requirements, and improvements with regard to security, confidentiality, and content integrity. IPv6 continues to be studied, and while it is in use, its pace of adoption is relatively slow. Its arrival is inevitable; however, the question is when.

Transport Layer Protocols

We turn our attention now to layer four, the transport layer. Two key protocols are found at this layer: the TCP and the *User Datagram Protocol*

(UDP). TCP is an ironclad, absolutely guaranteed service delivery protocol, with all of the attendant protocol overhead you would expect from such a capable protocol. UDP, on the other hand, is a more "lightweight" protocol, used for delay-sensitive applications like VoIP. Its overhead component is relatively light.

In TCP and UDP messages, higher layer applications are identified by *port identifiers*. The port identifier and IP address together form a *socket*, and the end-to-end communication between two or more systems is identified by a four-part complex address: the source port, the source address, the destination port, and the destination address. Commonly used port numbers are shown in Table 7-2.

Table 7-2
Commonly Used Port Numbers

Port #	Protocol	Service	Port #	Protocol	Service
7	TCP	echo	80	TCP	http
9	TCP	discard	110	TCP	pop3
19	TCP	chargen	119	TCP	nntp
20	TCP	ftp-control	123	UDP	ntp
21	TCP	ftp-data	137	UDP	netbios-ns
23	TCP	telnet	138	UDP	netbios-dgm
25	TCP	smtp	139	TCP	netbios-ssn
37	UDP	time	143	TCP	imap
43	TCP	whois	161	UDP	snmp
53	TCP/UDP	dns	162	UDP	snmp-trap
67	UDP	bootps	179	TCP	bgp
68	UDP	bootpc	443	TCP	https
69	UDP	tftp	520	UDP	rip
79	TCP	finger	33434	UDP	traceroute

The Transmission Control Protocol (TCP)

TCP provides a connection-oriented communication service across the network. It stipulates rule sets for message formats, establishing virtual circuit establishment and termination, data sequencing, flow control, and error correction. Most applications designed to operate within the TCP/IP protocol suite use TCP's reliable, guaranteed delivery services.

In TCP, the transmitted data entity is referred to as a *segment* because TCP does not operate in message mode: It simply transmits blocks of data from a sender and receiver. The fields that make up the segment, shown in Figure 7-39, are described in this section.

The *source port* and the *destination port* identify the originating and terminating connection points of the end-to-end connection as well as the higher-layer application. The *sequence number* identifies this particular segment's first byte in the byte stream, and since the sequence number refers to a byte count rather than to a segment, the sequence numbers in sequential TCP segments are not, ironically, numbered sequentially.

Figure 7-39
The TCP
header

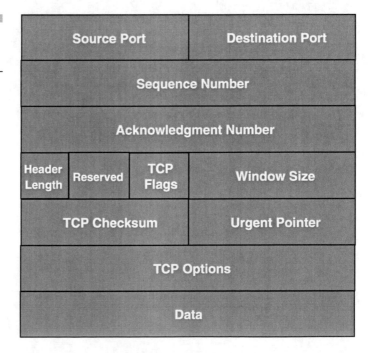

The *acknowledgment number* is used by the sender to acknowledge to the transmitter that it has received the transmitted data. In practice, the field identifies the sequence number of the next byte that it expects from the receiver. The *Data Offset field* identifies the first byte in this particular segment; in effect it indicates the segment header length.

TCP relies on a collection of *control flags*, which, in fact, do control certain characteristics of the virtual connection. They include an *Urgent Pointer Field Significant* (URG) which indicates that the current segment contains high-priority data and that the Urgent Pointer field value is valid; an *Acknowledgment Field Significant* (ACK), which indicates that the value contained in the Acknowledgment Number field is valid; a *Push Function* (PSH) *Flag*, which is used by the transmitting application to force TCP to transmit data immediately that it currently has buffered, without waiting for the buffer to fill; a *Reset Connection* (RST) *flag*, which is used to immediately terminate an end-to-end TCP connection; a *Synchronize Sequence Numbers* (SYN) *flag*, which is used to establish a connection, and to indicate that the segments carry the proper initial sequence number; and finally, a *Finish* (FIN) *flag*, which is set to request a normal termination of a TCP connection in whatever direction the segment is traveling.

The *Window field* is used for flow control management. It contains the value of the permitted *receive window size*, the number of transmitted bytes that the sender of the segment is willing to accept from the receiver. The *Checksum field* offers bit-level error detection for the entire segment, including both the header and the transmitted data.

The *Urgent Pointer field* is used for the management of high-priority traffic as identified by a higher layer application. If so marked, the segment is typically allowed to bypass normal TCP buffering.

The last field is the *Options field*. At the time of the initial connection establishment this field is used to negotiate such functions as maximum segment size and selective acknowledgement (SACK).

The User Datagram Protocol (UDP)

UDP provides connectionless service. And while "connectionless" often implies "unreliable," that is a bit of a misnomer. For applications that require nothing more than a simple query and response, UDP is ideal

IP Source Address			
IP Destination Address			
Unused	Protocol Type	UDP Datagram Length	
Source Port Number		Destination Port Number	
UDP Datagram Length		UDP Checksum	

because it involves minimal protocol overhead. UDP's primary responsibility is to add a port number to the IP address to create a socket for the application.

The fields of a UDP message (see Figure 7-40) are described here.

The *Source Port field* identifies the UDP port used by the sender of the datagram, while the *Destination Port field* identifies the port used by the datagram receiver. The *Length Field field* indicates the overall length of the UDP datagram.

The *Checksum field* provides the same primitive bit error detection of the header and transported data as we saw with TCP.

The Internet Control Message Protocol (ICMP)

The Internet Control Message Protocol (ICMP) is used as a diagnostic tool in IP networks to notify a sender that something unusual happened during transmission. It offers a healthy repertoire of messages, including *Destination Unreachable*, which indicates that delivery is not possible because the destination host cannot be reached; *Echo* and *Echo Reply*, used to check whether hosts are available; *Parameter Problem*, used to indicate that a router encountered a header problem; *Redirect*, which is used to make the sending system aware that packets should be forwarded to another address; *Source Quench*, used to indicate that a router is experiencing congestion and is about to begin discarding datagrams; *TTL Exceeded*, which indicates that a received datagram has been discarded because the Time-to-Live field (sounds like a soap opera for geeks, doesn't it?) reached 0; and finally, *Timestamp* and *Timestamp Reply*, which are similar to Echo messages except that they timestamp the message, giving systems the ability to a measure how long is required for remote systems to buffer and process datagrams.

The Application Layer

The TCP/IP application layer protocols support the actual applications and utilities that make the Internet — well, useful. They include the BGP, the DNS, the FTP, the HTTP, OSPF, the Packet Internetwork Groper (PING; how can you *not* love that name), the Post Office Protocol (POP), the Simple Mail Transfer Protocol (SMTP), the Simple Network Management Protocol (SNMP), the Secure Sockets Layer Protocol (SSL), and TELNET. This is a small sample of the many applications that are supported by the TCP/IP application layer.

Multiprotocol Label Switching (MPLS)

Before we talked about TCP/IP, we mentioned that we would discuss the evolving role of such technologies as TCP/IP and MPLS. As the network evolves, these two collections of protocols become extraordinarily important because the network is evolving at a rapid and radical pace.

When establishing connections over an IP network, it is critical to manage traffic queues to ensure the proper treatment of packets that come from delay-sensitive services, such as voice and video. To do this, packets must be differentiable, that is, identifiable so that they can be classified properly. Routers, in turn, must be able to respond properly to delay-sensitive traffic by implementing queue management processes. This requires that routers establish both normal and high-priority queues, and handle the traffic found in high-priority routing queues faster than the arrival rate of the traffic.

MPLS delivers QoS by establishing virtual circuits known as *label switched paths* (LSPs) which, in turn, are built around traffic-specific QoS requirements. An MPLS network, such as that shown in Figure 7-41, comprises *label switch routers* (LSRs), at the core of the network, and *label edge routers* (LERs), at the edge. It is the responsibility of the LERs to set QoS requirements and pass them on to the LSRs, which are responsible for ensuring that the required levels of QoS are achieved. Thus, a router can establish LSPs with explicit QoS capabilities and route packets to those LSPs as required, guaranteeing the delay that a particular flow encounters on an end-to-end basis. It's interesting to note that some industry analysts have compared MPLS LSPs to the trunks established in the voice environment.

MPLS uses a two-part process for traffic differentiation and routing. First, it divides the packets into *forwarding equivalence classes* (FECs) based on their quality of service requirements, and then maps the FECs to their next hop point. This process is performed at the point of ingress at the edge of the network. Each FEC is given a fixed-length "label" that accompanies each packet from hop to hop; at each router, the FEC label is examined and used to route the packet to the next hop point, where it is assigned a new label.

MPLS is a "shim" protocol that works closely with IP to help it deliver on QoS guarantees. Its implementation will allow for the eventual dismissal of ATM as a required layer in the multimedia network protocol stack. And while it offers a promising solution, its widespread deploy-

Figure 7-41
Schematic
representation
of an MPLS
network
showing LERs
and LSRs: The
LERs are at the
periphery of
the network;
the LSRs are in
the core.

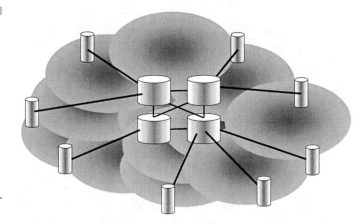

ment is still somewhat in the future because of the logistics of deployment. Its use, however, is growing at a rapid rate.

Chapter Summary

This concludes our long-winded coverage of transport solutions. Clearly they are remarkably diverse, but with that diversity and complexity comes an equally far-ranging degree of capability. And that capability is necessary, considering how diverse the market is — as we'll see in the following chapter on market segments.

Chapter 7 Questions

1. What is the difference between access and transport?
2. Explain the advantages and disadvantages of dedicated vs. switched transport facilities.
3. Draw and label the switching hierarchy.
4. How do frame relay and ATM differ?
5. "Even though frame relay is viewed as a transport technology, it is really an access solution." Explain.
6. What is the purpose of the ATM adaptation layer?
7. How do the ATM layer and the ATM adaptation layer differ?
8. Explain how addressing differs in ATM and frame relay.
9. What is the difference between single mode and multimode fiber?
10. How does an EDFA work?
11. What are the primary optical sources in use today?
12. Even though optical fiber is immune from electromagnetic interference, it still suffers from transmission impairments. What are some of them?

13.   Has anyone yet implemented broadband over power lines? (Use the Web to research your answer.)

14.   Draw and label the TCP/IP protocol stack.

15.   How do TCP and UDP differ? How are they each used and why?

16.   What is the difference between a label-switched router and a label edge router?

Telecom Market Segments

In this chapter we dissect the telecommunications marketplace and examine, in detail, the segments that it comprises. They include the following:

- The incumbent local exchange carriers (ILECs)
- The competitive local exchange carriers (CLECs)
- The interexchange carriers (IXCs)
- The cable multiple system operators (MSOs)
- The wireless providers
- The content providers
- The component manufacturers (including optoelectronics)
- The system manufacturers

We will discuss the characteristics of the players in each segment, the key challenges they face as the industry recovers from its recent meltdown, and the positioning moves they will make as they once again begin to redefine their role in the industry. We begin our discussion with the ILECs.

The ILECs

The ILECs have had a tumultuous ride since their inception in 1984, following the divestiture of AT&T. At that time, AT&T, through its negotiated settlement with the Justice Department, fought to retain control of its Western Electric manufacturing arm and spin off the local service providers because it was clear at the time that the real money was in hardware and that the local telephone companies would not be long for the world. In retrospect, perhaps they should have kept the local telephone companies (the Regional Bell Operating Companies, or RBOCs) and divested themselves of Western Electric. Amazing thing, 20-20 hindsight.

Today, of course, we know that the ILECs are in positions of significant market power. The four that remain—Verizon, SBC, Bellsouth, and Qwest—control 90 percent of the roughly 200 million access lines in the United States today. The remainder of these lines are served by CLECs —small independent telephone companies, cable providers, and wireless providers. Much to the chagrin of the ILECs, that 90 percent market share that they enjoy is declining at the rate of about 2.1 percent per year, and has been for the last couple of years, while "lineshare" for the other sectors has increased *incrementally*—but increased. It is critical

- 1963: 81 million subscriber lines in United States; 159 million worldwide.
- 1966: First optical cable used for voice transmission.
- 1976: First digital switch installed by AT&T.
- 1984: AT&T is divested; at time of divestiture it has more than one million employees and is worth $155 billion.
- 1988: First transatlantic optical cable is completed.
- 1989: 138 million subscriber lines in United States; 496 million worldwide.
- 1991: Bell Labs develops photonic (optical) switching capability.
- 2000: International voice traffic quadruples in 10 years to 132.7 billion minutes—but revenue only doubles to $70 billion.
- 2002: U.S. residential cable modem subscribers: 7.7 million; DSL: 4.4 million; data traffic passes voice at 200,000 TB/day; also in 2002, IP surpasses all other traffic in volume.

that the incumbent telephone companies stop focusing on numbers of access lines and preservation of minutes of use, and start focusing on *preservation of customers*. That is the only winning measure.

Contrary to popular belief, the most significant challenge faced by the ILECs is cable, not the threat posed by the CLECs. Cable companies are beginning to take on a more strategic role in the industry at large. After years of competing with satellite providers for "eyeshare," cable providers are finally beginning to enjoy price stability for the services they provide over their almost completely digital broadband networks. Furthermore, they are beginning to offer a variety of value-added services in addition to their traditional distributive entertainment content, such as high-speed cable modem Internet access and Internet-based telephony—both of which have become significant revenue producers as well as disruptive forces among their new competitors, the ILECs. Furthermore, the recently upgraded cable modem standards (DOCSIS 2.0) add significantly to the capabilities of broadband cable. These companies are now offering what has come to be known as the quadruple play—voice, video, data, and wireless—all packaged as a single, converged, and very lucrative bundle.

The fact is that the ILECs are under siege on multiple fronts. One of the critical questions that now comes up routinely in strategy discussions is this: Should the big three (Verizon, SBC, and Bellsouth) stop spending so much on capital outlay to build out their own networks and instead start spending their cash on the acquisition of preexisting assets? Companies like WorldCom, Cingular, Nextel, and Sprint PCS are, by many measures, undervalued at the moment; and because the ILECs have a recurring revenue stream, they have relatively easy access to capital compared to some of their competitors. One of the biggest challenges they face is that they have very high fixed costs and very low variable costs—which makes it difficult for them to make major adjustments in their cost base without drastic changes (like huge headcount reductions). The danger is that as they begin to lose access lines to competitors—such as cable MSOs, wireless providers, and CLECs—at what point do they lose critical mass and begin to collapse under the weight of their own infrastructure? Furthermore, as ARPU levels decline without a closely tied reduction in costs, the hole gets deeper.

Consider that between 1998 and 2002 alone Bellsouth, Verizon, and SBC spent $140 billion in CAPEX on their networks, a move that yielded less than one percent revenue growth. In 2003 they made combined profits of $20 billion on a collective market value of $240 billion. Clearly they want these numbers to increase. In 1998, ILEC CAPEX was 30 to 33 percent of revenues; today it is far lower, about 14 percent.

What steps have they taken to remedy the ongoing revenue shortfall? The first steps they took involved a direct attack on the FCC with a petition to reenter the long-distance market, referred to by regulators as Section 271 relief. Section 271 of the 1996 Telecommunications Act allows the remaining ILECs (Qwest, Verizon, SBC, and BellSouth) to enter the long-distance market if they can prove that they have sufficiently opened up their local markets to competition. As you know from your readings in the regulatory chapter, they have aggressively pursued this course of action; so, suffice it to say that the ILECs universally fought for long-distance relief in response to local telephony entry by their competitors and have in fact been granted large entry concessions. Verizon, for example, replaced Sprint as the third largest long-distance provider in terms of customer count. The originally stated reason for long-distance entry was incremental revenue, since the cost to an ILEC for in-region entry was near zero. However, while the ILECs doubled the number of long-distance customers they serve, the revenues from the long-distance lines of business they established declined 6 percent as the bandwidth glut and the magic of declining ARPUs took effect. Clearly, ILEC entry into long distance has bought them precious little if any margin relief as they watch their local services revenues decline.

So what must they do? There are a number of steps that service providers can take to shore up their falling revenue targets. They include acceleration of the circuit-to-packet migration, which facilitates convergence; marshalling a renewed focus on the importance of the metro marketplace; aggressive revamping of their overall network infrastructures; a much stronger focus on both element and network management; an aggressive examination of the services that customers actually want and a plan to deliver them cost effectively; and a strong focus on both enterprise and residence customer requirements.

Circuit-to-Packet

The circuit-to-packet migration plan stems from the convergence activity that is underway throughout the greater industry. There is no question that circuit-based services such as voice, ATM, and frame relay will be around for some time to come, but there is aggressive movement going on with regard to packet migration. Companies are rapidly developing and rolling out service packages that rely on IP, softswitch technology, and packet-based services, including voice. The VoIP marketplace for wireless, Centrex, and PBX service alone is enormous. Now that the industry is back on its feet and money is moving in, this trend will accelerate. VoIP is a critical success component and not simply because it helps the telco reduce costs. It is critical because it facilitates the delivery of a new suite of converged applications and services that represent new revenue streams for the telco. There is nothing more important than that.

Metro Markets

Because of the diverse, multiprotocol nature of the metro marketplace and the ongoing evolution of the corporation—from operating out of a single metro-based location to a distributed corporate architecture with offices scattered throughout customer locations—metro has become one of the fastest growing market sectors for service providers. And because optical technology has taken root in the metro rather effectively, with low-cost, multichannel DWDM technology offering lower cost solutions due to optical's ability to reduce the total number of network elements under management control, it garners attention. Today, ILECs control more than 60 percent of all metro spending; this is a significant percentage of the overall market and deserves the attention it attracts. Furthermore, the fundamental local-loop technology for the metro is Gigabit Ethernet, and to be even more granular, *switched* Gigabit Ethernet. Service providers can simply drop a fiber connection in the basement of a metro office building, terminate it on a high-volume Ethernet switch, and deliver a broad range of high-bandwidth services to multiple customers in the building.

Network Revamp

Today the typical ILEC operates a wide array of disparate and functionally unrelated networks: the PSTN, the IP network, the ATM network, the frame-relay network, the ISDN overlay, the DSL overlay, the wireless network, the high-speed Ethernet access and transport network, and so on. Recent technological advances such as Multiprotocol Label Switching (MPLS) and Resilient Packet Ring (RPR) allow for many of these network infrastructures to be converged onto a reduced set of protocols and physical networks, a move that dramatically simplifies the complex management task that ILECs face. This complexity translates into revenue barriers because of service delays, QoS issues, and billing discrepancies. Anything that can be done to simplify "the cloud" is a step in the proper direction because it reduces overall operating cost and stands a good chance of increasing customer service levels because of reduced network complexity.

The move to an IP infrastructure is clearly underway, but in many cases the telcos are reluctant to make big moves in that direction because of uncertainty over the business reasons for such a radical migration. In fact, there are relatively few valid business reasons for the migration to IP. The first is obsolescence: If a telco is facing a situation in which a switch in a central office is nearing obsolescence, then an IP overlay is probably worth considering. Or, if the telco is about to offer service in a greenfield (new, devoid of services) market, then IP is worth considering. Alternatively, if the service provider is facing a situation in which a switch is nearing capacity and will potentially require real estate expansion to accommodate the new equipment required, then IP is worth considering because of its reduced floor space requirements. Finally, if the service provider is operating in a region with a high degree of business penetration, or is in an area where fiber-to-the-home (FTTH) is being considered, then IP represents a serious advantage and should be considered as a key infrastructure component.

Management

Network and element management, like staff training, are often the first things cut from the budget and the last things implemented. Perhaps that's a bit harsh, but not by much. One of the loudest complaints voiced by customers, especially enterprise customers, is billing complexity. Customers claim to want a single, simple-to-understand bill[1] for all of their network services. And because of the diverse and logically disconnected nature of the network elements that come into play when provisioning customer services, the actual provisioning process is often slow, inaccurate, and expensive. Reduction of managed elements and the creation of an aggressively accurate and capable management interface must be of paramount importance to incumbent service providers.

Furthermore, there is ample evidence to suggest that the management systems operated by service providers since time began will soon be inadequate to handle the evolving demands of the services marketplace. Those systems are based on a model of predictable, recurring charges as well as a few nonrecurring charges every month. The market model, however, is beginning to embrace a more transaction-oriented billing model, which means that the OSS systems operated by the traditional telco will no longer suffice. It is critical that these companies actively evolve their internal systems to meet the needs of the changing customer base.

Another factor is physical deployment. The operating expenses incurred by ILECs to deploy DSL, for example, are inordinately high because it often requires multiple "truck rolls" (installation technician dispatches) to make it work. Not only is this expensive in terms of real dollars, it's frustrating for customers.

Services

Next on the wish list is an aggressive examination of the services that customers actually want and a plan to deliver them cost-effectively. Service providers are now putting into place data-mining and knowledge-management applications to help them better understand the evolving needs of their customers and make them more capable of responding to demands for service, perhaps even before the customer

---

[1]Personally, if I received a single bill for all of the telecommunications services I pay for every month, I'd probably go into cardiac arrest. But that's just me.

realizes the need. Most important is the aforementioned understanding:
At the risk of sounding simplistic, it is absolutely critical that service
providers offer services the customers actually want rather than the services that they think they want or that are based on the technologies
that happen to be available at the time.

Finally, related to the last item, it is important that service providers
carefully differentiate between enterprise and residence customer
requirements, as much for the commonalities of demand as for the differences.

So what are these additional services? Many believe that the key to
enhanced market success lies in a well-planned entry on the part of the
ILECs into the video services market, offering content and interactive
video-based services. This could also lead to enhanced success for broadband wireless technologies such as the Local Multipoint Distribution
Service (LMDS), WiMAX, and the Multipoint, Multichannel Distribution
Service (MMDS). Some service providers are already looking at content
delivery; consider, for example, Telus in western Canada. Telus was the
first service provider in North America to deploy a regionwide IP backbone, one of the first to offer IP-based voice services, and now offers Telus
TV, which is the equivalent of cable content delivered over DSL. Industry analysts continue to examine the viability of ILEC/satellite provider
alliances, which would give ILECs an enormous footprint for service
delivery and potential access to far-ranging content, and the satellite
company access to a large collection of potential subscribers. And while
this combination may not be ideal, it is worth considering in the future.
Consider SBC's recent interest in DirecTV. Today, 11 service providers
control 85 percent of the global telecommunications marketplace, and
ILECs are responsible for 85 percent of all equipment spending. So, they
are without question a force to be reckoned with.

ILEC Summary

So, what are the primary challenges facing the ILECs? First, they are
experiencing declining revenues brought on by wireless substitution,
cable incursion and broadband IP voice in both the enterprise and end
user sectors. Second, because of OPEX issues, they often lose money on
DSL deployment. Third, their debt load is high. Finally, uncertainty associated with regulatory decisions leaves them with a number of unan-
swered questions. Nevertheless, the market is currently theirs to lose
because they do own the bulk of the customers. However, they *must* refo-
cus their efforts on both cost reduction and customer-specific service

delivery if they are to hold on to their advantage, and must also develop the logical infrastructure required to offer, deploy, and bill for converged service packages.

We now move our attention to the much-maligned CLECs.

The CLECs

As we noted earlier, the CLECs came into existence in concert with the release of the Telecommunications Act of 1996, with plans to create a fully competitive market at the local loop level. Until that time, only the long-distance sector was fully competitive, with customers able to choose from among three major providers. The local loop was dominated by the ILECs. Since that time, the CLECs have carved out a reasonable piece of the market for themselves, supported by a certain degree of customer dissatisfaction with the ILECs, the promise of lower prices, and a number of favorable regulatory decisions including the Unbundled Network Element-Platform (UNE-P), which offered a generic switching and access platform to all entrants.

CLEC Challenges

Of course, the creation of a local competitive market has been fraught with difficulty. Many of the CLECs filed for bankruptcy because of inability to attract an adequate customer base, and while some analysts believe that CLECs may own as much as 65 percent of the medium enterprise market, their overall showing is still relatively small. In fact, while it is easy to conclude that CLECs are largely small startups that are giving fits to the ILECs, this is not true: Ironically, the largest CLECs in the United States have traditionally been AT&T and MCI, although recent regulatory decisions may force them to drastically reduce their footprint in the local marketplace.

For the most part, CLECs target the small-to-medium business market, which is largely underserved by the ILECs, whose primary focus is on the large business and residence sectors. Most CLECs resell services at a price that is 15 to 20 percent lower than the ILEC price and can thus attract a share of the market that is price sensitive. In fact, the numbers are significant: ILECs lease 20 million lines to resellers at an average of 40 percent of retail, although these numbers could change with the recent regulatory changes announced in recent months by the FCC—and which continue to be announced. The ongoing regulatory circus, combined with the burgeoning deployment of IP-based voice by various market sectors, will continue to make this game an exciting one for the CLECs. Stay tuned!

The Interexchange Carriers (IXCs)

There are four main long-distance players in the United States: AT&T, MCI, Verizon, and Sprint. These carriers face a number of seemingly insurmountable problems including rapidly increasing customer counts, growing minutes of use (MoU) per subscriber, and a rapidly declining dollars-per-transported-bit-per-mile figure, the combination of which is deadly. Needless to say, this explains their compelling argument for accelerated local service and broadband entry.

One bone of contention for the IXCs is that more than 40 percent of all IXC revenues are paid to ILECs as access charges. Access charges are the fees paid to ILECs by long-distance carriers for the right to use local exchange facilities for the origination or termination of traffic transported to or from one exchange to another by an interexchange carrier. And while some access charges are billed directly to the end user, most of them are paid by interexchange carriers—not an insignificant amount of money. IXCs hope that future regulatory decisions will address the magnitude of this fee. In fact, one reason that VoIP has become so popular among those deploying the service is that it is currently classified by the FCC as an information service rather than as a telecom service, which means that VoIP carriers are exempt from many of the regulatory tethers that bind traditional service providers.

The Cable Multiple System Operators (MSOs)

As we often hear, "There's the good news, and then there's the bad news . . ." In cable, the good news is that cable MSOs are taking two-thirds of all broadband data (ISP) customers and own two-thirds of the broadband consumer market. In fact, three out of five of the largest broadband providers are cable companies.

Furthermore, they are going after voice customers with VoIP, offering the service at an incremental fee plus video, a strategy that erodes the ILECs' base. In fact, because they are also content providers with plenty of mindshare, cable MSOs are a more serious competitive threat to ILEC local service dominance than the CLECs are. The CLECs, after all, do not, for the most part, go after residence customers; they are too difficult to provision en masse. Instead they go after the aggregated enterprise customers that reside in MTUs because provisioning them is a simpler and less costly process. In fact, many analysts predict that if cable MSOs can crack the code on delivery of voice and can centralize their management and repair functions, they could become the first choice for the delivery of integrated services. And today they seem to be doing this— and are just beginning to announce wireless as a secondary service option. In late December 2004, Sprint and Time Warner announced their intent to work together to give Sprint the ability to sell the "quadruple play" of voice, video, data, and wireless. This places them squarely in the game, offering a converged set of bundled services to customers for as single, low price.

The U.S. cable industry is taking its voice penetration strategy seriously. The industry as a whole has invested over $70 billion in network upgrades to accommodate the demand for high-speed data, and although cable companies have attracted significant numbers of voice customers away from ILECs, the ILECs have done very little in the way of competitive pricing or service expansion to combat cable—though SaskTel's MAX service is intriguing. The newly released DOCSIS 2.0 for high-speed data promises transmission at very high bandwidth—as high as 30 Mbps. Furthermore, the operating margins on cable modems are quite high—35 to 40 percent.

So what's the bad news? The bad news is that there is more to running a successful company than having impressive technology. The cable industry's debt load is enormous because they have all invested heavily to ensure market penetration and their ability to deliver the services demanded by customers. Adelphia, for example, the sixth largest cable company in the United States, has nearly $14 billion in current debt and approximately 5 million subscribers—which works out to a debt load of $2,800 per subscriber. And while this number seems large, it also seems (perhaps) manageable—until we begin to factor in other numbers. Subscribers spend approximately $800 per year on services. However, cable companies spend an average of $480 per year to provide routine service to each subscriber, which leaves $320 per subscriber. Furthermore, capital spent on routine maintenance averages roughly $100 per subscriber, leaving far less in the way of income against incurred debt. And to add insult to injury, this figure does not take into account in any way the capital required to perform ongoing (and required) network upgrades.

Cable analysts use a number of measures for assessing the financial health of cable companies, including operating income (EBITDA) as opposed to free cash flow; they also use return on capital as a good indicator. Today the number averages about 5 percent on a pretax basis, which is not a particularly impressive number.

The top five cable MSOs serve 81 percent of the overall U.S. market, while the top ten serve 95 percent. There is still significant competition for cable from satellite providers: Consider DirecTV and the many services it offers. Their primary market is distributed, commercial video service with 80 percent market penetration and very slow growth (roughly 1 to 2 percent per year). They primarily serve residential markets and are motivated to (1) enter the enterprise space and (2) diversify their line of product offerings. (Readers may have seen the frontal assault on cable posed by the satellite providers in the voracious pig commercials that were recently aired.)

Cable MSOs are not taking the attack idly, however. As the satellite industry pushes hard for customers, cable providers are striking back, keeping rates steady in markets that are most prone to attack by satellite providers and raising prices in markets where it is warranted. They are also relying on broadband Internet access and its high profit margins to bolster revenues and to provide a loudly proclaimed competitive advantage, since satellite providers cannot yet offer Internet access that provides equally satisfactory service. Others are aggressively advertising bundled service packages that offer distributive cable services, broadband Internet access, and telephony for a small additional fee.

Even though satellite prices are typically cheaper than cable for entertainment packages, the magnitude of the price difference is shrinking in many markets. As pay-per-view sporting events and other premium lineup components become more and more expensive, satellite providers are being forced to make the same price hikes that the cable providers have been forced to make to cover costs. And while there is a place for satellite in the provider lineup, its advantage as a big-footprint provider is rapidly being winnowed away as cable and telephony service provider penetration increase and provide fixed access to the same set of services.

The Wireless Providers

We covered the wireless sector in great detail earlier in the book, so we won't go into detail here. Suffice it to say that usage is up, but ARPU is dramatically down—not a healthy picture. Watch this sector for continued consolidation and buy-down: The recently announced merger of Nextel and Sprint is one of the first; it certainly won't be the last.

The Internet Service Providers (ISPs)

In reality it doesn't make sense to address the ISPs alone because they have been subsumed by a greater collection of service providers known collectively as the communications service providers (CSPs). These are broadly defined as companies that provision IP-based services including those based on wireline, wireless, and cable access technologies and that include Internet service providers (ISPs), applications service providers (ASPs), and both wireline and next-generation service providers.

These CSPs are launching bundled services by aggregating content through partnerships among the content providers, software providers, and network providers. By offering bundled service packages, they will enjoy the coveted first-mover advantage by selling a collection of services that are both desirable and that offer broad margins. The biggest challenge they face will be customer retention: To attract and retain them, win adequate market share, and drive profits, CSPs must develop and make the best advantage of customer care and billing systems. Without these systems, which provide the ability to not only bill but also to monitor the purchasing behavior of the customer and respond to it, CSPs cannot respond fast enough to maintain a leading position among their competitors. Key to their survival is the migration from being providers of traditional flat-rate, dial-up access to the Internet to being providers of richly contextual, just-in-time, user-specific services such as VoIP; fax over IP; voice over xDSL; video and audio content streaming; video e-mail; electronic gaming; and text, audio, and video instant messaging. Of course, these services are fundamentally dependent on the availability of broadband access because they are bandwidth intensive. The combination of rapid response and bandwidth availability are two critical components. A third, naturally, is critical mass in the market. Successful customer acquisition and retention strategies will be crucial in the medium to long term. Finally, provisioning these services in a way that makes them profitable is the last piece of the puzzle. Billing algorithms may need to change as this evolution occurs: Today most services are billed on a flat-rate basis; in the future a combination of flat rate and measured billing may become the norm.

If we assume that this model of access, application, and content is valid—that is, that successful companies in the CSP space will comprise some combination of these three elements—then we can identify and speak to the characteristics of each element.

Under the purview of the CSPs are found a collection of company types including the wireless content providers, traditional access providers, and application providers. In the paragraphs that follow we discuss each in turn.

Let's now examine the manufacturing sector, beginning with the component manufacturers responsible for semiconductor and optoelectronic devices.

The Component Manufacturers (Semiconductor and Optoelectronics)

As the lowermost layer in the technology market food chain, the semiconductor industry serves as the layer of silicon-based plankton that provides viability to the entire chain. As a result, it is typically the last layer to contract in a recession and the first to recover. And, true to form, that is precisely what we are currently seeing. Global semiconductor sales were roughly $200 billion for 2003, with major contributors including WiFi, automotive demand, screen (display) technology, LEDs, microprocessors, DVD, DRAM, nanotechnology devices, and smart cards. Today there are more than 1.5 billion smart cards in use, underscoring the great demand for this alternative technology. Similarly, wireless demand has driven attention to related technologies, such as improved battery life, multiprotocol support, and reduced space requirements.

Major players in the component sector include Intel, AMD, IBM, Agere, and Texas Instruments. Intel has long been the dominant player in the industry, with its functional tentacles extended into numerous successful ventures that span the breadth of the industry. AMD is the number two PC processor manufacturer after Intel. Interestingly, AMD started its life as an Intel "imitator," licensed to sell Intel-designed chips as a second supplier to the market. The relationship between the two companies eventually unraveled, and after a series of minimally successful products, AMD finally came out with the 1 GHz Athlon, which put the company on the map. Their success story has continued ever since, and today they are a true force to be reckoned with.

IBM has been successful in its own right in this space. Long known as a manufacturer of memory chips, IBM recently announced a line of 3D chipsets, made by combining high-density, high-performance layers into a coupled stack. This allows the embedded transistors to be placed closer to each other, which accelerates the process of processing.

A problem that the industry has often faced with 3D arrays is the heat required in the manufacturing process to add a second layer: The heat often degraded the first layer during the bonding process. The IBM method uses a lower temperature technique to bond the multiple layers to one another. In practice, a thin glass layer is applied to the face of the second layer, after which the other side is etched. The resulting wafer is 200 millionths of a millimeter thick, and is aligned and bonded to the first wafer. Because the resulting stacked wafer is so thin, it behaves as if the circuitry actually exists in a single layer. By using this 3D method, multiple functions can be combined in less "real estate," and processing speeds are dramatically improved.

As optical technology has matured and become a viable alternative to electrical transport in certain circumstances, a new set of challenges have arisen. One challenge that has always faced the component industry is the ability to build optical chipsets. Electrons will live happily in two-dimensional space, moving in a very controlled fashion along a designated conductor. Photons, on the other hand, are much more difficult to control. Attempts to build waveguide technology have succeeded using etched gratings, multiple fibers, and mirrors, but these solutions are by necessity quite large—a problem in a real-estate–constrained industry. New research into particle physics in Europe has yielded a small packet of energy called a (ready?) *surface plasmon polariton (SPP),* sort of a hybrid between electrons and photons. These SPPs are formed in the region between a metal surface and a dielectric surface, commonly found (indeed, designed into) chipsets. These SPPs can be manipulated in 2D space. The result is that etched wafers can now be built that move photons but that are as small as those used for electrons—that is, along the lines of traditional semiconductor designs. And while this technology is nascent, it will mature and become a part of the technological pantheon among component manufacturers.

Component Industry Challenges

Component manufacturers face major challenges in the months ahead. The downturn in the greater telecommunications industry resulted in massive revenue reduction, which in turn caused a buy-down of the number of manufacturers in the sector, particularly among those focused on building components for the much-anticipated, never-ending optical core transport buildout. In fact, 2002 was the first year in history in which the United States failed to be the largest market for semiconductors. Nevertheless, certain applications continue to drive sales among component manufacturers. Mobile phones, for example, drive 30 percent of global component sales and China consumes 25 percent of all mobile output. Other drivers, mentioned earlier, include automotive, display technology, wireless, and nanotechnology.

Another challenge is the ongoing commoditization (if that's a word) of the component industry. There are quite a few similarities between the histories of the American steel and component industries; it is worth a brief side trip.

Steel vs. Silicon: Parallel Stories

The steel industry has its roots in the iron industry, which started in North America as early as 1621 in Jamestown, Virginia. The industry grew over the years until the Civil War, when the shift from iron to steel took place. Steam engines made it possible to build massive factories capable of producing larger quantities of product, and over the years the industry became completely, vertically integrated.

In 1901, under the management genius of such visionaries as J. Pierpont Morgan and Elbert H. Gary, the United States Steel Corporation was formed as the largest industrial corporation on the planet. At that time its capitalization was $1.4 billion and the company controlled more than 60 percent of the American market. Both it and the industry it dominated continued to be major contributors to the nation's economy until after World War II, when union labor costs and aging technology conspired to send the domestic steel industry into a long, inexorable decline that ultimately resulted in its death. At the same time, steel manufacturing facilities began to appear offshore in countries like Korea and China, where labor costs were extremely low, technology was more modern, and product quality was high. Combined with a dramatic decline in transportation costs, these offshore advantages resulted in the decline and fall of the American steel industry. In the early 1970s a consolidation and buy-down of the industry began; by 1975, American steel production had declined by 37 percent. Even so, the industry still employed 500,000 workers at very high union scale wages. By the late 1980s, domestic steel production had rebounded, but the number of workers had declined dramatically. Furthermore, the nature of the product had changed radically: Instead of producing enormous lots of traditional steel products such as building materials, the foundries turned out small volume, highly specialized products with a very high profit margin.

In much the same way, segments of the component industry have found their way offshore. Many domestic manufacturers are selling off their large-footprint silicon foundries and are instead farming out the work to the lowest price, high-quality bidder, many of which are offshore. In Asia, government-subsidized foundries are being built with the stated intent of cornering segments of the global manufacturing work, including memory (DRAM). Micron is still the largest manufacturer of DRAM in the world, but large fabricators are on the horizon that will give them a run for their money.

One response to this commoditization is the move to "fabless manufacturing." Major component manufacturers have realized that the actual manufacturing process, which is highly labor and capital intensive, has reoriented their businesses. Instead of focusing on the actual manufacturing, they focus on the design of highly specialized, custom-designed chipsets for individual clients. In effect, they become the architects of the chip design and then hire a reliable manufacturer to actually build the physical component. Today, there are more than 350 fabless manufacturers in operation, and like the steel industry that learned to succeed through specialization and cost control, these companies are proving to be viable. By putting a customer's IP on a custom-designed device, a manufacturer can create a small product monopoly by offering customer-specific services instead of large-lot, capital-intensive manufacturing—and by focusing on accuracy, cost reduction and revenue enhancement as well.

Another successful direction for component manufacturers is the creation of more complex system-on-a-chip (SoC) devices. As the IC-to-FPGA-to-ASIC evolution has moved inexorably forward, the need to increase functionality while reducing footprint has also increased. The end-user device that, while physically shrinking also grows in functionality, has presented a new set of challenges to the manufacturers that must meet both of those design criteria.

We now turn our attention to the systems manufacturers.

The System Manufacturers

The system manufacturers—that is, the companies that manufacture complex network systems such as routers, switches, multiplexers, and other devices required in a large network—were hard hit by the downturn in the telecom industry. Because of their relative position in the food chain—they sell directly to the service providers—they were affected early in the process. As soon as end-user and enterprise spending began to decline in 2000, service providers put their capital expenditure plans on indefinite hold. Unfortunately, the systems manufacturers had based their manufacturing plans on continued service provider spending; so, when the spending stopped, manufacturers were left with a massive overhang of inventory—inventory, by the way, that ages rapidly and quickly becomes unusable. Between 1996 and 2000, total hardware spending grew 16 percent per year, from $24.5 billion in 1996 to $47.5 billion in 2000—an absolutely unsustainable level, as we now know. Between 2000 and 2001, spending fell to its pre-2000 levels, which on the surface appears to be a catastrophic response. Combined spending in 2001 fell roughly 14 percent to $40.1 billion, a significant decline. The bubble burst in 2001 and the industry declined into a multiyear nuclear winter from which the industry is just now thawing.

Expansion and contraction of capital-intensive markets like telecom, however, are normal events. What made this particular contraction so painful was the immense overgrowth and overexpectation that took place prior to the collapse. As we described earlier, service providers were buying network infrastructure at top dollar because (1) the money was available, (2) the growth was expected to go on forever, and (3) they had no choice—they had to keep up with the competitors that were engaged in the same practice.

The decline of the hardware market was protracted and financially bloody. Cisco's market cap declined from $555 billion in March 2000 (at which point it was the most valuable company on the planet) to $74 billion at the end of 2002, with an attendant share price decline from $82 to $9.50. Similar declines can be charted for any of the major players—Lucent, Nortel, Alcatel, Fujitsu, Marconi, none were immune. The core transport glut resulted in a precipitous decline in the optical sector as well. Consider the fate of JDS Uniphase, a premier manufacturer of optical components, which recorded the single greatest loss at the time in corporate history—$51 billion. Manufacturers as a whole saw a 26 percent decline in revenues year-over-year, and new orders were down more than 50 percent in the same period. And while capital spending in the United States showed an admirable 25 percent growth rate, average revenues grew only 25 percent, while profits dropped precipitously and return on equity faltered. Between 1996 and 2000, capital expenditures in telecommunications rose from $41 billion to $110 billion—but return on capital fell 50 percent. Furthermore, since 1984, the domestic consumer price index is up 73 percent and local communications call volume is up 71 percent, but long-distance prices are *down* 35 percent—not a good sign for service providers, particularly given the capital-intensive nature of telecomm infrastructure. Even the venerable PC market saw a downturn; for 2003, growth declined from an expected level of about 13 percent to slightly less than 10 percent.

Amidst all this darkness there has been good news, however. While optical transport revenues declined precipitously during the postbubble disaster, the revenues from SONET and SDH (optical transport infrastructure) sales climbed. By the end of 2005 the optical metro marketplace, a strong market sector, is expected to roughly double, while the total market for IP-based VPN services will grow to approximately $14.7 billion by the beginning of 2006 because of the proliferation of mobile workers, improved mobile technology, security concerns, and changing work models.

Technology Indicators

Several technology sectors show promise and are recovering more rapidly than others. They are discussed in the following sections.

Softswitch

Softswitch technology is designed to support next generation networks that rely on packet-based voice, data, and video communications technologies and that can interface with a variety of transport technologies including copper, wireless, and fiber. One goal of the softswitch concept is to functionally separate network hardware from network software. In traditional circuit-switched networks, hardware and software are dependent on one another, resulting in what many believe to be an unnecessarily inextricable relationship. Circuit-switched networks rely on dedicated facilities and are designed primarily for delay-sensitive voice communications. The goal of softswitch is to bring about dissolution of this interdependent relationship where appropriate.

In concert with the evolution of the overall network, the softswitch concept has evolved to address a variety of network payload types. More and more, telecommunications networks are utilizing IP as a fundamental protocol, particularly in the network backbone. This backbone is also making VoIP services viable as an alternative to circuit-switched voice.

There is more to softswitch, however, than simply separating the functional components of the network. Another key goal is to create an open service-creation development environment so that application developers can create universal products that can be implemented across an entire network. Part of the evolution will include the development of call control models that will seamlessly support data, voice, and multimedia services. We will discuss more specific examples of this functional evolution in the last section of this report.

The result of widespread softswitch deployment will be the creation of a switching model that does not have the same restrictions that plague circuit switches, such as intelligent network triggers, application invocation mechanisms and complex service logic.

This functional distribution will result in faster and more-targeted feature development and delivery and significantly lower service delivery costs. Softswitches will be architecturally simpler, operationally efficient, and less expensive to operate and maintain.

A number of corporations have now focused their efforts on the development of softswitch products. Both Lucent and Nortel announced softswitch products as early as 1999 in the form of the 7R/E (Lucent) and the Succession (Nortel). Today, they have been all but abandoned because the market simply wasn't ready to bridge the gap between the circuit-switched Bellheads and the IP bit-weenies, although both companies have now come roaring back into the fray with Succession (Nortel) and iMerge (Lucent). Today, however, the market is once again growing, and ferociously. Companies like Taqua Systems are beginning the process of reinvigorating the softswitch marketplace. A key target sector is the small rural telephone company. These companies are targets because they can take advantage of the incremental-deployment nature of the softswitch, and because they are one of the few segments that have money earmarked for capital growth.

Softswitch was originally intended as a local-switch replacement technology to facilitate the circuit-to-packet migration. The economy, however, has been problematic, and progress has not been as aggressive as the sector would like. In truth, softswitches have been used to replace lines, not wholesale switches, and while the incumbents are not buying, they are shopping with RFPs and RFQs—a good indicator of later movement. To date the main applications that softswitch solutions have addressed are enterprise applications such as PBX management, IP Centrex, VoIP, and wireless LANs.

The Server Market

There is an evolution underway in the server market that is pulling the players away from the stand-alone server model in favor of a new design. The next evolutionary stage is called the *blade server.* A blade server is a complete server on a board that plugs into a common chassis and shares power, cooling, backplane cabling, and network access with other blades. It is designed to simplify the process of server management for IT staff and to reduce the cost of shared infrastructure.

The second evolutionary stage is called a *brick server.* A brick server allows IT staff to interchange memory, processors, disks, and other elements on an individual blade, a model that allows for highly modular plug-and-play capability. Several manufacturers are already going down the blade-to-brick path including IBM, Dell, HP, and Sun.

The Content Providers

This is what it's all about: Designing and building the content that customers will access from a wide array of both stationary and mobile devices. Consider these numbers: 98 percent of all American households have TV; 69.4 percent of them have cable. Furthermore, there are 187 national cable channels. Ninety-one percent of American households have VCRs, while a slightly smaller (but equally respectable) number have DVD players. And just to provide fuel for that entertainment fire, 500 feature films emerge each year on 37,000 screens. And it isn't just television and movies: According to consultancy IDC, U.S. wireless gaming alone will grow to $71.2 million in 2007. And that's just the wireless segment!

The key content providers have well-known names: AOL Time-Warner, Google, Sony, Vivendi, Terra-Lycos, Yahoo!, and Microsoft. Their activities in the marketplace illustrate that they understand all too well that the key to success is not only having a broadband access and transport network available, but having content to transport that customers will pay for. We noted earlier that 3G has not been particularly successful, largely because of the dearth of applications that can effectively take advantage of 3G's considerable technological capabilities. What seems to be largely missing in the market is an understanding that customers by and large are not looking for a new killer application: What they *are* looking for is a killer new way to access existing applications, such as (in addition to good-quality voice) instant messaging, e-mail, Internet access, and gaming. The addition of simple multimedia applications such as video clips and photograph sharing are nice additions, but far from killer apps. Of course, many of the services delivered over networks are based on cultural demand. In South Korea, SK Telecom offers 15 channels of satellite TV service to mobile phones for about $4 per month. While slick, that's probably not an application that would generate large revenues in North America, in spite of the fact that companies like Sprint have announced TV broadcast plans for mobile devices.

These applications imply that the handset market must evolve to accommodate them. As we noted previously in the wireless discussion, there is considerable activity underway as manufacturers scramble to define the nature of evolving services and the capabilities that handsets and other mobile devices must offer. A few companies are already moving into the multimedia handset space: Microsoft recently acquired Vicinity Corporation for $96 million; the company's products allow them to deliver maps and directions to mobile devices. Based on VoiceXML, call centers are among the firm's major targets.

Originally less than willing to sell Windows to the mobile market, Microsoft is now making a concerted effort to penetrate that world. Concerned about a repeat of the Microsoft monopoly that occurred in the PC world, however, mobile manufacturers quickly pulled together following Microsoft's Windows CE entrée and created Symbian, a software consortium tasked to create a Windows-like mobile phone OS that would *not* be dominated by Microsoft. Symbian members include Nokia, Motorola, Siemens, SONY/Ericsson, Panasonic, and Samsung. Currently, the Symbian organization dominates the market with roughly 80 percent of the market because of its numerous licensees.

In response to Symbian's position, Microsoft did an end-run. Instead of trying to sell to the major manufacturers that already had a stake in Symbian (although Samsung has licensed both platforms, as well as Palm's), Microsoft went directly to the mobile network operators and contract manufacturers that buy large volumes of handsets with plans to

influence them. More than 25 percent of all handsets are contract manufactured and can therefore be influenced by Microsoft. And while
Microsoft has succeeded in some cases, they have met with challenges in
others.

One barrier that Microsoft faces is that the existing economic mobile
model leans heavily in favor of the large manufacturers. They produce
millions of handsets, while contract manufacturers sell hundreds of
thousands, resulting in devices that tend to be more expensive. Some
analysts believe that Microsoft's attempted end-run will fail, but it's still
too early to tell. Microsoft may be able to appeal to corporate clients if
they craft a way to integrate mobile and desktop applications to create a
seamless service continuum. The key point here is service: Technology,
while important, is not the lead act in this play. Services, applications,
solutions, and value represent the lead theme, and content is the principal cast of characters.

Closing the Revenue Gap

The triad of content, carriage, and computing represents a strong
weapon in the war against the widening cost-revenue gap. The existence
of multiple service markets (enterprise, consumer) dictates that there
are multiple judgments that define their value in each segment. For
enterprise customers, network management and professional services
rank well in the hierarchy because they provide a technology-to-services
transition strategy. For residence customers, the main item identified is
broadband access. Other so-called value-added services include video,
music, games, simple wireless videoconferencing, and encryption. Note
that these are all content-based services. On the other hand, value-added
services (by the market's definition of value) include back-office integration, online payment and consolidated billing, content delivery (photos
and video, primarily), and messaging (instant messaging and SMS).
SMS, the GSM-equivalent of IM, is an enormously lucrative service offering: It is more than 1,000 times more lucrative in terms of dollars per
transported bit than the nearest competitive service. Furthermore, the
margins on traditional value-added services—customarily, those provided through the magic of the SS7 signaling network and driven by
database-intensive caller ID capability—can be as high as 95 percent,
one reason voice is still a lucrative product offering in spite of the fact
that the price point for the service is declining. In this industry, we have
often observed that the killer application for voice networks is data,

referring to the revenue-rich nature of these services. They cost comparatively little to deliver, yet offer good revenue returns. That's why service providers count on them.

So if data is the killer application for voice networks, what's the killer application for data networks? The entire industry is scrambling to build IP networks, to add broadband access in any of a variety of forms, yet every economic study ever done shows that the revenue generated by the data transported over service provider networks accounts for a small fraction of their actual revenue—as little as 5 percent according to some studies. The answer, not surprisingly, is voice—in addition to a growing array of potentially lucrative data-dependent applications. What is interesting, however, is that as we stated earlier, the marketplace is not looking for the next great killer application—what they are looking for is a killer new way to access existing, utilitarian applications such as voice, e-mail, instant messaging/short messaging, Web access, content delivery, security (hence the success of wireless VPNs), caching, and storage. Furthermore, instant messaging is not an application targeted exclusively at teens and preteens: Industry reports indicate that nearly 100 million enterprise employees rely on some form of IM for intracorporate communications, and many corporations have seen a measurable decline in e-mail volume as a result of growing IM usage. The phenomenon is decidedly human: Most employees, upon receipt of an e-mail message, decide whether they must respond immediately or can delay the response until later. IM users, on the other hand, psychologically feel an obligation to respond immediately to the appearance of an instant message—as if the sender knows they are in their cubicle. The result of this is that information flows more quickly, resulting in an acceleration of decision making. There is a downside, however: IM users are scattered across a handful of providers including AOL, Microsoft, and Yahoo! Interoperability among these providers has always been a problem, but a few companies, including Verisign, are stepping up to the challenge of creating an IM interoperability gateway and putting it into service.

Summary

The added function of managing value-added services is a critical differentiator: Consider that AT&T's professional services business is now a $4 billion product line. Consider also that the true professional services firms (Deloitte, for example) are in enormous demand by telecom industry players to help them navigate their way through the stormy waters of market recovery, product positioning, competitive positioning, and value delivery. One key to success is the ascendancy of the seven-layer value chain that I described in an earlier book. Because of positive reader response, I repeat that discussion here in a somewhat amended form.

In the early 1960s, American psychologist Abraham Maslow proposed his now-famous five-layer hierarchy of basic human needs. In it, Maslow ranked the basic human needs: physiological demands; security and safety concerns; love and feelings of belonging; competence, prestige, and esteem; self-fulfillment; and finally, curiosity and the need to understand the surrounding world. Maslow concluded that these basic requirements had to be fulfilled before the higher level, more cognitive demands could be fulfilled.

Interestingly, there is a similarly structured hierarchy-of-needs model within the domain of telecommunications, particularly as it relates to the activities of network designers, service providers, and manufacturers of components and network devices. This hierarchy, comprises a seven-layer model with these identified layers, starting from the bottommost layer: features; functions; benefits; applications; services; solutions; and finally, at the top, value.

It's interesting to observe that all of these terms appear in marketing, sales, and technical documentation, and all are important. However, the priority they are given and the qualities they each represent deserve attention.

The dictionary defines the seven terms as follows:

- *Feature*: A prominent or distinctive aspect, quality, or characteristic
- *Function*: The action for which (something) is particularly fitted or employed
- *Benefit*: Something that promotes or enhances well-being; an advantage
- *Application*: The act of putting something to a special use or purpose
- *Service*: An act of assistance or benefit to another or others
- *Solution*: The method or process of solving a problem
- *Value*: Worth in usefulness or importance to the possessor; utility or merit

There are two ways to approach the use of this seven-layer network value hierarchy. One is from the bottom up, starting with features and passing through the various layers of the model; the other is to start with value at the top and work downward. Both are valid, but must be utilized appropriately. Let's consider the functional differences between the seven layers.

Features, functions, and *benefits* are characteristics that define the technical capabilities and defining parameters of a device or service. "A prominent or distinctive aspect, quality, or characteristic," the definition of a feature, clearly speaks to such a thing as a physical footprint (the space a device occupies in a central office), the amount of heat it generates while in operation, the amount of electricity it consumes, backplane capacity, and component redundancy. Function, "the action for which (something) is particularly fitted or employed," describes the technical inner workings of a device or software module that result in some sort of operational value for the purchaser of the product. A benefit, "something that promotes or enhances well-being; an advantage," provides precisely that: an advantage that the product conveys to the customer; although it is typically interpreted as a technical rather than a market advantage, such as the ability to perform a hot swap of a component.

Moving up the stack, we come to *application,* "The act of putting something to a special use or purpose." An application refers to the manner in which a product is actually used—usually the reason that the customer purchased the item in the first place and the first occurrence of possible value to the consumer. *Service,* "an act of assistance or benefit to another or others," refers to the process of converting what is usually a generic application set into a more focused, almost customized treatment to address a specific set of needs for a client. *Solutions,* defined as "methods or processes used to resolve a problem," carry the specificity of services to the next level, addressing the customer's specific business requirements and often presenting a product directed as much at the customer's customer as at the actual customer.

Finally, we come to *value,* something that is useful or important to the possessor and offers utility or merit. Value is the most critical of the seven elements, because it is universal and timeless. Value, custom-defined in the mind of each customer, is a personal, specific, and difficult-to-quantify essence unique to every client. The other six—feature, function, benefit, application, service, and solution—are relatively static and time dependent. In other words, an application that is timely, useful, and effective today may not be six months from now. A solution that resolves today's vexing customer problem may not do so this time next year. Value, however, is a constant, and has little to do with technology: It has everything to do, however, with a discrete knowledge and understanding of what makes the customer tick. There is one additional characteristic that must be addressed: While the service provider or system manufacturer may be in a position to define the characteristics of the lower seven layers, only the customer can define value. A vendor who understands this sublime fact will have a significant advantage over those who do not.

In the days when bandwidth held sway and was a tremendous money generator, the lower layers represented a cash cow ripe for milking. Today, however, with the perceived glut of bandwidth that is available, the price of bandwidth has plummeted to near zero—a frightening reality for those companies that have traditionally made their fortunes through the sale of bits-per-second. Today, the big money lies at the top of the stack, closer to the customer, a place where unique, specially designed network products can be positioned on a customer-by-customer basis. The traditional service providers—the LECs, cable companies, and IXCs—are scrambling to establish a toehold that will allow them to climb the stack, to move out of the primordial network ooze into the lofty heights of content and services. Of course, the ILECs and cable MSOs are in an enviable position: They touch the customers with their networks, and whoever is closest to the customers has the greatest influence over them and the infrastructure deployment decisions they are likely to make. The combination of network provisioning and content is unbeatable, and is fast becoming a major focus for converged providers. The ability to provision an end-to-end network as well as deliver application content is a powerful combination. Thus, the service providers' interest in moving up the food chain is a valid one.

So what does all this mean? It means that the winners in this game to acquire and keep customers will be those companies that understand the importance of providing highly targeted value based upon a discrete understanding of the drivers behind every customer. Technology, expressed as features, functions, benefits, and so on, is still critically important. However, the people who care about that level of technical detail are not generally the people who make buying decisions or write checks for product purchases. Value takes on many different forms and must be expressed appropriately to each audience. Technology bells and whistles certainly matter—I would never suggest otherwise—but there is much more to the capability equation. The characteristics that represent value to a technician responsible for installing, operating, and maintaining a network are different than those that provide value to a network designer, a beneficiary of the services that the network delivers, or the end user who must make a buying decision about a highly capital-intensive acquisition. Thus, the manner in which each product is represented to the client must be customized to an appropriate degree if the sale is to be successful.

Chapter Summary

In this chapter we explored the nature of the telecom industry: Who are the players, what are their roles, and perhaps most important, what is the future that awaits them? In the final chapter, we delve into the past before we examine the future, looking at the now infamous telecom bubble.

Chapter 8 Questions

1.   For each of the major industry sectors, list at least three players.
2.   What advantages do CLECs have over ILECs?
3.   Use the Web to research Section 271 Relief. Is it successful?
4.   Why is the metro sector such an important market?
5.   What are the biggest challenges facing the ILECs?
6.   Should the industry just get it over with and "revest" itself?
7.   What are the biggest challenges facing the cable sector?
8.   Why were the component manufacturers least affected by the burst bubble?
9.   What lessons do we learn about the component industry by studying the steel industry?
10. What is softswitch and why is it important?

CHAPTER **9**

Final
Summary

As I write this final chapter for the second edition of *Telecom Crash Course*, I am sitting on a lounge chair in the backyard of my South African colleague Roy Marcus, founder of the Da Vinci Institute, listening to the exotic calls of South African weaver birds in the trees and monstrous frogs in the pond. It is November, and therefore midsummer; today it's well over 100 degrees. I've spent the last few months working closely with Telkom, the incumbent, monopoly service provider for South Africa, helping them prepare for the arrival of a truly competitive market in their country, which has begun to happen following the government's recent announcement of second carriers in the country. In many ways it is reminiscent of AT&T in late 1983, just before divestiture. In other ways, it is actively, dynamically, dramatically different. There is no question that things will change here in the coming months as competition takes root at the local level and introduces chaos into the game; on the other hand, there are few places I have been in this world where there is so much hope, so much promise, so much *vibrancy* in the market. These are capable, clever, truly entrepreneurial people. South Africa specifically and Africa in general are powerful harbingers of capability, and telecommunications will play a key role in the continent's economic development and success in coming years. From the shocking warmth of Soweto, to the power of modern business in the big cities, to the innovative strength seen inside telecom-equipped shipping containers scattered about the African countryside, this is what it's all about.

As I mentioned earlier in the book, a few years ago I was asked by the World Bank to deliver a television broadcast from their studio in Washington, D.C., to 50 (or so) viewers in four Sub-Saharan African countries. The topic was "The Internet and Its Use as a Business Tool." The session was highly interactive and exciting, and I answered as many questions as time allowed while online. After the fact I was contacted by a gentleman in Kenya named Edwin N., who asked if I could offer him some assistance. He wanted to set up a Web site that would serve as a central contact nexus for information about Africa. Tourists wishing to visit the continent could come to his site to look for country-specific data about tours, safaris, hotels, restaurants, and so on. Over the course of several months I helped Edwin put together the HTML for his Web site, and as soon as it was online and working we slowly drifted apart. A year later, I received a letter in the mail (ironically, a letter, not an e-mail) from him. It is reproduced in Figure 9-1.

There was a time in my life when I measured my productivity each day by how many inches I could reduce my inbox (this was back when

Figure 9-1
Letter from
Edwin
Ngorongo
(Reprinted with
permission)

13 November 2000

2187 King Edward Road
Nairobi, Kenya

Dear Professor Shepard:

Perhaps you will remember me from the World Bank seminar you conducted for Africa.

I wish to thank you from the bottom of my heart for **helping** me start my Internet business in Kenya. It was not easy: I **had to** take out a large business loan from the World Bank for the equal of $100 U.S., but I believe that I will be able to pay **it** back within 18 months. Already I have made sales of my magazine in the US and Europe and because of your help my family **is** eating well and my children are smiling. God bless you and the Internet.

Asante Sana, rafiki wako Edwin

Most sincerely,

Yours,

Edwin

Edwin Ngorongo

inboxes involved massive quantities of paper) before leaving for home. When I received Edwin's letter I read it, then read it again. It made me cry, and it made me realize that telecom has the fundamental ability to change lives. It can create wealth, bring people together, channel hopes for a better future, and make children smile.

Since receiving the letter from Edwin I have had the opportunity to watch a woman—a mother, in a small rural village—cautiously pick up a telephone for the first time, dial a number, wait a few seconds, and talk to a son who she hadn't seen or spoken to in ten years. It's simply too costly to travel, and until recently there were no phones anywhere near her village. Ten years ago the child had moved to the city in search of work and had not yet been able to return. Today, the mother and son can talk daily if they wish. Again, I cried—and I smiled.

It doesn't get any better than that. In fact, Edwin's letter is framed on the wall behind my desk. It bears a small metal plaque that says, "Never forget why you do what you do."

The same promise holds true in the developed world. In the wake of the millennial meltdown, the companies that make up the technology industry struggle to redefine themselves as players in a game in which the rules have all changed; intense, fierce competition leaps from the shadows, and customers, once meek and malleable, are now hardened, jaundiced, and wary. The freedoms that once prevailed have disappeared: Manufacturers must accept that they can no longer build *everything* for *everybody* and must partner with others, including competitors, to succeed—and must even pass on some products if they fail to offer lucrative financial returns. Service providers must redefine the very meaning of their name in a constant struggle to anticipate what customers expect in the way of service—and all that the word implies. And the industry supremacy that so many legacy firms enjoyed for so very long is changing as new companies in often surprising industries rise to the top and begin to catch the eye of the consumer.

Welcome to the technology marketplace in the twenty-first century. It doesn't get much more exciting than this, and the *sturm und drang* is far from over. The companies that make up this remarkable industry seem to be floating on a vast sea of uncertainty at the moment but, like soap bubbles, are being pulled slowly and inexorably toward each other, coalescing into new shapes. A year from now this industry will look very different, as powerful tectonic forces reshape it. Financial recovery, consolidation, fickle customers, and companies looking to be the first to offer the next new thing will all play parts in this passionate dance of technologies.

There are those who believe that the bubble has burst, the market has calmed, and the road to recovery is straight and true. I am reminded, however, of a quote from my friend and colleague Bob Camp, who splashed figurative cold water on the other directors of a board on which we served with this memorable quote. As we prematurely celebrated a business victory that had not yet materialized, he smiled and said, "At the moment of victory, tighten your helmet strap."

More to come. Buckle up.

INDEX

D